TECHNOLOGY, MANAGEMENT AND SYSTEMS OF INNOVATION

To Michelle

Technology, Management and Systems of Innovation

Keith Pavitt

R.M. Philips Professor of Science and Technology Policy and Director of Research, SPRU, University of Sussex, UK

Edward Elgar

Cheltenham, UK • Northampton MA, USA

Published by
Edward Elgar Publishing Limited
Glensanda House
Montpellier Parade
Cheltenham
Glos GL50 1UA
UK

Edward Elgar Publishing, Inc.
136 West Street, Suite 202
Northampton
Massachusetts 01060
USA

A catalogue record for this book
is available from the British Library

Library of Congress Cataloguing in Publication Data
Pavitt, Keith.
    Technology, management and systems of innovation / Keith Pavitt.
    Includes bibliographical references.
    1. Technological innovations—Economic aspects. 2. Strategic planning. 3. High technology industries. I. Title.
HC79.T4P38 1999
658.5'14—dc21   99–12825
                                                                          CIP

ISBN 1 85898 874 8

Printed and bound in Great Britain by MPG Books Ltd, Bodmin, Cornwall

Contents

Acknowledgements

I would like to thank Edward Elgar for encouraging me to prepare this volume, Giovanni Dosi and Nick von Tunzelmann for commenting on the Introduction, and Dymphna Evans for helping with the practicalities of printing and marketing. Many thanks also to Susan Lees, who prepared and read the manuscripts, and dealt with matters of copyright, with her usual efficiency, dedication, and powerful mixture of firmness and good humour.

Since I came to SPRU in the early 1970s, I have been fortunate in at least four respects. First, I have always had wise advice on what to do – and what not to do – from Chris Freeman inside SPRU, and Dick Nelson outside. In particular, both have always encouraged me and others to strike out in new directions.

Second, I have worked in an environment of continuous exchange of ideas, of debate and of mutual support, involving colleagues and post-graduate students. My debts are too many to mention, but I owe a great one to Pari Patel, who has enabled us to develop and exploit the science and technology indicators made possible by improvements in information technology; and an important one to Joe Townsend, who was a pioneer in compiling a database on significant innovations in the UK, long before it became fashionable to do so.

Third, I have been fortunate in obtaining continuous and long-term financial support for my research, and for colleagues with similar interests, first from the Leverhulme Trust and then from the Economic and Social Research Council.

Fourth, and most important, my wife Michelle has been with me throughout, supporting me and my work, and ensuring that the many other important things in life get proper care and attention. It is to her that this book is dedicated.

The author and publishers also wish to thank the following who have kindly given permission for the use of copyright material.

Beech Tree Publishing for: 'The Objectives of Technology Policy', *Science and Public Policy*, **14** (4), August 1987, 182–8.

Blackwell Publishers for: 'The Size Distribution of Innovating Firms in the UK: 1945–1983', with M. Robson and J. Townsend, *The Journal of Industrial Economics*, **XXXV** (3), March 1987, 297–316.

California Management Review for: 'What We Know about the Strategic Management of Technology', **32** (3), Spring 1990, 17–26; 'Multi-technology Corporations: Why They Have "Distributed" Rather Than "Distinctive Core" Competencies', with O. Granstrand and P. Patel, **39** (4), Summer 1997, 8–25.

Cassell for: '"Chips" and "Trajectories": How Does the Semiconductor Influence the Sources and Directions of Technical Change?', in *Technology and the Human Prospect*, R. MacLeod (ed.), 1986, Frances Pinter, 31–54.

Elsevier Science for: 'Sectoral Patterns of Technical Change: Towards a Taxonomy and a Theory', *Research Policy*, **13**, 1984, 343–73; 'The Continuing, Widespread (and Neglected) Importance of Improvements in Mechanical Technologies', with P. Patel, *Research Policy*, **23**, 1994, 533–45; 'The Technological Competencies of the World's Largest Firms: Complex and Path-dependent, but Not Much Variety', with P. Patel, *Research Policy*, **26**, 1997, 141–56; 'What Makes Basic Research Economically Useful?', *Research Policy*, **20** (2), April 1991, 109–19.

Journal of International Business Studies for: 'Large Firms in the Production of the World's Technology: An Important Case of "Non-globalisation"', with P. Patel, **22** (1), 1991, 1–21.

National Academy of Sciences for: 'National Policies for Technical Change: Where Are the Increasing Returns to Economic Research?', *Proceedings of the National Academy of Sciences*, **93**, November 1996, 12693–700.

Oxford University Press for: 'Uneven (and Divergent) Technological Accumulation among Advanced Countries: Evidence and a Framework of Explanation', with P. Patel, in *Technology, Organization and Competitiveness: Perspectives on Industrial and Corporate Change*, G. Dosi, D.J. Teece and J. Chytry (eds), 1998, 289–317; 'Technologies, Products and Organization in the Innovating Firm: What Adam Smith Tells Us and Joseph Schumpeter Doesn't', *Industrial and Corporate Change*, **7** (3), September 1998, 433–52.

Introduction

Framework

This book reprints thirteen of the papers and book chapters that I have published over the past fifteen years. They do not derive from a thought-out research strategy based from the outset on well-established theoretical foundations. Life in a multi-disciplinary, policy-oriented research institute – which gets most of its funding through competing continuously for programmes, projects and contracts – does not leave much time for such deliberations, and even less, any inclination. They derive instead from a central concern of the Science Policy Research Unit (SPRU): to improve understanding of the nature and sources of the technological knowledge that under-lies the continuous technical change in contemporary society, and of the implications for public policy and corporate management. This central concern provides the necessary coherence for this book.

I came to SPRU in the early 1970s with a thorough training in engineering, a smattering of economics, and nearly ten years' experience in policy analysis in what is now the Directorate for Science, Technology and Industry (DSTI) at the Organization for Economic Cooperation and Development (OECD). After this heavily problem-oriented experience, I was disappointed to find that established theory did not help me much in understanding and solving the problems of science and technology policy. In particular, the then prevailing assumptions in mainstream economics – that technological knowledge comes into the business firm, either in the form of easily usable information, or embodied in machinery – did not explain where and why the knowledge was originally produced, nor why so many firms found it necessary to perform their own research and development (R&D) activities.

Fortunately for me (and others), more useful styles of enquiry were opening up, pioneered mainly by economists dissatisfied by mainstream explanations of technical change. Christopher Freeman had helped pioneer the empirical 'innovation studies' which became the hallmark of SPRU.[1] His research was and is always theoretically embedded, especially in the writings of Adam Smith, Karl Marx, Joseph Schumpeter and other 'classicals', all of whom had important things to say about the nature and sources of technical change.[2] Together with Sidney Winter, Richard Nelson (Nelson and Winter, 1982) built on Schumpeterian foundations to develop an evolutionary theory of economic change, at the centre of which is the purposive and continuous search by business firms for improvements in perfor-mance (that is, technical change). And Nathan Rosenberg's historical studies began to unravel the importance of both the socio-economic and the cognitive factors shaping the accumulation of scientific and technological knowledge.[3] He also pointed to the neglect in many social sciences of the *sources* – compared to the *impact* – of technical change, thereby identifying a field of enquiry that differen-tiates SPRU's research from that of a conventional economics or management department.[4] The extent of my intellectual debt to these writers will be apparent from the contents of this book.

These contents also reflect choices, and the exclusion of most of what I have written on specific issues of science and technology policy, in particular, on the innovative performance of the UK economy,[5] on government policies towards innovation,[6] on innovation systems in the former centrally planned economies,[7] on technological accumulation in developing countries,[8] and on science and technology in the European Union.[9]

With even greater regret, I have not included any paper concerned exclusively with methods of measuring scientific and technological activities.[10] We know from the writings of Nathan Rosenberg (1974) and Derek de Solla Price (1984) that improvements in techniques of measurement have often been at the origin of major theoretical and practical advances in the sciences, by answering questions that could not be answered beforehand, and (equally important) by raising important new questions outside established theory that needed answering. Advances in our understanding of technological change over the past thirty years have often resulted from improvements in systematic data on R&D, patents, significant innovations, scientific papers and citations.[11] As will become apparent, one of the foundations of this book is new and better data on scientific and technological activities.

The other foundation is that technological knowledge is much more than codified information. It is knowing how to solve the complex problems in designing, developing, testing and making artefacts that work, and that are useful. The complexity is of two kinds. First, in their design and use, most artefacts are made up of numerous components and subsystems whose interactions are often non-linear and therefore impossible fully to predict. This helps explain the following *cognitive* dimensions of technological activities.

- The continuing dominance in corporate technological activities of expensive experiments to test the performance characteristics of prototypes and pilot plant, compared to much less expensive experiments to develop scientific understanding. If theory were in itself a strong guide to practice, cost-conscious capitalist firms would presumably spend less on prototypes and more on theory.[12]
- The central importance of tacit knowledge in interpreting the performance of complex artefacts and knowing how and where to search for improved performance.
- The incremental, step-by-step nature of corporate search for improved technical performance, given the difficulties in interpreting the effects of simultaneous changes in many dimensions, and the costly risks associated with large step changes.

The second dimension of complexity is that a wide and increasing range of fields of specialized knowledge are being mobilized to take artefacts from their initial design to their widespread use. This helps explain the importance of the *organizational* dimensions of technological activities.

- Within business firms, ensuring the assimilation of the relevant fields of technological knowledge, through linkages with the wider knowledge communities,

and the capacity within the firm to experiment and learn across cognitive and functional boundaries.
- In the regional, national and international environments, establishing competencies, institutions and incentive structures that ensure the development and diffusion of technological knowledge. Given the nature of the problem in hand, some of these will involve market institutions and transactions, but others will involve public funding and professional networks of development and exchange.

Combining both kinds of complexity – cognitive and organizational – defines two further properties of technological activities.

- The specialized and differentiated nature of corporate technological and related organizational knowledge, determined by their principal product markets. What business firms have been able to do in the past defines and constrains what they can hope to do in the future.
- The advantages of physical proximity in innovative activities that involve tacit knowledge, uncertainty, and coordinated experimentation across functional and disciplinary boundaries.

Thus, most of this book is about the cognitive and institutional dimensions of technological changes. I have very little to say about incentive structures, not because they are unimportant, but because others (mainly economists) have written extensively about them, and I have nothing new to add. Like Nelson (1990) and others before him, I am persuaded that innovation and change are difficult, painful and uncertain; and that they therefore thrive in general in a framework of pluralistic competition and experimentation. But I am similarly persuaded that even the most powerful incentive structures cannot overcome cognitive and organizational limits. I can offer the most entrepreneurial person aged over 18 $1 million if he or she proves able to speak Chinese fluently after one year. If he or she has never learned any other language than English since birth, I am pretty sure to win my bet, even if I increase it to $10 million.

Main findings
The book is divided into three sections reflected in its title: the nature of technological knowledge, the particular characteristics of the management of innovation (especially in large firms), and the systems of innovation within which the knowledge and the firms are embedded.

The nature of technology
Chapter 1 is based on my Inaugural Lecture as R.M. Phillips Professor of Science and Technology Policy at the University of Sussex in 1987.[13] I start with a theme that is now quite widely accepted, namely, that useful technological knowledge cannot be reduced either to scientific knowledge, or to freely available information. I explore the implications of these distinctions for a number of theoretical and practical dimensions for science and technology policy. For example, the output of basic research may be a 'public good' but it is not a 'free good', and research skills are

an important component of such output; all business firms need a problem-solving capacity, even if they are imitating rather than innovating; the separation of R&D from production in Soviet-style countries is harmful to technical progress.

Chapter 2[14] is an early attempt to systematize our understanding of intersectoral differences in the sources and directions of technical change, based on the major pioneering exercise in collecting data on significant British innovations, begun by Christopher Freeman in the early 1970s, and continued by Joe Townsend into the early 1980s. Case studies had shown clearly that patterns of technical change differed markedly amongst industries, in terms of the relative importance of firms of different sizes, of suppliers and customers, and of formal R&D activities. The challenge was therefore to develop a meaningful system of putting firms and sectors into reasonably homogeneous categories; otherwise the possibilities for learning from empirical study and experience would be severely limited. The new data enabled this to be done.[15] Amongst other things, the conclusions show that different types of sectors approximate to different models found in the economics literature, and that Adam Smith's threefold classification of the sources of technical change – producers and users of machinery, and 'philosophers' (that is, scientists) – is still a useful beginning.

The proposed taxonomic categories have held up reasonably well since then to independent, statistical validation.[16] In addition, they have been used in some scholars' analyses of structural change in national economies, emerging from intersectoral linkages and dynamic learning,[17] and they have been the basis for analysing differences in corporate innovation strategies between sectors (see Chapter 4, below). Their weakness is the high degree of variance still found within each category, and the fact that some firms can simultaneously belong to more than one of them. It also proved necessary to add a fifth category in the early 1990s, with the rapid and revolutionary growth of new sources of software technology in the services sector.[18]

One important and complex element in sectoral patterns of technical change is the link between producers and users of capital goods (including control instrumentation and software), where user firms often make important contributions to technological advance that they make available to their suppliers. One of Nathan Rosenberg's major contributions has been to explain these processes.[19] It is therefore appropriate that Chapter 3[20] was written for his *Festschrift*, since Pari Patel and I were able to confirm statistically one of his major insights into the development of the capital goods industry: namely, that small, specialized suppliers of capital goods often emerge through a process of *vertical disintegration* as a result of *technological convergence*, where experience in building and operating machinery in part of a production system in one sector (for example, boring a hole in a metal tube in gun-making) can be applied in a number of others (for example, making sewing machines). At the same time, we were able to show that R&D statistics are a very bad reflection of the continuing importance of improvements in mechanical technologies. We also found a puzzle that needed explaining: why do large firms in all sectors continue to make inventions and innovations in capital goods when they often do not produce them, and when the barriers to entry to do so are low? We suggest that the low barriers to entry enable large firms to make incremental and low-cost improvements in their production methods. A fuller explanation emerges in Chapter 6, below.

Managing the innovating firm

The five papers on managing the innovating firm build on a long-established SPRU tradition, and especially on its early work identifying factors distinguishing success from failure in the management of innovations.[21] Most economists have not been interested in conceptualizing the activities and processes inside firms that initiate and implement major changes and result in improvements in products, processes and organization. Even evolutionary economics has not got much beyond the conceptually important, but difficult to operationalize, notion of firm-specific 'routines'. The papers here were all written with the belief that new insights into the nature of the innovating firm – and into the problems of its management[22] – will emerge from our improving knowledge of the nature of technology, and from the wider range of quantitative and qualitative knowledge on the nature and sources of innovation in different industries.

Thus, Chapter 4[23] identifies the following key characteristics of innovative activities in large firms: differentiated and cumulative paths of development, uncertainty in outcomes, in an organizational context of professional and functional specialization. It concludes that management in even large firms are heavily constrained by their size and their principal activities in their choices of technology strategy, that continuous learning across professional and functional boundaries is essential, and that conventional methods of project appraisal and organization discourage effective innovation.

In Chapter 5,[24] Pari Patel and I use data on the patenting activities of the world's largest firms to confirm that firms are heavily constrained in their choice of innovative activities by the products that they make and their country of origin. There remains significant room for managerial choice and discretion in the *volume* of innovative activities. But the *directions* (that is, technological fields) of search – and the consequent ability to exploit promising new technological opportunities – are heavily constrained by what they do and what they know now. Contrary to a widely held view amongst evolutionary economists, large firms within a given sector show hardly any differences amongst themselves in their directions (that is, mix of fields) of technological search. Significant differences amongst firms in directions (that is, mix of fields) of search can only be observed *between* sectors.

Diversity in directions of search can be observed *within* large firms – not in the sense of differences in the *mix* of technological fields, but the *wide variety* of fields in which large firms are active. This is a central theme in Chapter 6,[25] written with Ove Granstrand and Pari Patel. Both aggregate patenting statistics and case studies of Ericsson, Rolls-Royce and the development of opto-electronics in Japanese firms all show that firms are more diversified in their technological activities than in their product range: for example, a much larger number of the world's largest firms are active in making inventions in computing technology than actually develop, make and sell computers.[26] Furthermore, this corporate technological diversity is increasing over time, together with the number of external alliances involving technological exchanges, and the increases in expenditures on R&D.

These trends go dead against much of the conventional management wisdom of the 1990s, which says that firms should concentrate their resources on their fields of advantage and outsource the rest. We explain these observed patterns of corporate behaviour by two factors: first, firms need to have a capacity to coordinate changes in their own products and processes, with changes made by their suppliers of machinery, components and subsystems; second, firms need to evaluate – and if necessary absorb and master – the increasing range of potentially useful technologies that are emerging over time. The main analytical and policy conclusion is that we should not confuse technologies with products.

These themes are taken up again in Chapter 7,[27] which explores more systematically the theoretical and practical implications of the evidence presented in Chapters 5 and 6. I argue that the key features of the large innovating firm are better understood through the increasing division of labour in knowledge production, as foreseen by Adam Smith, than through the creative destruction of established firms, as described by Joseph Schumpeter. The evidence shows that the R&D laboratories of contemporary large corporations are well able to identify, absorb and master emerging fields of technological knowledge. As a consequence, notions of 'competence-destruction' and of 'paradigm shifts' in technologies should be treated with caution. The main managerial difficulties arise in large corporations in matching organizational systems of coordination (for example, between R&D and other corporate functions) and control (for example, financial and strategic) with changing technological opportunities.

The themes addressed in Chapter 8[28] are less concerned with cognitive and organizational characteristics within firms than with the interactions between especially large firms and the national systems of innovation within which they are embedded. Pari Patel and I use the patenting activities of the world's largest firms to demonstrate the major differences between fields of technology in the degree of concentration of corporate technological activities (the two highest are in petroleum and semiconductors). We also show that, amongst the major OECD countries, international differences in the intensities and rates of increase of national innovative activities cannot be explained by the relative importance of large firms – whether national or foreign – in total innovative activities.

However, the major finding of the paper is that the innovative activities of the world's largest firms are amongst the least internationalized of their functions. Although the validity of this conclusion based on patent data was criticized at the time, it has since been corroborated almost exactly by an OECD study using R&D data.[29] The reasons for the domestic concentration of innovative activities were first identified by Vernon more than 30 years ago (1960, 1966): namely, the advantages of physical agglomeration in combining the specialized knowledge and other inputs necessary for the launching of major new products and production processes. Pari Patel (1995, 1996) has since extended and refined the analysis, without detecting any major shift, although we have recently argued that the privileged links between the innovative activities of large firms and the public science base of their home countries are increasingly under strain.[30]

Systems of innovation
The blind spot of many economists in ignoring what goes on inside innovating firms is often matched by the blind spot of many management specialists in ignoring what goes on outside. As we have already hinted above, managements are heavily constrained in their choices of innovation strategy by what have come to be called the

'systems of innovation' in which they are embedded. Chapter 9,[31] written with Mike Robson and Joe Townsend, confirms earlier work by industrial economists showing the importance of sectoral systems of innovation in determining the firm size distribution of innovative activities. It uses systematic data on significant innovations in the UK from 1945 to 1983 to confirm the dominance of large firms in the chemical and electrical–electronic industries shown by official statistics on R&D activities. However, it shows a much larger (about ten times greater) share of innovative activities in firms with fewer than 1,000 employees, principally in the mechanical and instruments industries where innovations emerge from activities not officially designated as R&D, either because they are described in the firm under a different name (for example, 'design'), or because they are undertaken informally or on a part-time basis. Sectors where small firms have relatively large shares of innovations also have relatively large shares from user sectors. The data also show a consistent reduction over time in the size (in terms of employment) of 'innovating units' (that is, divisions, strategic business units). Some commentators (but not the original authors) made much of an apparently significant reduction in the share of large firms from 1975 to 1983. A secondary analysis of the data by Tether et al. (1997) casts doubt on any such trend.

Chapter 10[32] was written for Chris Freeman's *Festschrift* and addresses his major contribution to our understanding of the emergence of new technological systems,[33] and especially of what was then called the micro-electronics revolution. Whilst acknowledging the revolutionary nature of such a technology, it reflects a certain scepticism about the extent of the 'discontinuities' and the 'creative destruction' generated by revolutionary technologies.[34] It concludes that 'creative destruction' is no more prevalent than 'creative accumulation', since revolutionary new technologies either grow out of old technologies, or must be combined with changes in them, in order to develop useful artefacts. Like much of the writing in the 1980s, the focus of attention is micro-electronics in manufacturing and not software in services where the real technological revolution is probably taking place (see Chapter 13, below).

Chapter 11,[35] written with Pari Patel, is an ambitious attempt to define, measure, explain and predict trends in the national systems of innovation of the advanced OECD countries that have such a strong influence on what even the largest domestically based firms undertake by way of innovative activities. It argues that the notion of 'national systems of innovation' usefully encompasses the national investments in 'intangible capital' that complement those in tangible capital as the foundations of national improvements in productivity and output. Proxy measures of intangible capital – R&D, patenting, education of the workforce – all show important international differences in the sectoral composition of technological activities, with no signs of convergence. We explain the international differences in the persistent sectoral patterns of strengths and weaknesses by differences in local inducement mechanisms; and in the volume and rate of increase of aggregate innovative activities by differences in systems of corporate governance (that is, in management and financial control), and in the level and composition of workforce education. We conclude that, given the international difference in investments in intangible capital and their causes, economic convergence amongst OECD countries is unlikely, and divergence is possible – even probable.

Since publication, these conclusions have been criticized by Verspagen (1996)[36] who argues that there is no convincing statistical evidence showing international divergence in technological activities. In addition, events have contradicted some of our predictions. The increasing lead of Germany and Japan over the United States, in business-funded R&D and international patenting, has gone into reverse, and the United States has replaced Japan amongst commentators as the exemplary economy for the future. And South Korea and Japan are now considered part of the 'East Asian Crisis' rather than the 'East Asian Miracle'. With the benefit of hindsight, our conclusions about future divergence were too strong. However, what we said about methods of measuring national systems of innovation, and about the causes and persistence of international differences is, in my view, still valid. And it is too soon to remove Germany, Japan and Korea from the list of possible exemplars of innovative activities for the future.

I next address another important dimension of 'systems of innovation', namely, the links between the largely privately funded applications-oriented technological activities performed by business firms, and the largely publicly funded activities in academic research. By the beginning of the 1990s, the public consensus that governments should fund academic research, and leave the choice of what to fund mainly to academic researchers was breaking down, for two reasons. First, the theoretical justification for public funding – that the output of academic research was costly-to-produce but costless-to-reproduce 'information' which should be made freely available to everybody – broke down in a multi-country world in which such information could in principle be used by 'free-rider' countries who spent nothing on producing it. Second, public expenditure in most OECD countries on basic research had reached a scale (0.2–0.4 per cent of GDP) where questions were inevitably being asked about its usefulness.

In Chapter 12,[37] I argue that the nature of the usefulness of academic research is widely misunderstood. The links with technological practice are more varied and complex than is often assumed, in particular involving the transfer of tacit knowledge through personal contacts and personal mobility. For this reason, the specific economic benefits of any particular piece of basic research are very difficult – if not impossible – to predict, but most are likely to be geographically localized in nature.[38]

In the final Chapter 13,[39] I return after ten years to one of the themes of Chapter 1, namely, that policy prescriptions depend on underlying assumptions. In the intervening period, the so-called 'new trade' and 'new growth' theories – each giving a central role to endogenously generated technical change in trade and growth – had become accepted in the economics mainstream. However, these theories still assumed that firms have perfect information and foresight, and that implementation is costless. They continue to ignore or neglect the compelling empirical evidence that knowledge is more than information, and that firms are organizations that generate learning and change, as well as allocating resources in response to market signals. I argue that these differences in assumptions lead to vastly different policy recommendations for the public support of basic research, and for the encouragement of corporate innovative activities, with less emphasis on incentives compared to competencies and institutions. I further argue that emphasis in quantitative and policy analysis on manufacturing R&D as the main or only source of innovations neglects the contributions to innovative activities of small firms, as well as the growing contributions made by firms in service sectors to the development of software technology.

Future directions of research

What do the above results suggest for future directions of research on the nature, sources and policy implications of technological change?

On the *nature of technology*, major contributions can be expected from what the cognitive sciences can tell us about experimentation and learning. The challenge will be to find useful linkages to the specific characteristics of experimentation and learning in technological change: in particular, the specialized nature of its component parts, institutional and functional boundaries, professional networks, and the growing importance of Information and Communication Technology.

Improved understanding will also depend on progress in the *measurement* of technological activities. One danger is that the improvements in databases on scientific and technological activities will greatly increase the emphasis given to the professionally safe option in academic careers: namely, regression analysis to test established theories. This would be a pity, since – as was pointed out at the begining of this Introduction – new theories emerge from new data. In this context, old challenges remain, especially in finding systematic measurements of deliberate technological learning processes in firms and countries behind the world technological frontier. New challenges have emerged: exploring the influence of firm-level indicators of technological activity on corporate performance; and measuring the increasingly important software technology-generating activities of firms in service sectors.[40] And major opportunities for improving our understanding of professional networks are emerging from advances in bibliometric methods and data.

In the *management of innovation*, we still need a more precise and practical understanding of both the firm-specific 'routines' associated with innovation, and the factors determining what might be called the 'knowledge boundaries' of the firm, given that products and firms are incorporating an increasing range of technologies. Similarly, the important notion of the 'coevolution of technology and organization' needs a firmer conceptual and empirical basis, especially in the definition and measurement of organizational change.

Two dimensions of *systems of innovation* also can expect much attention. The first is the effects of ICT and biotechnology on technological systems, where there has been some success already in establishing multi-disciplinary research programmes. The second is the contrast and comparison of different national systems of innovation, where public discussion too easily gets entangled with ideological and nationalist competition between different forms of capitalism.[41] There is here a crying need for careful and detached empirical research to answer the following type of question. To what extent are the US achievements in Silicon Valley and biotechnology the products of flexible private institutions, or of massive public investments in the underlying technologies? Is the lag in German performance in software and biotechnology, compared to the United States, increasing or declining? Are the Japanese institutions and practices that allowed rapid technological catch-up inappropriate for future development?

Links with theory and practice

As pointed out at the beginning, the largely empirically based research agenda discussed above has certainly benefited from advances in the evolutionary theory of economic change. However, its effects on developments in evolutionary theory itself have been diminishing over time, probably because practical problem-solving has been displaced by formal theory and model-building as the main drivers in its development: hence the ever-present danger of the continuing co-existence of 'numbers without theory' with 'theory without numbers'.

Its influence on private and public policy is difficult to assess because – as with academic research in the natural sciences and engineering – such influence is often diffuse and roundabout. One is therefore forced to speculation and anecdote. On the former, we might expect the incorporation of research findings into management textbooks and journals to have some influence on management practice. On the latter, our analyses of practical benefits of academic research have led public policymakers to give greater weight to the important contribution of research training.

Similarly, public policymakers increasingly acknowledge the importance of technology and innovative activities for national economic performance. But in debating and devising policies, the emphasis remains on the incentive structures that are central to mainstream economics, rather than on the competencies and the institutions that embody them, which are central to evolutionary economics. This will change only when public officials are more familiar with evolutionary economics.

Notes

1. Many of these studies are embodied in the latest version of *The Economics of Industrial Innovation* (Freeman and Soete, 1997).
2. And still relevant things – see, for example, Chapters 1, 2 and 7 of this book.
3. See, for example, Rosenberg (1963, 1974).
4. Although SPRU staff and doctoral students have made major contributions to understanding the effects of technological activities on national economic performance. See in particular, Soete (1981), Fagerberg (1987, 1988) and Dosi et al. (1990).
5. Patel and Pavitt (1987b); Pavitt (1980, 1992, 1993a, 1993b, 1994, 1995, 1996b).
6. Pavitt and Walker (1976).
7. Hanson and Pavitt (1987); Pavitt (1997a).
8. Bell and Pavitt (1993).
9. Patel and Pavitt (1987a); Pavitt (1972, 1998c); Sharp and Pavitt (1993); Robson et al. (1998).
10. Patel and Pavitt (1996); Pavitt (1982, 1988, 1998a); Pavitt and Patel (1988).
11. See, for example, Soete (1981) and Fagerberg (1987, 1988) who made pioneering use of patent statistics to demonstrate the importance of investments in technological activities in explaining international differences in productivity and export performance; and Narin et al. (1997) similarly used citations in patents to confirm the relevance of high quality, publicly funded academic research to practical applications.
12. Some writers argue that advances in theoretical understanding, coupled with those in information technology, are considerably strengthening the predictive links from basic understanding to technological practice (Ergas, 1994; Arora and Gambadella, 1994; Cowan and Foray, 1997). Others argue that such advances mainly improve the efficiency of prototyping activities (Nightingale, 1997). My position is that, in situations of complexity as defined above, theoretical understanding will never be a sufficient guide to technological practice.
13. Originally published in a revised form as Pavitt (1987).
14. Originally published as Pavitt (1984).
15. A similar exercise at the time was undertaken as the so-called 'Yale Survey', based on a detailed and systematic survey of the experience of R&D directors in large US corporations (see Klevorick et al., 1995). The results were consistent with those of Chapter 3.
16. See, for example, Cesaretto and Mangano (1992), Evangelista et al., (1997) and Arundel et al. (1995).
17. See Guerrieri and Tylecote (1994) and Laursen (1996).

18. See Chapters 4 and 13, below.
19. See Rosenberg (1963). Eric von Hippel (1988) has also made major contributions to our understanding.
20. Originally published as Patel and Pavitt (1994a).
21. See, for example, Rothwell et al. (1974).
22. See, for example, Tidd et al. (1997).
23. Originally published as Pavitt (1990).
24. Originally published as Patel and Pavitt (1997).
25. Originally published as Granstrand et al. (1997).
26. An earlier analysis of less detailed data on significant innovations in the UK suggested that this might be the case. See Pavitt et al. (1989). See also Gambardella and Torrisi (1998).
27. Originally published as Pavitt (1998b).
28. Originally published as Patel and Pavitt (1991).
29. See OECD (1997).
30. See Patel and Pavitt (1999).
31. Originally published as Pavitt et al. (1987).
32. Originally published as Pavitt (1986).
33. Sometimes called 'technological family', 'technological system', 'technological paradigm'. See Soete (1986).
34. See also Chapters 4, 6 and 7 in this volume.
35. Originally published as Patel and Pavitt (1994b).
36. See also our response (Pavitt and Patel, 1996).
37. Originally published as Pavitt (1991).
38. On this basis, I formulated a revised case and guidelines for the public funding of basic research (Pavitt, 1995, 1997b). I also argued – without success – against the UK government's 'Foresight' programme (Pavitt, 1992, 1993a, 1993b, 1994), but – perhaps with more success – in favour of continuing, large-scale public expenditure on academic research (Pavitt, 1996b and Kealey, 1996a, 1996b. See, also, David, 1997).
39. Originally published as Pavitt (1996a).
40. The unsatisfactory rate of progress in collecting systematic data on software-generating activities suggests that academics and statisticians are like military generals, in being better at dealing with the technologies of the past, rather than with those of the future.
41. See Albert (1992).

References

Albert, M. (1992), *Capitalism against Capitalism*, Whurr, London.
Arora, A. and A. Gambardella (1994), 'The changing technology of technological change: general and abstract and the division of innovative labour', *Research Policy*, **23**: 523–32.
Arundel, A., G. van de Paal and L. Soete (1995), *Innovation Strategies of Europe's Largest Industrial Firms* (PACE Report), MERIT, University of Limbourg, Maastricht.
Bell, M. and K. Pavitt (1993), 'Technological accumulation and industrial growth: contrasts between developed and developing countries', *Industrial and Corporate Change*, **2**: 157–210.
Cesaretto, S. and S. Mangano (1992), 'Technological profiles and economic performance in the Italian manufacturing sector', *Economics of Innovation and New Technology*, **2**: 237–56.
Cowan, R. and D. Foray (1997), 'The economics of codification and the diffusion of knowledge', *Industrial and Corporate Change*, **6**: 595–622.
David, P. (1997), 'From market magic to calypso science policy. A review of Terence Kealey's *The Economic Laws of Scientific Research*', *Research Policy*, **26**: 229–55.
Dosi, G., K. Pavitt and L. Soete (1990), *The Economics of Innovation and International Trade*, Wheatsheaf, Hemel Hempstead.
Ergas, H. (1994), 'The new face of technological change and some of its consequences' (mimeo), Center for Science and International Affairs, Harvard University, 15 March.
Evangelista, R., G. Perani, F. Rapiti and D. Archibugi (1997), 'Nature and impact of innovation in manufacturing industry: some evidence from the Italian innovation survey', *Research Policy*, **26**: 521–36.
Fagerberg, J. (1987), 'A technology gap approach to why growth rates differ', in C. Freeman (ed.), *Output Measurement in Science and Technology: Essays in Honour of Y. Fabian*, North-Holland, Amsterdam, 33–45.
Fagerberg, J. (1988), 'International competitiveness', *Economic Journal*, **98**: 355–74.
Freeman, C. and L. Soete (1997), *The Economics of Industrial Innovation*, Pinter, London.
Gambardella, A. and S. Torrisi (1998), 'Does technological convergence imply convergence in markets? Evidence from the electronics industry', *Research Policy*, **27**: 445–63.

Granstrand, O., P. Patel and K. Pavitt (1997), 'Multi-technology corporations: why they have "distributed" rather than "distinctive core" competencies', *California Management Review*, **39**, Summer: 8–25.

Guerrieri, P. and A. Tylecote (1994), 'National competitive advantages and microeconomic behaviour', *Economics of Innovation and New Technology*, **3**: 49–76.

Hanson, P. and K. Pavitt (1987), *The Comparative Economics of Research Innovation in East and West: A Survey*, Vol. 25, *Fundamentals of Pure and Applied Economics* (eds J.M. Montias and J. Kornai), Harwood Academic Publishers, Chur, London, Paris, New York, Melbourne.

Hippel, E. von (1988), *The Sources of Innovation*, Oxford University Press, Oxford.

Kealey, T. (1996a), *The Economic Laws of Scientific Research*, Macmillan.

Kealey, T. (1996b), 'You've got it all wrong', *New Scientist*, 29 June, 23–6.

Klevorick, A.K., R. Levin, R. Nelson and S. Winter (1995), 'On the sources and significance of inter-industry differences in technological opportunities', *Research Policy*, **2**: 185–205.

Laursen, K. (1996), 'Horizontal diversification in the Danish national system of innovation: the case of pharmaceuticals', *Research Policy*, **25**: 1121–37.

Narin, F., K. Hamilton and D. Olivastro (1997), 'The increasing linkage between US technology and public science', *Research Policy*, **26**: 317–30.

Nelson, R. (1990), 'Capitalism as an engine of progress', *Research Policy*, **19**: 193–214.

Nelson, R. and S. Winter (1982), *An Evolutionary Theory of Economic Change*, Belknap, Cambridge, MA.

Nightingale, P. (1997), 'Knowledge and technical change: corporate simulations and the changing innovation process', DPhil Thesis, University of Sussex.

OECD (1997), *Internationalisation of Industrial R&D: Patterns and Trends*, DSTI/IND/STP/SWP/NESTI(97)2, Paris, October.

Patel, P. (1995), 'The localised production of global technology', *Cambridge Journal of Economics*, **19**: 141–53.

Patel, P. (1996), 'Are large firms internationalizing the generation of technology? Some new evidence', *IEEE Transactions on Engineering Management*, **43**: 41–7.

Patel, P. and K. Pavitt (1987a), 'Is Western Europe losing the technological race?', *Research Policy*, **16**: 5–32.

Patel, P. and K. Pavitt (1987b), 'The elements of British technological competitiveness', *National Institute Economic Review*, **4** (122): 72–83.

Patel, P. and K. Pavitt (1991), 'Large firms in the production of the world's technology: an important case of non-globalisation', *Journal of International Business Studies*, **22**: 1–21.

Patel, P. and K. Pavitt (1994a), 'The continuing, widespread (and neglected) importance of improvements in mechanical technologies', *Research Policy*, **23**: 533–46.

Patel, P. and K. Pavitt (1994b), 'Uneven (and divergent) technological accumulation among advanced countries', *Industrial and Corporate Change*, **3**: 759–87.

Patel, P. and K. Pavitt (1996), 'Patterns of technological activity: their measurement and performance', in P. Stoneman (ed.), *Handbook of the Economics of Innovation*, Blackwell, Oxford: 14–51.

Patel, P. and K. Pavitt (1997), 'The technological competencies of the world's largest firms: complex and path-dependent, but not much variety', *Research Policy*, **26**: 141–56.

Patel, P. and K. Pavitt (1999), 'National systems of innovation under strain: the internationalisation of corporate R&D', in R. Barrel, G. Mason and M. Mahony (eds), *Productivity, Innovation and Economic Performance*, Cambridge University Press, Cambridge (forthcoming).

Pavitt, K. (1972), 'Technology in Europe's future', *Research Policy*, **1**: 210–73.

Pavitt, K. (ed.) (1980), *Technical Innovation and British Economic Performance*, Macmillan, Basingstoke.

Pavitt, K. (1982), 'R and D patenting and innovating activities: a statistical exploration', *Research Policy*, **11**: 33–51.

Pavitt, K. (1984), 'Sectoral patterns of technical change: towards a taxonomy and a theory', *Research Policy*, **13**: 343–73.

Pavitt, K. (1986), 'Chips and trajectories: how does the semiconductor influence the sources and directions of technical change?', in R. Macleod (ed.), *Technology and the Human Prospect*, Frances Pinter, London: 31–54.

Pavitt, K. (1987), 'The objectives of technology policy', *Science and Public Policy*, **14**, August: 182–8.

Pavitt, K. (1988). 'Uses and abuses of patent statistics', in A.F.J. van Raan (ed.), *Handbook of Quantitative Studies of Science and Technology*, Elsevier, Amsterdam: 509–36.

Pavitt, K. (1990), 'What we know about the strategic management of technology', *California Management Review*, **32**, Spring: 3–26.

Pavitt, K. (1991), 'What makes basic research economically useful?', *Research Policy*, **20**: 109–19.

Pavitt, K. (1992), 'Britain's excellence in basic research is no longer guaranteed', *Times Higher Education Supplement*, 23 October, p.19.

Pavitt, K. (1993a), 'Back to basic', *Science and Public Affairs*, Winter: 9–10.

Pavitt, K. (1993b), 'Viewpoint: what's always missing in UK Government White Papers on Science and Technology', *Social Sciences*, Issue 20, ESRC.

Pavitt, K. (1994), 'Try business class research world-wide', *Times Higher Educational Supplement*, 18 November, p. 140.

Pavitt, K. (1995), 'Backing basics: basic research should not just depend on what industry needs now', *New Economy*, **2** (2): 71–4.

Pavitt, K. (1996a), 'National policies for technical change: where are the increasing returns to economic research?', *Proceedings of the National Academy of Sciences*, Issue 23, 12 November, Washington DC.

Pavitt, K. (1996b), 'Road to ruin', *New Scientist*, 3 August, **2041**: 32–5.

Pavitt, K. (1997a), 'Transforming centrally planned systems of science and technology: the problem of obsolete competencies', in D. Dyker, *The Technology of Transition*, Central European Press, Budapest, 43–60.

Pavitt, K. (1997b), 'Academic research, technical change and government policy', in J. Krige and D. Pestre (eds), *Science in the Twentieth Century*, Harwood Academic, Amsterdam, 143–58.

Pavitt, K. (1998a), 'Do patents reflect the useful research output of universities?', *Research Evaluation*, **7**: 105–11.

Pavitt, K. (1998b), 'Technologies, products and organisation in the innovating firm: what Adam Smith tells us and Joseph Schumpeter doesn't', *Industrial and Corporate Change*, **7** (3): 433–52.

Pavitt, K. (1998c), 'The inevitable limits of EU R&D funding', *Research Policy*, **27**, 6: 559–68.

Pavitt, K. and P. Patel (1988), 'The international distribution and determinants of technology activities', *Oxford Review of Economic Policy*, **4**: 35–55.

Pavitt, K. and P. Patel (1996), 'On uneven technological development amongst countries: a response to Bart Verspagen', *Industrial and Corporate Change*, **5**: 903–4.

Pavitt, K., M. Robson and J. Townsend (1987), 'The size distribution of innovating firms in the UK: 1945–1983', *The Journal of Industrial Economics*, **XXXV**, March: 297–316.

Pavitt, K., M. Robson and J. Townsend (1989), 'Technological accumulation, diversification and organisation in UK companies, 1945–83', *Management Science*, **35**: 81–99.

Pavitt, K. and W. Walker (1976), 'Government policies towards industrial innovation: a review', *Research Policy*, **5**: 11–97.

Price, D de Solla (1984), 'The science/technology relationship: the craft of experimental science and policy for the improvement of high technology innovation', *Research Policy*, **13**: 3–20.

Robson, M., K. Pavitt and J. Townsend (1988), 'Sectoral patterns of production and use of innovations in the UK: 1945–1983', *Research Policy*, **17**: 1–14.

Rosenberg, N. (1963), 'Technological change in the machine tool industry, 1840–1910', *Journal of Economic History*, **23**: 414–46.

Rosenberg, N. (1974), 'Science, invention and economic growth', *Economic Journal*, **84**: 333.

Rothwell R., C. Freeman, A. Horsley, P. Jervis, A. Robertson and J. Townsend (1974), ' SAPPHO updated – project SAPPHO phase II', *Research Policy*, **3**: 258–91.

Sharp, M. and K. Pavitt (1993), 'Technology policy in the 1990s: old trends and new realities', *Journal of Common Market Studies*, **31**: 129–51.

Soete, L. (1981), 'A general test of the technological gap trade theory', *Weltwirtschaftliches Archiv*, **117**: 638–60.

Soete, L. (1986), 'Technological innovation and long waves: an inquiry into the nature and wealth of Christopher Freeman's thinking', in R.Macleod (ed.), *Technology and the Human Prospect*, Frances Pinter, London: 214–38.

Tether, B., I. Smith and A. Thwaites (1997), 'Smaller enterprises and innovation in the UK: the SPRU innovations database revisited', *Research Policy*, **26**: 19–32.

Tidd, J., J. Bessant and K. Pavitt (1997), *Managing Innovation: Integrating Technological, Market and Organizational Change*, Wiley, Chichester.

Verspagen, B. (1996), 'On technological convergence amongst countries: a comment on Patel and Pavitt', *Industrial and Corporate Change*, **5**: 897–901.

Vernon, R. (1960), *Metropolis, 1985*, Harvard University Press, Cambridge, MA.

Vernon, R. (1966), 'International investment and international trade in the product cycle', *Quarterly Journal of Economics*, **80**: 190–207.

PART ONE

TECHNOLOGICAL
KNOWLEDGE

[1]

The nature of technology
Keith Pavitt

It is not useful, either in theory or in practice, to equate technology with science or with information. Many empirical studies show that most technology is specific, complex, partly tacit, and cumulative in its development. As a consequence, key questions for analysis and policy are, first, the nature and the extent of the interactions between science and technology; and, second, the accumulation in firms of technological knowledge, necessary not only for new and original inventions and applications, but also for effective assimilation of science and technology developed elsewhere.

I shall be using the term 'technology' to encompass both physical artefacts themselves, and the person-embodied knowledge to develop, operate and improve them.

The continuous development and diffusion of technology is widely recognized to be one of the key distinguishing features of modern and modernizing societies. It has become central to major political and moral debates: about war and peace, about life and death, and about the future survival of the species and the planet.

I shall address an issue which is central to most of these debates: namely, the development and diffusion of often rather prosaic technologies in industry, agriculture and services; technologies associated with economic and social change, with international competition, and with increases in measured gross national product per head. Such a focus inevitably reflects my values which, for what it is worth, are that continuous improvements in the quality of life depend – amongst other important things – on keeping up with, or getting closer to, world best practice in these technologies; and that the experience of the UK over the past 25 years is an ample justification for this position.

Fortunately, the relevance of what I shall have to say does not require agreement with this view. But I suspect that we nearly all would agree that effective action to guide technology towards whatever objectives we consider desirable requires an accurate understanding of its nature, sources and determinants.

There are perhaps two groups who might argue that such understanding is not necessary. The first consists of the so-called technological determinists, who often

This is an edited version of Professor Keith Pavitt's inaugural professorial lecture, delivered at the University of Sussex on 23 June 1987.

He would like to thank Richard Nelson and Geoffrey Oldham for comments and suggestions which helped greatly in preparation.

argue or assume that technology develops and diffuses strictly according to its own logic, and that society can (and does) simply adapt to its requirements. In places like Sussex University, technological determinists are generally in bad odour, and considered to be intellectually and morally suspect. I would simply add that their arguments fail the empirical test, since

- a high proportion of developed technology does not get diffused but is rejected on economic and social grounds,
- much technology is continually being adapted in the light of economic and social constraints, and
- any given technology allows some variations in organizational forms in exploiting it.

There is also a second group that might not see much point in understanding the nature and determinants of technology. Its members would tend to argue that both the rate and the direction of technological change are very sensitive and responsive to economic and social signals, be they market prices or planning targets. Provided that markets, or plans, or social and political relations more broadly, are got right, the appropriate technologies will inevitably follow. I suspect that these (what might be called) 'socio-economic' determinists are in fact more important in the UK in influencing action and attitudes than their technological counterparts. In their reasoning, the development and diffusion of technology are assumed to be cheap, whereas they are not.

In any event, since technology is not strictly determined by either of its dynamic, or by economic and social forces, wise action to influence it depends, amongst other things, on the degree of understanding of how these and other factors interact. Results emerging from empirical research are adding to this understanding, and are influencing policy. I believe that they should also influence underlying theory, models and their assumptions. Keynes' comforting dictum that practical men are influenced by academic scribblers is one persuasive reason why academics should get it right. Another is that academics might ultimately be ignored if they get it wrong; being ignored is not a healthy state for an academic community.

In this context, I shall argue that assumptions equating technology with either science or information are misleading, and can result in unbalanced or inaccurate policy prescriptions. My concerns are not particularly original. They were expressed in the 1960s by Derek de Solla Price (1965), in the 1970s by Nathan Rosenberg (1976), and in the 1980s by David Mowery (1983).

Science and technology
Analysis of the links between science and technology goes back a long time. In Chapter 1 of *The Wealth of Nations*, Adam Smith identifies 'philosophers or men of speculation' as one of the three major sources of technology, together with both producers and users of machinery – a categorization that still has its relevance today.

In the 1830s, Alexis de Tocqueville predicted a rosy and growing future for science in the modernizing society that he observed in the USA. Its application would create considerable opportunities for profit, so that business demand would grow for applied scientists and for the institutions that trained them. Basic research, unconstrained by preoccupations of immediate application, would be necessary if profitable opportunities for application were not to dry up.

Events since then have on the whole confirmed de Tocqueville's prediction: the relative contribution of science to technology has been increasing. For example, Christopher Freeman and his colleagues (1982) have shown the pervasive influence across all sectors of the economy of the diffusion of technologies growing out of basic research in chemistry and physics. Rosenberg (1985) has described the economically important economic contributions made by the application of rather elementary chemistry and other sciences in sectors like steel and food processing. Numerous writers have also shown the major contributions made by technology to science, mainly through improved or radically new instrumentation.

With the growth of industrial research and development (R&D) departments in the 20th century, and the large-scale recruitment of university trained scientists by them, the debate has become more complicated. When scientists working in US General Electric, in Dupont, in the Bell laboratories or in EMI, win Nobel Prizes, is the distinction between science and technology useful any more?

Such events help explain why some social scientists have recently questioned the analytical usefulness of distinguishing between the context of science and of technology. Thus, it has been suggested that the analytical tools of the sociology of science can readily be transferred to technology (Pinch and Bijker, 1984, 1986). Two distinguished economists, Partha Dasgupta and Paul David (1986), have argued that the essential difference between science and technology is that the former produces public and published knowledge, whilst the latter produces private and often unpublished knowledge. They thereby implicitly define science as what goes on in universities and related institutions, and technology as what goes on in business firms.

Similar definitions underlie most of the studies to which I shall refer later, as well as the recent UK policy report, entitled *The Science Base and Industry*, published jointly by the Advisory Council for Applied Research and Development (ACARD) and the Advisory Board for the Research Councils (ABRC). Dasgupta and David's is a useful definition, although it leaves academic engineering in an ambiguous position (that I return to later).

However, Dasgupta and David also assert that much of the content of science and of technology has become indistinguishable, and they cite molecular biology, biochemistry and solid state physics as examples. By doing so, I think that they have fallen into a trap, common in studies of science and technology policy, of generalizing from the particular, in what are in fact very heterogeneous activities. This was understandable when science and technology policy studies were mainly of particular cases, but has become less so with the accumulation of this case material, and (perhaps more important) with the development of a variety of statistical data banks on various aspects of scientific and technological activity.

These show that the content of science is in general different from that of technology, that the nature and extent of interaction between the two varies considerably across industrial sector, and that these interactions are important subjects for analysis and policy.

To begin with, aggregate statistics for countries in Western Europe and North America all show very different types of activities being carried out in universities and in business firms. In universities 90 per cent of research activity is defined as basic and applied research: in other words, knowledge aiming at general applicability. In the firms, up to a quarter is also basic and applied research, but three-quarters or more is the development and testing of prototypes and related production systems: in other words, knowledge related to the specific products and the specific production processes that firms eventually hope to commercialize.

Clearly, the significant proportion of basic and applied research undertaken in industrial firms can and does sometimes lead to very fundamental results, and in some sectors the contents of science and of technology may well be very similar. Whether and where they are similar can now be assessed relatively systematically through the examination of citations in patent documents, which are the most extensive public record of codified technology. Patents citing other patents reflect technology building on technology: patents citing journals reflect technology building on science.

This field of analysis has been pioneered largely by Francis Narin and his colleagues in the USA (Carpenter, 1983; Carpenter et al., 1981; Narin, 1982; Narin and Noma, 1986). The main conclusions emerging from their analysis is that technology builds largely on technology, but that the rate of interaction with science varies considerably amongst scientific fields and amongst technologies.

Between biotechnology patents and biomedical research, the links between the two are at present very strong, with the former using scientific results just as up to date as the latter – and a recent case study of a development in monoclonal antibodies comes to a very similar conclusion (Mackenzie et al., 1987). In this type of sector, at least, the assumption that science and technology are very close may well be correct, and concern about the private encroaching on the public, to the long-term detriment of both, entirely justified.

However, biotechnology may well be the limiting case, since a comprehensive survey of citation patterns shows enormous variations amongst technologies in 'science dependence', as measured by the frequency of patent citations to journal papers: on average, a pharmaceutical patent cited journals 36 times more frequently than a patent in transportation. Chemical and biochemical patents are the most frequent in their citations of journals, the majority of which relate to basic research. Electronics and electrical patents are the next most frequent, but with a higher proportion of citations in journals reporting the results of applied research and engineering. Machinery patents have most of their citations to applied engineering, whilst instrumentation patents cite over the whole range of journals.

As might be expected, chemical patents refer mainly to journals in chemistry, biology and medicine; electrical and electronics patents to journals in physics and engineering; and mechanical patents to those in engineering.

Similar differences amongst industrial sectors and scientific fields emerge from a recent survey of 650 directors of industrial R&D in the USA (Nelson, 1986; Levin et al., 1984). They expected a higher proportion of biochemical and chemical knowledge emerging from academic research to find direct applications, than of knowledge from other disciplines.

However, the survey also found that, across a wide range of industries, the perceived value of universities was less for the content of their research, than for the training that it gave to future industrial researchers. This pattern is reflected in a Science Policy Research Unit (SPRU) study of British radio-astronomy. It concluded that the main economic benefit of such research has been the skills that it has given to Doctoral and Masters students who eventually work in industry: in particular, skills in computing and electronics, and in the ability to define complex problems, to communicate effectively and to work in a team (Irvine and Martin, 1980).

Another important benefit of such postgraduate training emerges from other studies of the information sources used by technical problem-solvers in firms (Gibbons and Johnson, 1974; Rothwell et al., 1974; Allen, 1977). In nearly all cases, the sources of such information are varied (a point that I shall return to later). What the university-trained scientist or engineer brings to technical problem-solving is not just substantive and methodological skills, but also the rich and informal network of professional contacts which can be called upon to help to solve problems: science and engineering graduates involved in application are part of a larger intelligence system, involving their former teachers and colleagues.

Given these characteristics, the relations between the differentiated and interrelated systems of science and technology are bound to be of concern to policy-makers at a number of levels. In the British context, they point to the following conclusions.

First, the arguments that Britain could usefully carry a smaller 'burden' of the world's freely available scientific knowledge begins to look as thread-bare and wrong-headed as those earlier arguments about the white man's imperial burden (which Britain was sometimes said to be carrying for the benefit of the world). As we have seen, the world's basic research cannot be applied by users without costs, comprising the costs to firms of employing graduate scientists and engineers, and the costs to governments of providing the academic infrastructure, including post-graduate training and research. If a government decides to run down the infrastructure, industrial firms will have to provide it themselves or, as hinted by the recently retired chairman of ICI (*The Economist*, 18 April 1987), they will move their core activities to places where an adequate infrastructure is provided.

Second, any policy must recognize that the nature of the complementarities between science and technology varies considerably amongst sectors of application, in terms of the direct usefulness of academic research results, and of the relative importance attached to such results and to training. In addition to direct transfers of knowledge, any evaluation of the science base for industry should include academic engineering, research training, and an assessment of the effectiveness of informal networks between academic research and places of application.

These networks can probably best be traced and assessed through the employment patterns of scientists and engineers outside academia. In the short term, this will be difficult to do, given the poor state into which British statistics on scientists and engineers have fallen since the late 1960s. In the meantime, we should be asking ourselves whether research funded by the Research Councils is sufficiently linked to postgraduate training, since such links are essential for the formation of these informal networks (Hague, 1986).

Third, we should be aware that the complementarities of the science and technology systems mean that the efficiency of the whole does not necessarily result from making the science system more like the technology system. There has been considerable pressure in the past few years to make the British science system more applied in its objectives. This should not be allowed to be taken too far. As we have seen, de Tocqueville rightly predicted a growing division of labour and inter-dependence between basic research and training, on the one hand, and technological activities on the other. He also predicted that neglect of the former would eventually destroy the latter.

Finally, we should be aware that some British science has been very relevant to application. In a paper recently presented at the British Government's Department of Industry, Narin and a colleague (1987) identify the most frequently cited US patents of British origin in the period from 1975 to 1982. Since frequency of citation in other patents turns out to be a pretty good indicator of an invention's usefulness, it is relevant to note that the top patent by far, with nearly 100 citations, was not granted to a British firm, but to the National Research Development Corporation (now called the British Technology Group), and resulted from publicly-funded research by the Agricultural Research Council into synthetic pyrethrin insecticides (Barclay and Fottit, 1987; Davies, 1980).

It is also relevant that the second most frequently cited British patent was for work on liquid crystal materials and devices carried out at Hull University.

British-based firms have been active in converting these examples of relevant science into commercial technology. Unfortunately, there is more general evidence that the relevance of British science is often better perceived by foreign firms. There are at least two cases on the Sussex campus – my home ground – where foreign firms have been more active in the commercial exploitation of the University's academic research than British firms. The Medical Research Council has gone on record to regret the lack of prominence of British firms, compared to their foreign counterparts, in exploiting the results of the Council's research in scientific instrumentation and specialized patient care (Select Committee on Science and Technology, 1983).

In other words, the British problem should be seen not as 'too much science' but as 'too little technology', a point to which I shall return later.

Experience and doing

Given the central importance of technology in firms for the use of outputs from the science base, it is necessary to delve more deeply into the nature of technology.

One influential tradition of analysis assumes that technology has the properties of information, in being much more costly to produce than to transmit and to use. Often related to this has been the assumption that technological knowledge can be more or less completely codified in the form of patents, blueprints, operating manuals, and the like.

Empirical analysis suggests that neither of these assumptions are true: most technology is specific, complex, often tacit, and cumulative in its development. These characteristics have important implications for the ways in which national and international systems for the development, the evaluation and the diffusion of technology can and do work.

In market economies, technology is specific in two senses: it is specific to firms, where most technological activity is carried out; and it is specific to products and processes, since most expenditure is not on research, but on development and production engineering, after which knowledge is also accumulated through experience in production and use – or what has come to be known as 'learning by doing' and 'learning by using'.

This combination of activities reflects the essentially pragmatic nature of most technological knowledge. Although a useful input, theory is rarely sufficiently robust to predict the performance of a technological artefact under operating conditions, and with a high enough degree of certainty, to eliminate the costly and time-consuming construction and testing of prototypes and pilot plant.

One of the oldest technologies – civil engineering – still has difficulties in foreseeing all important contingencies on the basis of past experience, design studies and computer simulations. And even the most science-based and science-dependent of all technologies – pharmaceuticals and pesticides – was described recently in *The Economist* as 'a highly empirical business' (1 February 1987), involving the development and screening of a vast range of synthetic compounds, the full range of biological effects of which cannot be completely predicted from knowledge of their molecular structure.

The problem is even more severe in the design and development of complex machinery and production systems, involving multiple objectives and multiple constraints, and the combination of a variety of technologies and materials. In this context, the essence of engineering skill and the engineering profession is the ability to make things work, by drawing upon and combining technology from a variety of sources, which is presumably why academic engineering teaches the range of subjects that it does.

This inherent complexity of technology has implications for both its development and its diffusion. The first is uncertainty and the inability to make an early and accurate assessment of either the performance or the utility of an innovation. Predictions made before development turn out to be unerringly inaccurate; and after initial development, there are often considerable possibilities for improvement, as a result of learning from production and use. Amongst competing innovations, it is therefore often difficult to know precisely when a technological race has ended, and (without much hindsight) who has won or lost it.

The second implication is that the assimilation of technologies developed in other organizations (or countries) is rarely costless, because the specificity of technology means that assimilating organizations have to undertake technological modifications and obtain additional knowledge, if they are to make it work. This is well illustrated in the detailed analysis by the Economic Council of Canada (de Melto et al., 1980) of the characteristics and costs of more than 200 innovations commercialized in Canada, and about a quarter of which involved agreements to assimilate technology from other firms, often in other countries. The magnitude of licence payments by users to firms supplying technology were in fact much smaller than the sums spent by the user firms themselves on technological activities.

The study concludes as follows 'the importation of technology by agreement or arrangement with other firms . . . acts as a strong substitute for research spending but has little effect on the relative proportion of development spending required to launch the . . . innovations . . .' (p. 35).

The relatively high cost and long time, necessary in most sectors for imitators to imitate, is reflected in innovating firms' assessment and use of the patent system as a barrier to imitation. Although patents are sought by firms in most industries, they are considered as the most important barrier against imitation only in very few, particularly in fine chemicals, where product innovations are expensive to develop, but often cheap to replicate. In most other industries, the costs of imitation are in themselves a sufficient barrier against imitation to justify the costs and risks incurred by the innovating firms themselves (Bertin and Wyatt, 1986; Levin et al., 1984; Mansfield et al., 1981; Wyatt, 1985).

In other words, there are very few free technological lunches. Even borrowers of technology must have their own skills, and make their own expenditures on development and production engineering; they cannot treat technology developed elsewhere as a free, or even a very cheap, good. The lessons for both developed and developing countries are obvious. And like his predecessors, Mr Gorbachev is no doubt pondering the effects of the continuing separation in the USSR of most development and production engineering competence from operating firms, on the Soviet Union's lack of technological dynamism (Hanson and Pavitt, 1987).

The second consequence of the complexity of technology is paradoxical: although (as we have seen) imitation is rarely cheap, it can only equally rarely be prevented completely. Of course, innovating firms are always trying to protect their lead, through patent protection, through secrecy, and/or through a variety of other means.

However, a number of empirical studies show that they hardly ever succeed. Precisely because technology is complex, and therefore difficult to define completely and precisely, it is possible to invent round existing patent protection, and firms that want to generally succeed in doing so pretty quickly: according to an American study, about 60 per cent of them within four years (Mansfield et al., 1981). Furthermore, since most new technology is embodied in products, one major method of imitation – much neglected in the academic literature – is what is called reverse engineering: taking other firms' product innovations to pieces to understand how they work and how they are made (De Melto et al., 1980; Levin et al., 1984; Nelson, 1986).

As with the assimilation of outside technology, neither inventing round other people's patents, nor reverse engineering, are cheap. They require considerable in-house investments in development, production engineering, and even research, and they sometimes result in imitations that are better than the original.

As we can see, this is a world very different from that where, technology having the properties of information, innovation is expensive but imitation virtually costless. We find instead that innovation and imitation are often indistinguishable, both in their inputs and their outputs, and that it is difficult to decide which is the 'best' of a number of competing innovations. Under such circumstances, some pluralism and variety in technological activities is likely to be beneficial.

Another justification for in-house or indigenous technological activity emerges from the second key characteristic of technology, namely, its cumulative nature. Partly as a result of its specificity and complexity, technological activity in firms tends to build out incrementally from what they know already, even when they are seeking major changes or breakthroughs – a characteristic already remarked upon in the 19th century by Marx. Technological search and choice in firms are therefore constrained by what they (or firms they purchase or join with) have already learned, and their technological activities tend to follow a 'trajectory'.

There is at present much research at the Science Policy Research Unit (SPRU) on these so-called trajectories, at the level of firms, of countries, and of major technologies. There are also links with other research groups to explore their theoretical implications: in particular, the development of evolutionary theories, that have as their central characteristics the diversity and adaptation of business firms in a dynamic and uncertain environment (Nelson and Winter, 1982).

I want to spell out briefly one of the implications of the existence of such trajectories that has recently been explored by the economist, Joseph Stiglitz (1986). He had already been joint author of a paper in 1969 that began to explore the consequences of what he called localized technological change, and I am only sorry that it has been 17 years before he has returned to the subject.

He argues in his recent paper that, if technical change is specific and cumulative, choices about technology should reflect not only what is available at present, but also expected developments in future. A decision to choose the cheapest alternative now will be short-sighted if an immediately more expensive technology offers greater opportunities for improvement in future, in terms of both cost and the development of more new products.

In his paper, Stiglitz is concerned mainly with the problems of developing countries, but what he says is of potentially wider applicability. For example, many technological decisions in British firms might be myopic, because of short-term financial pressures, or of an inability to assess seriously the future rate and direction of technological change. More generally, consider the typical choice facing either a firm or a country on how best to assimilate some successful technology first developed elsewhere. Whilst the immediate lower cost alternative may be the purchase of a licence, it may be that reverse engineering and independent redevelopment will offer greater benefits in future, in terms of accumulated knowledge, and the generation of a stream of competitive innovations.

The reality of such choices is supported in the study of Canadian innovations, mentioned above, and in a study of the Brazilian software industry recently completed at SPRU by Fatima Gaio.

The final key characteristic of technology that I want to mention briefly is its partial tacitness: in other words, it cannot be completely codified. Learning through experience, example and training is therefore an essential feature of technological accumulation, and this is reflected in the greater volume of technology trade amongst companies and countries in know-how than in patent rights. Neglect of the tacit element in technology can lead to policies and practices that turn out to be unproductive. For example, in the 1960s and 1970s, government policies in some third world countries set out to reduce the costs and the time periods of licensing agreements for technology purchases from the advanced countries. This sometimes led to the neglect of the personal contacts, the training and the experience, essential for effective technology transfer, and to disappointing results in recipient firms (Bell and Scott-Kemmis, 1985).

Technological capacity in firms

There is, I suggest, one major policy conclusion that emerges from this discussion, namely, the central importance of technological knowledge and activities embedded within firms, and necessary not just for new ideas and innovations, but also to enable effective assimilation of knowledge – technological and scientific – from outside. R&D expenditures, and other indicators of technological activity in business enterprises, therefore reflect not just the ability to get ahead, but also the capacity to keep up.

In this context, I am obliged to draw attention to the low level and rate of growth of British industrial R&D compared to other OECD countries, a subject that received considerable coverage and debate earlier this year. In a paper that Pari Patel and I are preparing for the 1987 meeting of the British Association for the Advancement of Science, we show the considerable changes that have taken place over the past 20 years in the sectors of British technological strength and weakness; in particular, the relative improvement compared to the rest of the world in the chemical-based sectors, with spectacular growth in pharmaceutical products; and at the same time, the relative decline in engineering – mechanical, electrical and (above all) electronic.

In major sectors like chemicals, electronics and aerospace, British growth and decline can be attributed to the technological policies and commitments of not much more than a handful of companies, the comparative study of which will be rich in future lessons for both business strategy and national policy.

References

Advisory Council for Applied Research and Development, and the Advisory Board for the Research Councils (1986), *The Science Base and Industry* (London, HMSO).

Allen, T. (1977), *Managing the Flow of Technology* (MIT Press).

Barclay, G. and M. Fottit (1987), British Technology Group, personal communication.

Bell, M. and D. Scott-Kemmis (1985), 'Technology import policy: have the problems changed?', and 'Technological dynamism and technological content of collaboration', *Economic and Policy Weekly*, vol. XX, nos 45, 46, 47, pages 1975–2004.

Bertin, G. and S. Wyatt (1986), *Multinationales et Propriété Industrielle* (Paris, Presses Universitaires de France).

Byatt, I. and A. Cohen (1969), *An Attempt to Quantify the Economic Benefits of Scientific Research* (London, HMSO).

Carpenter, M. (1983), 'Patent citations as indicators of scientific and technological linkages', Annual Meeting of the American Association for the Advancement of Science.

Carpenter, M., F. Narin and P. Woolf (1981), 'Citation rates to technologically important patents', *World Patent Information*, volume 3, Detroit, 30 May.

Clark, N. (1987), 'Similarities and differences between scientific and technological paradigms', *Futures*.

Dasgupta, P. (1986), *The Economic Theory of Technology Policy* (London, Centre for Economic Policy Research).

Dasgupta, P. and P. David (1986), 'Information disclosure and the economics of science and technology', in G. Feiwel (ed.), *Essays in Honour of K. Arrow* (Macmillan).

Davies, D. (1980), personal communication.
Dosi, G. (1982), 'Technological paradigms and technological trajectories', *Research Policy*, **11**(3), 147–62.
The Economist (1987), 'In search of a cure', 7 February.
The Economist (1987), 'Britain's university challenge', 18 April.
Eliasson, G. (1986), 'Innovative change, dynamics market allocation and long-term stability of economic growth', Working Paper No. 156, The Industrial Institute for Economics and Social Research, Stockholm.
Fagerberg, J. (1987), 'A technology gap approach to why growth rates differ', *Research Policy*, **16**(2, 3 and 4).
Freeman, C. (1977), 'Economics fo research and development', in I. Spiegel-Fosing and D. de Solla Price, *Science, Technology and Society* (Sage).
Freeman, C. (1986), 'The case for technological determinism' (mimeo), Brighton, Science Policy Research Unit.
Freeman, C., J. Clark and L. Soete (1982), *Unemployment and Technical Innovation: A Study of Long Waves and Economic Development* (London, Frances Pinter).
Gibbons, M. and R. Johnston (1974), 'The roles of science in technological innovation', *Research Policy*, **3**(3), 220–43.
Hague, D. (1986), Evidence submitted to the House of Lords Select Committee of Science and Technology, on behalf of the Economic and Social Research Council. *Civil Research and Development*, Vol. II (HMSO, London).
Hanson, P. and K. Pavitt (1987), *The Comparative Economics of Research, Development and Innovation in East and West* (mimeo), Brighton, Science Policy Research Unit.
Irvine, J. and B. Martin (1980), 'The economic effects of big science: the case of radio-astronomy', in Proceedings of the International Colloquium on Economic Effects of Space and Other Advanced Technologies, Strasborg, 28–30 April (Ref ESA SP 151, September 1980).
Kay, J. and C. Llewellyn Smith (1985), 'Science policy and public spending', *Fiscal Studies*, **6**.
Keynes, J. (1935), *The General Theory of Employment, Interest and Money* (Harcourt Brace and Company).
Levin, R., A. Klevorick, R. Nelson and S. Winter (1984), 'Survey research on R&D appropriability and technological opportunity. Part 1: Appropriability', (mimeo, Yale University).
Mackenzie, D. and J. Wajcman (1985), 'Introductory Essay', in *The Social Shaping of Technology* (UK, Open University Press).
Mackenzie, M., A. Cambrosia and P. Keating (1987), 'The Commercial Application of a Scientific Discovery; The Case of the Hydridoma Technique' (mimeo) (Crest, University of Quebec at Montreal).
Madeuf, B. (1984), 'International technology transfers and international technology payments: definitions, measurements and firm's behaviour', *Research Policy*, **13**(3), 125–40.
Mansfield, E., M. Schwartz and S. Wagner (1981), 'Imitation costs and patents: an empirical study', *Economic Journal*, **91**, 907–18.
Martin, B. and J. Irvine (1983), 'Assessing basic research: some partial indicators of scientific research in astronomy', *Research Policy*, **12**.
de Melto, D., K. McMullen and R. Wills (1980), *Innovation and Technological Change in Five Canadian Industries*, Discussion Paper No. 176, Ottawa, Economic Council of Canada.
Mowery, D. (1983), 'Economic theory and government technology policy', *Policy Sciences*, **16**, 27–43.
Narin, F. (1982), 'Assessment of the Linkages between Patents and Fundamental Research', Workshop on Patents and Innovation Statistics, OECD (DSTI/SPR) 82, 36.
Narin, F. and E. Norma (1985), 'Is industry becoming science?', *Scientometrics*, **7**.
Narin, F. and D. Olivastro (1987), *Identifying Areas of Strength and Excellence in UK Technology* (CHI Research, 1050 King's Highway North, Cherry Hill, NJ 08034, USA).
Nelson, R. (1959), 'The simple economics of basic scientific research', *Journal of Political Economy*.
Nelson, R. (1986), 'Institutions supporting technical advance in industry', *American Economic Review*, **76**. See also 'The Generation and Utilisation of Technology: A Cross Industry Analysis', (mimeo).
Nelson, R. and S. Winter (1982), *An Evolutionary Theory of Economic Change* (Belknap).
Patel, P. and K. Pavitt (1987), 'The Technological Activities of the UK: A Fresh Look', ESRC DRC Discussion Paper No. 49, Brighton, Science Policy Research Unit.
Pavitt, K. (1980), (ed.), *Technical Innovation and British Economic Performance* (Macmillan).
Pavitt, K. (1981), 'Technology in British Industry: A Suitable Case for Improvement', in C. Carter (ed.), *Industrial Policy and Innovation* (London, Heinemann).
Pinch, T. and W. Bijker (1984), 'The social construction of facts and artefacts: or how the sociology of science and the sociology of technology might benefit each other', *Social Studies of Science*, **14**,

399–441. See also (1986), 'Science, relativism and the new sociology of technology: reply to Russell', *Social Studies of Science*, **16**, 347-60.

de Solla Price, D. (1965), 'Is technology historically independent of science? A study in statistical historiography', *Technology and Culture*, **6**.

Rosenberg, N. (1976), 'Problems in the Economist's Conceptualization of Technological Innovation', *Perspectives on Technology* (Cambridge University Press).

Rosenberg, N. (1982), *Inside the Black Box: Technology and Economics* (Cambridge University Press).

Rosenberg, N. (1985), 'The Commercial Exploitation of Science by American Industry', in K. Clark, R. Hayes and C. Lorenz (eds), *The Uneasy Alliance: Managing the Productivity/Technology Dilemma* (Harvard Business School).

Rosenberg, N. and C. Frischtak (1985), *International Technology Transfer: Concepts, Measures and Comparisons* (Praeger).

Rothwell, R. et al. (1974), 'SAPPHO updated - Project SAPPHO, Phase 2', *Research Policy*, **3**(3), 258-91.

Select Committee on Science and Technology (1983), *Engineering Research and Development* (London, House of Lords, volumes I, II and III, HMSO).

Select Committee on Science and Technology (1986), *Civil Research and Development* (London, House of Lords, volumes I, II and III, HMSO).

Smith, Adam (1985 edition), *The Wealth of Nations* (G. Routledge and Sons).

Soete, L. (1981), 'A general test of technological gap trade theory', *Weltwirtschaftliches Archiv*, **117**(4), 638-60.

Stiglitz, J. (1986), 'Learning to learn, technological change, and economic and social structure', (mimeo, Princeton University).

de Tocqueville, A. (1963 edition), *De La Démocratie en Amérique* (abridged edition, Paris, Union Générale d'Editions).

Wyatt, S. (1985), 'Patents and multinational corporations: results from questionnaires', *World Patent Information*, volume 7.

Sectoral patterns of technical change: Towards a taxonomy and a theory

Keith PAVITT *

Science Policy Research Unit, University of Sussex, Brighton BN1 9RF, UK

Final version received January 1984

The purpose of the paper is to describe and explain sectoral patterns of technical change as revealed by data on about 2 000 significant innovations in Britain since 1945. Most technological knowledge turns out not to be "information" that is generally applicable and easily reproducible, but specific to firms and applications, cumulative in development and varied amongst sectors in source and direction. Innovating firms principally in electronics and chemicals, are relatively big, and they develop innovations over a wide range of specific product groups within their principal sector, but relatively few outside. Firms principally in mechanical and instrument engineering are relatively small and specialised, and they exist in symbiosis with large firms, in scale intensive sectors like metal manufacture and vehicles, who make a significant contribution to their own process technology. In textile firms, on the other hand, most process innovations come from suppliers.

These characteristics and variations can be classified in a three part taxonomy based on firms: (1) supplier dominated; (2) production intensive; (3) science based. They can be explained by sources of technology, requirements of users, and possibilities for appropriation. This explanation has implications for our understanding of the sources and directions of technical change, firms' diversification behaviour, the dynamic relationship between technology and industrial structure, and the formation of technological skills and advantages at the level of the firm, the region and the country.

* The following paper draws heavily on the SPRU data bank on British innovations, described in J. Townsend, F. Henwood, G. Thomas, K. Pavitt and S. Wyatt, *Innovations in Britain Since 1945*, SPRU Occasional Paper Series No. 16, 1981. The author is indebted to Graham Thomas and to Sally Wyatt who helped with the statistical work, to numerous colleagues inside and outside SPRU for their comments and criticisms, and to Richard Levin and two anonymous referees for their detailed and helpful comments on a longer and more rambling earlier draft. The research has been financed by the Leverhulme Trust, as part of the SPRU programme on innovation and competitiveness.

Research Policy 13 (1984) 343–373
North-Holland

1. Introduction

1.1. Purpose

The subject matter of this paper is sectoral patterns of technical change. We shall describe and try to explain similarities and differences amongst sectors in the sources, nature and impact of innovations, defined by the sources of knowledge inputs, by the size and principal lines of activity of innovating firms, and by the sectors of innovations' production and main use.

It is recognised by a wide range of scholars that the production, adoption and spread of technical innovations are essential factors in economic development and social change, and that technical innovation is a distinguishing feature of the products and industries where high wage countries compete successfully on world markets [55]. However, representations of the processes of technical change found in economics are in many respects unsatisfactory. According to Nelson:

> In the original neo-classical formulation, new technology instantly diffuses across total capital. In the later vintage formulation, technology is associated with the capital that embodies it and thus adoption of a new technique is limited by the rate of investment. [29]

Whilst such assumptions may be convenient or useful in macro-economic model building and analysis, they have – as Nelson [29] and Rosenberg [42] have pointed out – two important limitations. First, they make exogenous the production of technology and innovations. Second, they do not reflect the considerable variety in the sources, nature and uses of innovations that is revealed by empirical studies and through practical experience.

Such formulations of technical change are not

therefore very useful for analysts or policy makers concerned with either the nature and impact of technical change at the level of the firm or the sector, or with R&D policy at the level of the firm, the sector or the nation. Hence, the importance, we would argue, of building systematically a body of knowledge – both data and theory – that both encompasses the production of technology, and reflects sectoral diversity. The following paper is a contribution to this objective.

1.2. The data base

What makes is possible is data collected by Townsend et al. [60] on the characteristics of about 2000 significant innovations, and of innovating firms, in Britain from 1945 to 1979. The methodology, results and limitations are spelt out fully in the original publication. Suffice here to say that:

(1) Innovation is defined as a new or better product or production process successfully commercialised or used in the United Kingdom, whether first developed in the UK or in any other country.

(2) Significant innovations were identified by experts knowledgeable about, but independent from, the innovating firms; information about the characteristics of the innovations was collected directly from the innovating firms.

(3) The sample of innovations covers three and four digit product groups accounting for more than half the output of British manufacturing. At the two digit level, the sectoral distribution of innovations is similar to that measured by numbers of patents, but is not to that measured by expenditures on R&D activity. In concrete terms, this reflects a slight over-representation of innovations in mechanical engineering and metals; a considerable over-representation in instruments and textiles; a slight under-representation in chemicals and electronics; and a considerable under-representation in aerospace. [1]

(4) Experts in different sectors defined the threshold of significance at different levels, which means that our sample of innovations cannot be used to compare the volume of innovations

amongst sectors. However, it can be used to compare patterns of innovative activity within sectors, where the results are consistent with other independent sources of data on innovative activities in the UK and elsewhere (see [36]).

(5) The data measure significant innovations introduced into the UK. They do not measure significant world innovations, nor do they capture the incremental and social innovations that often accompany significant technical innovations. We shall assume that the data on significant innovations are the visible manifestations of deeper processes, involving incremental and social, as well as significant, innovations. We shall also assume that, although the pattern of innovative activities in the UK does have some distinctive features [2], what we are measuring on the whole reflects patterns in most industrial countries, rather than the specific characteristics of the UK.

1.3. Approach and structure

Given the nature of the problem as posed in subsection 1.1, and of the large data base as described in subsection 1.2, the reader might legitimately expect a paper that is largely econometric in nature: an alternative model of technical change to neoclassical ones would be proposed and formalised, and a series of statistical tests would be carried out, that discriminate between the explanatory powers of the competing models. However, this will not be the approach followed, for reasons that go beyond the intellectual propensities and professional limitations of this particular author. Although the statistical data are more comprehensive and systematic than any others previously assembled on innovations, the sample still has a number of limitations. As we have seen, it covers just one half of manufacturing, so important gaps remain. For purposes of statistical analysis, it can be grouped into 11 sectoral categories at the two digit level, and into 26 categories at the three and four digit level. Statistical data on other sectoral properties often cannot be conveniently assembled into the same categories and for the same time periods. We were therefore faced with a choice between "creating" data to make any regressions econometrically more convincing, or making for-

---

[1] For the number of innovations produced in each two digit sector, see table 2, column 3. For the three to four digit sectors included in the sample, see table 1.

[2] See, for example [34;35].

mal statistical analysis a minor part of the paper. We chose the latter approach, although tentative econometric analysis is described in the Appendix to this paper, and discussed in section 4.

This approach has the advantage of allowing the patterns of the statistical data to be compared to the mind's eye with the rich range of sectoral and firm studies of technical change that have accumulated over the past 25 years. Given that no obvious model of sectoral patterns of technical change emerges from previous theoretical writings, such direct and visual comparisons turned out to be particularly useful.

We present and discuss the main features of the data in section 2, and compare them with some prevailing theoretical assumptions. In section 3, we suggest a taxonomy of sectoral patterns of innovative activity, and a theoretical explanation, that are consistent with the data. In section 4, we explore some of the analytical implications of such a theory, and in section 5 we suggest further research that should be done.

2. Sectoral patterns of innovation

2.1. Analysis of the data

The information contained in the data bank describes characteristics of significant innovations and of innovating firms. In this paper, we shall be using information on the institutional sources of the main knowledge inputs into the innovations, on the sectors of production and of use of the innovations, and on the size and the principal sectors (or product groups or lines) of activity of the innovating firms.

Sources of the main knowledge inputs into the innovations were identified by asking the sectoral experts and the innovating firms to identify the type of institution that provided up to the three most important knowledge inputs into each innovation. This information provides a basis for assessing the relative importance in providing such knowledge, of the innovating firms themselves, of other industrial firms, and of institutions providing public knowledge, such as universities and government laboratories. This is done in subsection 2.2.

Information on the sectors of production of innovations comes from the sectoral experts, and

on sectors of use from the innovating firms [3]. We define innovations that are used in the same sectors as those in which they are produced (e.g. direction reduction of steel) as *process* innovations, and those that are used in different sectors (e.g. the Sulzer Loom) as *product* innovations. Such information provides what can be considered as the technological equivalent of an input/output table. It shows how intersectoral patterns of production and sale of goods is reflected in intersectoral transfers of technology. It is strictly equivalent in purpose, if not in method, to the table compiled recently for the USA by Scherer [51]. It is discussed in subsection 2.3.

Information on the size and principal sector of activity of innovating firms was provided by the firms themselves, and sometimes checked through other sources. Size is measured in terms of total world employment, and (for the innovations in the period from 1969 to 1979) also of employment in the UK. Such information allows comparisons of the size distribution of innovating firms amongst sectors, over time, and in comparison to other indices of economic activity.

Information on the principal activity of innovating firms allows comparisons, amongst sectors and over time, of the degree to which firms produce innovations outside their principal sector of activity, and to which innovations in sectors are produced by firms with their principal activity elsewhere. Such comparisons can be seen as the equivalent for technology of comparisons of firms' diversification in output, employment or sales. Patterns of size and of "technological diversification" of innovating firms are analysed in subsection 2.4.

It is to be noted that each innovation in the data base is attributed three numbers in the Standard Industrial Classification, or Minimum List Heading, as it is called in the UK: (1) the sector of production of the innovation; (2) the sector of use of the innovation; (3) the sector of the innovating firm's principal activity. We are therefore able to construct an (as yet incomplete) three-dimensional matrix encompassing links amongst sectors in the production and use of innovations, and in the sectoral patterns of "technological diversification" of innovating firms. Such a construct enables us to

---

[3] When an innovation found a use in more than one sector, we defined the main user sector as the sector of use.

compare sectors in terms of:

(1) The sectoral *sources* of technology *used* in a sector: in particular, the degree to which it is generated within the sector, or comes from outside through the purchase of production equipment and materials.

(2) The institutional *sources* and *nature* of the technology *produced* in a sector: in particular, the relative importance of intramural and extramural knowledge sources, and of product and process innovations.

(3) The *characteristics* of *innovating firms*: in particular, their size and principal activity.

Such comparisons have been made systematically by the author, at the two and the three to four digit level, in the preparation of this paper. They were essential for an evaluation of the empirical validity of prevailing models of technical change, and *a fortiori* for working out the sectoral taxonomy and theory proposed in section 3. However, they will not be reproduced in comprehensive detail since they are long, tedious and sometimes potentially confusing. We shall instead present statistical material mainly at the two digit sectoral level, although we shall also refer to some patterns at the three to four digit level.

Suffice to say here that a central feature in our search for a taxonomy and an explanatory theory was the classification of innovations in each sector according to whether or not the sectors of production, of use, and the principal activity of the innovating firm, are the same. There are five possible combinations:

Category 1: sectors of production, use, and principal firm activity are all the same: e.g. a process innovation by a steel making firm. (MLH [4] 311)

Category 2: sectors of production and principal firm activity are the same, but different from sector of use: e.g. a specialised firm making textile machines (MLH 335), designing a new textile machine (MLH 335) for use in the textile industry (MLH 411).

Category 3: sectors of principal firm activity and of use of the innovation are the same, but different from the sector of production of the innovation: e.g. a shipbuilding firm (MLH 370) develops a special machine tool (MLH 332), for use in building ships (MLH 370).

Category 4: sectors of production and use of the innovation are the same, but different from that of the firm's principal activity: for example, a firm principally in general chemicals (MLH 271) develops a process innovation in textiles (MLH 411).

Category 5: sectors of production of the innovation, of its use, and of the firm's principal activity are all different: for example, a firm principally in electronic capital goods (MLH 367) develops and produces an innovation in instrumentation (MLH 354.2) for use in making motor vehicles (MLH 381).

In the particular examples given above, the categories are the same at the two digit as at the three to four digit level. But in some cases they are not. For example, a firm in general chemicals (MLH 271), producing an innovation in pharmaceuticals (MLH 272), for use in medical services (MLH 876) will fall into category 5 at the three digit level, and category 2 at the two digit level.

2.2. Institutional sources of main knowledge inputs

As we have already pointed out, experts could allocate up to three institutional sources of knowledge inputs for each innovation. All provided one such source, about 40 percent provided two sources, but only 3 percent provided three sources.

The results at the three to four digit level are summarised in table 1. Only about 7 percent of the knowledge inputs comes from the public technological infrastructure (higher education, government laboratories, and research associations). The highest proportion is reached in a number of electronics sectors, but even here it is never as much as 25 percent. On the other hand, 59 percent came from within the innovating firms themselves, and about a third from other industrial firms.

These data have a number of imperfections. Given that they were collected mainly from industrial experts, and that only about 1.5 sources were identified for each innovation, they underestimate the contribution made by the public technological infrastructure to person-embodied knowledge and to essential background knowledge for the innovations. [5] More generally, the distribu-

---

[4] MLH = Minimum List Heading.
[5] See Gibbons and Johnston [14] for an excellent analysis of these sources.

Table 1
Distribution of knowledge inputs into significant innovations, according to institutional source

Sector [a]	Source of knowledge inputs (%) [b]			Number of observations
	Intra-firm	Other firm	Public Infrastructure	
Food (211-229)	53.4	44.6	2.0	101
Pharmaceuticals (272)	62.8	37.2	0	129
Soap and detergents (275)	60.0	40.0	0	30
Plastics (276)	40.4	55.2	4.4	114
Dyestuffs (277)	68.1	30.5	1.4	69
Iron and steel (311)	47.7	44.9	7.4	149
Aluminium (321)	68.0	28.0	4.0	50
Machine tools (332)	64.1	29.8	6.1	231
Textile machinery (335)	61.2	36.6	2.2	278
Coal-mining machinery (339.1)	52.3	31.6	16.1	199
Other machinery (339.4 + 339.9)	59.1	36.6	4.3	115
Industrial plant (341)	51.6	41.9	6.5	31
Instruments (354.2)	61.6	25.2	13.2	440
Electronic components (364)	48.2	37.1	14.7	170
Broadcasting equipment (365)	64.4	33.9	1.7	59
Electronic computers (366)	50.6	33.3	16.1	81
Electronic capital goods (367)	67.2	9.7	23.0	113
Other electrical goods (369)	60.8	35.3	3.9	51
Shipbuilding (370)	47.9	43.8	8.2	73
Tractors (380)	78.7	21.3	0	47
Motor vehicles (381)	69.3	29.7	1.0	101
Textiles (411-429)	67.3	32.7	0	110
Leather goods and footwear (431/450)	44.4	48.1	7.4	54
Glass (463)	48.2	44.6	7.1	56
Cement (464)	62.5	33.3	4.2	24
Paper and board (481)	66.7	28.2	5.1	39
Other plastics (496)	55.8	41.9	2.3	43
Other	–	–	–	56
Total	58.6	34.0	7.4	3013

[a] Numbers in brackets refer to the appropriate Minimum List Heading.
[b] Each row adds up to 100 percent.

tion of knowledge sources in this kind of study depends heavily on the definitions and time perspectives of the data collected. [6] In spite of these imperfections, the distribution of knowledge

sources in table 1 is not dissimilar to that found in other studies. [7].

Given that innovating firms evaluate their own knowledge contributions at nearly 60 percent of the total, we cannot realistically assume that there exists a generally available and applicable stock or pool of knowledge, where each firm – being very

[6] See, for example, the classic US controversy at the end of the 1960s: the Hindsight and Traces studies arrived at very different conclusions about the contribution of basic research to industrial innovation. For a comparison, see Pavitt and Wald [39].

[7] See Langrish et al. [21], and Gibbons and Johnston [14].

small in relation to the total stock or pool – can gain much more from drawing on the pool, rather than by adding to it. The concept of the general "pool" or "stock" of knowledge misses an essential feature of industrial technology, namely, the firm-specific and differentiated nature of most of the expenditures producing it. In Britain and elsewhere, about three-quarters of all expenditures on industrial R&D is on "D", and an equivalent sum is spent on testing and manufacturing start up. [8] The purpose of these expenditures is to mobilise skills, knowledge and procedures in the firm in order to commercialise specific products and production processes, with the characteristics of operation, reliability and cost that satisfy user needs. Specificity is an essential feature of innovations and innovative activity in capitalist firms – both in terms of functional applications, and of the ability of the innovating firm to appropriate the relevant knowledge for a period of time.

This feature is missed in any simple equation of "technology" with "information." Whilst it may be reasonable to describe *research* and *invention* as producing "information" that is quickly and easily transmitted, [9] it is grossly misleading to assume that *development* and *innovation* have similar properties. Given their specific characteristics, the costs of transmission from one firm to another can be high, even in the absence of legal protection or secrecy in the innovating firm [7;33;57]. As Nelson [30] has recently argued, technological knowledge has both proprietary and public aspects, although table 1 and other studies suggest that the former outweigh a latter.

These features are missed in some representations of technology in a production function. According to Salter:

> ...the production function concept ... could refer either to techniques which have been developed in detail, or to techniques which are feasible in principle but have not been developed because the necessary economic pressures are absent. [48, p.26]

Salter plumps for the latter and, in doing so, makes exogenous to his analysis most of the innovative (i.e. development and post-develop-

ment) activities of industrial firms. As Rosenberg [42] has pointed out, most firms do not (and in the light of the above discussion cannot) have information on a full and complete range of alternative techniques. The assumption that most technological knowledge is or could be publicly available and generally applicable has little foundation in reality.

2.3. Sectoral patterns of production and use of innovations

As already described above, the innovation data base compiled by Townsend et al. [60] describes sectoral patterns of production and use of innovations in the UK. On the basis of a different method, Scherer [51] has compiled similar information for the USA. He obtained detailed data on the sectoral allocations of R&D resources in more than 400 large US firms in the 1970s. On the basis of examination of the patenting activity of these firms, he was also able to attribute the "output" of this R&D to sectors of use. Scherer's work covers more than 40 US sectors of production and use. The data collected by Townsend et al., on the other hand, cover small and medium sized, as well as large firms, but not all sectors. Most important for the purposes of this paper, both studies show comparable results in sectoral patterns of production and use of technology. [10]

Following Scherer, we define as product innovations those innovations that are used *outside* their sector of production, and *process* innovations as those that are used *inside* their sector. [11] Both studies confirm the prevalence of *product* innovations which accounted for 73.8 percent in the USA, according to Scherer, and 75.3 percent in the UK, when sectors are defined at the three to four digit level, and 69.6 percent when defined at the two digit level.

---

[8] For a recent review of empirical findings on the total costs of innovation, see Kamin et al. [19].
[9] See the classic paper by Arrow [3].

[10] See Pavitt [36].
[11] This definition is not strictly the same as product or process innovation at the level of the firm. Thus, what is a product innovation for the firm will be a process innovation for the sector, when the firm's innovation is purchased and used in the same sector; conversely, a process innovation in the firm will be a product innovation for the sector, when the firm produces and uses its capital goods. However, for the firm, as well as the sector, product innovation predominates. See Townsend et al. [60, tables 9.1 and 9.2].

Table 2
Innovations produced and used in two digit sectors

Innovations used in sector		Sector [a]	Innovations produced in sector	
Percentage produced in sector	Number		Number	Percentage that are product innovations
(1)	(2)	(3)	(4)	(5)
52.9	68	III Food and drink	65	44.7
60.5	71	V Chemicals	251	82.9
60.7	130	VI Metal manufacture	137	42.3
68.1	169	VII Mechanical engineering	662	82.7
38.4	60	VIII Instrument engineering	332	93.1
80.8	107	IX Electrical and electronic engineering	339	60.1
32.2	90	X Shipbuilding	52	44.1
37.6	221	XI Vehicles	128	35.2
16.2	377	XIII Textiles	91	32.9
60.0	45	XIV&XV Leather and Footwear	34	26.5
46.1	63	XVI Bricks, Pottery, glass and cement	72	85.0
na	823	Other	61	na
41.9 [b]	2224	Total	2224	69.6

[a] Roman numerals refer to the appropriate Order Headings.
[b] For the 1401 innovations in the sample that are attributed a sector of use.

Scherer's more complete and comprehensive data for the USA show a clear difference in the production and use of innovations between manufacturing and the other sectors of the economy (i.e. agriculture, mining, service industries, private and public services). For manufacturing as a whole, the ratio of production to use of technology is about 5.3 to 1. Outside manufacturing it is about 0.1 to 1, and the proportion of all the technology used outside manufacturing that is generated there amounts to less than 7 percent. In other words, manufacturing produces most of the innovations that get used in other parts of the economy.

However, manufacturing itself is far from homogeneous in patterns of production and use of innovations. Table 2 shows at the two digit level, the relevant characteristics of those sectors of British manufacturing for which we have a satisfactory sample of innovations. Column 5 shows the percentage of all innovations produced in each sector that are purchased and used in other sectors: in other words, the percentage of product innovations. These are relatively most important in instruments, mechanical engineering, chemicals, building materials (mainly glass and cement) and electrical and electronic engineering, whilst process innovations predominate in leather and footwear, textiles, vehicles, metal manufacture, shipbuilding and food and drink. Data at the three to four digit

level show that all the mechanical engineering product groups covered in the survey are strongly orientated towards product innovations whilst, within the chemical and the electrical/electronic sectors, there are two product groups with high percentages of process innovations: soaps and detergents, and broadcasting equipment.

Column 1 in table 2 shows the percentage of innovations used in each sector that are produced in the same sector: in other words, the degree to which each sector generates its own process innovations. [12] They show that most two digit sectors of manufacturing in the sample make a significant contribution to developing their own process technologies. The main exception is textiles, which is heavily dependent on innovations from other sectors.

Finally, a comparison between columns 4 and 2 of table 2 shows the differences between production and use of innovations in each sector. Production is greater than use in chemicals, mechanical engineering and instruments, and electrical/electronic products. The two are roughly in balance in industries characterised by continuous process

---

[12] Column 2 shows 823 innovations produced in the identified sectors of manufacturing but used elsewhere. Unlike Scherer, we cannot in this context usefully allocate these innovations to user sectors, since we do not yet have a sample of innovations produced by these sectors of use.

technology (i.e. food and drink, metal manufacture, building materials), whilst more innovations are used than produced in sectors characterised by assembly operations (i.e. shipbuilding and vehicles). These assembly industries also draw on a wider range of sectors for their process technologies than do those characterised by continuous process technology.

How does this pattern of production and use of innovations compare with the "vintage" model of technical change, which assumes that all technology is capital-embodied and enters the economy through investment? In his original formulation of this model, Salter [48] was very well aware of its limitations. He recognised the importance of innovations in capital goods, and of product innovations, but made them exogenous. He also stated that other assumptions made it "highly simplified" (p. 64): for example, that technical change involves no cumulative effects from one generation of capital equipment to another, or that "best practice" performance is clearly defined and instantly reached.

Nonetheless, Salter's assumptions do reflect the reality of most of the economy, namely non-manufacturing, where technical change comes mainly through the purchase of equipment, materials and components from manufacturing. Within manufacturing, it also reflects accurately the sources of process innovations in the textile industry. However, his characterisation of the sources of technical change at the more modern end of manufacturing industry is less satisfactory, in three respects.

First, whilst it may be conceptually correct in certain economic models to assume – as Salter does – that improvements in the performance of capital goods (i.e. product innovations) are equivalent to the relative cheapening of capital goods (i.e. process innovations), such an assumption is misleading about the directions and sources of technical change in the capital goods sector. Innovative activities are in fact heavily concentrated on product innovation: no amount of process innovation in, for example, the production of mechanical calculators would have made them competitive with the product innovations resulting from the incorporation of the electronic chip.

Second, Salter's model assumes that process innovations come to user sectors already developed. However, we see in table 2 that a significant

proportion of the innovations used in modern manufacturing are developed and produced in the innovating sectors themselves. It is worth dwelling a bit on one of the possible reasons why. We know from the research of Gold [15], Sahal [47] and others that two of Salter's simplifying assumptions are false: in continuous process and assembly industries, there is in fact cumulative learning, and "best practice" performance is rarely easily defined or quickly reached. The same design, engineering and operating skills that enable rapid learning are also capable of making innovations, particularly in production equipment. In other words, sectors with complex and expensive process technologies devote considerable technical resources to ensuring that equipment is used efficiently and continuously improved.

Third, and more generally, the production of all innovations is made exogenous to Salter's model. Before suggesting in section 3 a framework that makes such production endogenous, we shall describe characteristics of innovating firms in different sectors.

2.4. Characteristics of innovating firms: Size and technological diversification

Table 3 summarises the main features of the size distribution of innovating firms in different sectors. Columns 7–9 classify them according to the principal sector of activity of the innovating firm. This classification shows a relatively big contribution by small firms (1–999 employees) in mechanical and instrument engineering, textiles, and leather and footwear; and by large firms (10,000 and more employees) in the other sectors. This sectorally differentiated pattern is very similar to that emerging from a study of significant innovations and innovating firms undertaken for the USA. [13]

Columns 1–3 of table 3 show the size distribution of innovating firms according to the sector of the innovations, rather than the principal sector of the innovating firms' activity. In sectors where large firms predominate, the two size distributions are very similar. However, in mechanical and instrument engineering and in textiles, both the number of innovations and the relative contribu-

---

[13] See [20]. A comparison between the two sets of results is made in [60, table 5.3].

Table 3
Distribution of Innovations by firm size [a] and by sector

By sector of innovation				Sector [b]	By sector of firm activity			
Percentage distribution [c]			Number of innovations		Number of innovations	Percentage distribution [c]		
10,000+	1000–9999	1–999				10,000+	1000–9999	1–999
(1)	(2)	(3)	(4)	(5)	(6)	(7)	(8)	(9)
72.3	10.8	17.0	65	III Food and drink	78	79.5	7.7	12.8
74.9	16.8	8.4	251	V Chemicals	290	82.4	7.9	9.6
63.5	31.4	5.1	137	VI Metal manufacture	143	62.9	32.8	4.2
35.2	30.5	34.3	662	VII Mechanical engineering	536	24.3	36.9	38.8
41.0	16.6	42.4	332	VIII Instrument engineering	187	24.6	21.4	54.0
66.4	15.9	17.7	339	IX Electrical and electronic engineering	343	65.9	12.2	22.0
57.7	38.5	3.8	52	X Shipbuilding	89	61.8	34.8	3.3
70.3	18.0	11.7	128	XI Vehicles	158	72.2	20.3	7.6
56.0	30.8	13.2	91	XIII Textiles	77	35.1	40.3	24.7
11.8	20.6	67.6	34	XIV&XV Leather and footwear	50	44.0	18.0	38.0
70.8	18.1	11.1	72	XVI Bricks, pottery, glass and cement	87	74.7	16.1	9.1
–	–	–	112	Other	227	–	–	–
53.2	21.9	24.9	2265	Total	2265	53.2	21.9	24.9

[a] Measured by number of employees.
[b] Roman numerals refer to the appropriate Order Headings.
[c] Rows add up to 100 percent.

tions of large firms are bigger when classified by sector of innovation, than when classified by the principal sector of activity of the innovating firm. In other words, a relatively large number of in-

novations are produced in these sectors by relatively large firms with their principal activities in other sectors.

Table 4 shows that for the sample as a whole,

Table 4
The distribution of innovations produced outside innovation firms' principal two-digit activities

Innovations in other sectors by firms with principal activities in the sector		Sector [a]	Innovations in the sector by firms with principal activities In other sectors	
%	Number		Number	%
(1)	(2)	(3)	(4)	(5)
30.8	78	III Food and drink	65	17.0
26.5	290	V Chemicals	251	15.2
34.3	143	VI Metal manufacture [b]	137	31.4
(37.0)	(119)		(93)	(19.4)
16.0	536	VII Mechanical engineering	662	32.1
19.8	187	VIII Instrument engineering	332	54.6
23.8	343	IX Electrical and electronic engineering	339	23.0
58.4	89	X Shipbuilding	52	28.9
33.5	158	XI Vehicles	128	18.0
24.7	77	XIII Textiles	91	36.3
50.0	50	XIV&XV Leather and footwear	34	26.5
32.4	87	XVI Bricks, pottery, glass and cement	72	18.1
–	227	Other	102	–
31.5	2265	Total	2265	31.5

[a] Roman numerals refer to the appropriate Order Headings.
[b] Percentages between brackets refer to Iron and steel only.

31.5 percent of the innovations are produced by firms with their principal activities in other two digit sectors. Column 5 shows that a relatively large proportion of innovations in mechanical and instrument engineering and textiles are produced by firms with their principal activities elsewhere (32.1, 54.6 and 36.3 percent respectively), whilst column 1 shows that firms with their principal activities in mechanical and instrument engineering and in textiles produce a relatively small proportion of innovations in other sectors (16.0, 19.8 and 24.7 percent respectively).

Column 1 also shows the sectors where firms principally in them produce a proportion of innovations in other sectors that is above or round about the average: food and drink, metal manufacture, shipbuilding, vehicles, leather and footwear, and building materials. This is in contrast with firms principally in chemicals, or in electrical and electronic products, neither of which produce relatively high proportions of innovations beyond their two digit sector (26.5 and 23.8 percent respectively). Similarly, a relatively small proportion of innovations in these two sectors are produced by firms principally in other sectors (15.2 and 23.0 percent respectively).

This pattern suggests, amongst other things, that a relatively high proportion of innovations in mechanical and instrument engineering are produced by firms typified by continuous process and assembly production, such as metal manufacture, shipbuilding and vehicles. A more detailed examination of the data base confirms that this is the case. Innovations in two fundamentally important sectors of production technology – mechanical and instrument engineering – are therefore made both in relatively small specialised firms in these sectors, and in relatively large firms in continuous process and assembly industries.

One question springs to mind, when examining the data in tables 3 and 4: to what extent are the intersectoral differences in the size distribution of innovating firms, and in their patterns of technological diversification, similar to those found in the size distribution and patterns of sectoral diversification, in terms of sales, output and employment? Given the gaps in the data in the UK censuses of production, it is not possible to provide a straightforward answer to this question. Certainly, there are similarities: small firms makes a relatively greater contribution to net output and em-

ployment in mechanical and instrument engineering than in the other two digit sectors in our sample; and over time, both the increasing contribution to the production of innovations of firms with more than 10,000 employees and the constant share of firms with less than 200 employees, are reflected in trends in both output and employment.

The similarities are at first sight far less apparent in patterns of diversification. A comparison with Hassid's analysis [17], based on data from the UK census of production, shows that diversification at the two digit level is considerably less in net output than it is in the production of innovations: 14.0 percent in 1963 and 16.9 percent in 1968, compared to 31.5 percent for the whole period from 1945 to 1979. Neither is there any close relationship across sectors between the degree to which firms principally in them diversify into other sectors in net output, and in the production of innovations.

However, there is a similarity in the sectors into which firms diversify: a comparison of table 4 above with Hassid's data [17, table 3] shows that, in terms of both the production of innovations and the net output, mechanical and instrument engineering are sectors where relatively large contributions are made by firms principally in other sectors, whilst relatively small contributions are made in food, chemicals, electrical and electronic engineering, and vehicles by such firms.

Taking these comparisons further will need much more time and space, and will not be done in this paper. Our contribution here hopefully will be to enrich the ways in which such comparisons will be interpreted and explained. In particular, we intend to go beyond explanations of sectoral patterns of production of innovations simply in terms of sectoral industrial structures. Even if there turned out to be perfect statistical correlations across sectors between firm size and sectoral patterns of output, on the one hand, and firm size and sectoral patterns of production of innovations, on the other, it would be wrong to interpret the latter simply as causal consequences of the former. This would neglect the causal links running from the latter to the former: that is, from diversification in the production of innovations to diversification in output, and from the production of innovations to firms growth and firm size.

Most of the empirical studies of patterns of

diversification do in fact refer to the notion of "technological proximity" in explaining diversification in output [4;16;17;46;62]; our analysis and explanation will try to give some additional empirical and theoretical content to this notion. Similarly, a number of writers have recently stressed the causal links running from innovation to firm size [23,32]; we shall begin to explain, amongst other things, why high rates of innovation do not necessarily lead to heavily concentrated industries. Before doing this, however, we propose in section 3 how and why patterns of technological development and innovation differ amongst sectors.

3. Towards a taxonomy and a theory

3.1. The ingredients

Two central characteristics of innovations and innovating firms emerge from section 2. First, from subsection 2.2 it is clear that most of the knowledge applied by firms in innovations is not general purpose and easily transmitted and reproduced, but appropriate for specific applications and appropriated by specific firms. We are therefore justified in assuming, like Rosenberg [42], that, in making choices about which innovations to develop and produce, industrial firms cannot and do not identify and evaluate all innovation possibilities indifferently, but are constrained in their search by their existing range of knowledge and skills to closely related zones. In other words, technical change is largely a cumulative process specific to firms. What they can realistically try to do technically in future is strongly conditioned by what they have been able to do technically in the past.

The second characteristic is, of cource, variety. From subsections 2.3 and 2.4, it emerges that sectors vary in the relative importance of product and process innovations, in sources of process technology, and in the size and patterns of technological diversification of innovating firms. Nonetheless, some regularities do begin to emerge. In subsection 2.3, we can see a whole class of sectors where – as in vintage models – technical change comes mainly from suppliers of equipment: nonmanufacturing and traditional sectors of manufacturing like textiles. We also ssee that the other manufacturing sectors make a significant contribu-

tion to their process technology. However, whilst firms in assembly and continuous process industries tend to concentrate relatively more of their innovative resources on process innovations, those in chemicals, electronic and electrical engineering, mechanical engineering, and instrument engineering devote most of these resources to product innovation.

In subsection 2.4, we see that sectors making mainly product innovations can be divided into two categories. First, firms principally in the chemicals and electronic and electrical sectors are relatively big, they diversify relatively little beyond their two digit category in producing innovations, and they produce a relatively high proportion of all the innovations in the two sectors. Second, firms principally in mechanical engineering and instrument engineering are relatively small, they diversify technologically relatively little beyond their two digit category, and they make a smaller contribution to all the innovations in the two sectors, given the important contribution made by relatively large user firms, particularly those in sectors typified by assembly and continuous process production.

In subsections 3.2–3.5 below, we shall try to categorise and explain these characteristics: in other words, to propose a taxonomy and a theory of sectoral patterns of technical change. Ideally, these should be consistent with the data so far presented. They should also be capable of further empirical refinement and test, given the inadequacies of the data at present available, and in particular of using what is mainly static, cross-sectional data as the basis for a theory that is essentially dynamic.

In our proposed taxonomy and theory, the basic unit of analysis is the innovating firm. Since patterns of innovation are cumulative, its technological trajectories will be largely determined by what is has done in the past in other words, by its principal activities. Different principal activities generate different technological trajectories. These can usefully be grouped into the three catogories, that we shall call supplier dominated, production intensive, and science-based. These different trajectories can in turn be explained by sectoral differences in three characteristics: sources of technology, users' needs, and means of appropriating benefits. The three categories, the differing technological trajectories, and their underlying causes are

Table 5
Sectoral technological trajectories: Determinants, directions and measured characteristics

Category of firm	Typical core sectors	Determinants of technological trajectories			Technological trajectories	Measured characteristics			
		Sources of technology	Type of user	Means of appropriation		Source of process technology	Relative balance between product and process innovation	Relative size of innovating firms	Intensity and direction of technological diversification
(1)	(2)	(3)	(4)	(5)	(6)	(7)	(8)	(9)	(10)
Supplier dominated	Agriculture; housing; private services traditional manufacture	Suppliers; Research extension services; big users	Price sensitive	Non-technical (e.g. trademarks, marketing, advertising, aesthetic design)	Cost-cutting	Suppliers	Process	Small	Low vertical
Production intensive — Scale intensive	Bulk materials (steel, glass); assembly (consumer durables & autos)	PE suppliers; R&D	Price sensitive	Process secrecy and know-how; technical lags; patents; dynamic learning economies	Cost-cutting (product design)	In-house; suppliers	Process	Large	High vertical
Production intensive — Specialised suppliers	Machinery; instruments	Design and development users	Performance sensitive	Design know-how; knowledge of users; patents	Product design	In-house; customers	Product	Small	Low concentric
Science based	Electronics/electrical; chemicals	R&D Public science; PE	Mixed	R&D know-how; patents; process secrecy and know-how; dynamic learning economies	Mixed	In-house; suppliers	Mixed	Large	Low vertical / High concentric

* PE = Production Engineering Department.

summarised in table 5. Before discussing them in greater detail, we shall identify briefly the three traditions·of analysis on which the taxonomy and the theory are based.

First, there are analysts who have deliberately explored the diversity of patterns of technical change. In particular, Woodward [69] has argued that appropriate organisational forms and mixes of skills for manufacturing firms are a function of their techniques of production, which she divided into three: small batch production and unit production, large batch and mass production, and continuous process production. Our proposal is in the same spirit but, whilst it has some common elements, its focus is different: encompassing product as well as process changes, and linkages with suppliers, customers and other sources of technology. Already in the 18th century, Adam Smith was aware of diversity in the sources of technical change, and of its dynamic nature; as we shall soon see, he identified many elements of our proposed taxonomy in Chapter One of *The Wealth of Nations* [54].

Second, there is the work of Penrose [41] on the nature of firms' diversification activities, and the importance of their technological base. Recent French writings, exploring the notion of *filière*, are in the same tradition [58], as is the work of Ansoff [2] and others on business strategy, and the recent contribution by Teubal [59] on the nature of technological learning.

Third, a number of analysts have explored the cumulative and dynamic nature of technical change: for example, Dosi [8], Freeman et al. [12], Gold [15] Nelson and Winter [31;32], Rosenberg [42;43] and Sahal [47]. From their research has emerged the notion of "technological trajectories," namely, directions of technical development that are cumulative and self-generating, without repeated reference to the economic environment external to the firm.

Nelson has gone further and suggested a framework for explaining technological trajectories [20]. He has argued that it any institutional framework, public or private, market or non-market, technical change requires mechanisms for generating technical alternatives; for screening, testing and evaluating them; and for diffusing them. In the Western market framework, the rate and direction of technical change in any sector depends on three features: first, the sources of technology; second, the

nature of users' needs; third, the possibilities for successful innovators to appropriate a sufficient proportion of the benefits of their innovative activities to justify expenditure on them.

For our purposes, there can be a number of possible sources of technology. Inside firms, there are R & D laboratories and production engineering departments. Outside firms, there are suppliers, users, and government financed research and advice. Similarly, users' needs can vary. For standard structural or mechanical materials, price is of major importance one certain performance requirements are met. For machinery and equipment used in modern and interdependent systems of production, performance and reliability will be given a higher premium relative to purchase price. In the consumer sector – as Rosenberg [41] and Gershuny [15] have pointed out – modern equipment is used extensively for "informal" household production. However, compared to their equivalents in the formal economy, purchase price will have a higher premium relative to performance, given that household systems of production are relatively small scale, with little technical interdependence, and with weak pressures of competition from alternative production systems.

The methods used by successful innovators to appropriate the benefits of their activities compared to their competitors will also vary. [14] For example, process innovations can be kept secret; some product innovations can be protected by natural and lengthy technical lags in imitation (e.g. aircraft), whilst others require parent protection (e.g. pharmaceuticals); and both product and process innovations may be difficult to imitate because of the uniqueness of the technological knowledge and skills in the innovating firm.

These ingredients are summarised in table 5, where column 1 defines the categories of firm, column 2 enumerates typical core sectors for such firms, columns 3–5 describe the determinants and the nature of the technological trajectories of the firms, and columns 7–10 identify some of the measured characteristics of these trajectories. We shall now go on to describe and discuss them in more detail.

---

[14] For more detailed discussion, see Taylor and Silberston [46] Scherer [50] and von Hippel [64–66].

3.2. Supplier dominated firms

Supplier dominated firms can be found mainly in traditional sectors of manufacturing, and in agriculture, housebuilding, informal household production, and many professional, financial and commercial services They are generally small, and their in-house R&D and engineering capabilities are weak. They appropriate less on the basis of a technological advantage, than of professional skills, aesthetic design, trademarks and advertising. Technological trajectories are therefore defined in terms of cutting costs.

Supplier dominated firms make only a minor contribution to their process or product technology. Most innovations come from suppliers of equipment and materials, although in some cases large customers and government-financed research and extension services also make a contribution. Technical choices resemble more closely those described in Salter's vintage model, the main criteria being the level of wages, and the price and performance of exogenously developed capital goods.

Thus, in sectors made up of supplier dominated firms, we would expect a relatively high proportion of the process innovations used in the sectors to be produced by other sectors, even though a relatively high proportion of innovative activities in the sectors are directed to process innovations. According to Scherer's data on the sectoral patterns of production and use of technology in the USA [51, table 2], the following sectors have such characteristics: textiles; lumber; wood and paper mill products; printing and publishing; and construction; in other words, precisely the types of sectors predicted by our taxonomy and theory. [15]

With our data on innovating firms in the UK, we are able to identify these and other characteristics of supplier dominated firms (as well as those of production intensive and science-based firms, described in subsections 3.3 and 3.4 below). Table 6 shows clearly the supplier dominated characteristics of textile firms. Before describing them, we shall define precisely the content of each of the columns of table 6, since tables 7, 8 and 9 present similar figures for the other categories of firms:

Column 1 defines the principal two digit sector of activity of the innovating firms.

Column 2 gives the percentage of innovations used in the sector that are produced by innovating firms principally in the sector. [16] It shows the degree to which firms in the sector develop their own process technology.

Column 3 shows the percentage of innovations produced by firms principally in the sector that are used in other sectors: in other words, the percentage of product innovations. [17]

Column 4 shows the size distribution of innovating firms principally in the sector. These figures are identical to those in columns 7, 8 and 9 of table 3.

Column 5 gives more detail on the nature of innovating firms' innovations outside their principal sector of activity. It breaks down the figures of column 1, table 4 between "vertical" and "concentric/conglomerate" technological diversification. These terms are taken from the writings of Ansoff [2] on business strategy. The "vertical" figure is the percentage of the innovations produced by innovating firms, that are outside the innovating firms' principal sector of activity, but used within the innovating firms' sector: it reflects the relative importance of technological diversification into the equipment, materials and components for their own production. The "concentric/conglomerate" figure is the percentage of the innovations that are both produced and used outside the principal sector of the innovating firms' activities: it reflects the relative importance of technological diversification into related and unrelated product markets.

Column 6 shows the origins of all the innovations in the sector, broken down between those produced by firms principally in the sector, those both produced and used by firms principally producing outside the sector (i.e. users of the output of the sector), and those from other sources. The figure in the first sub-column of column 6 adds up to 100 percent with the figure in column 5 of table 4.

---

[15]   Scherer's data are incomplete for agriculture and for services, which we would predict to have similar characteristics.

[16]   This percentage is not identical to the one in column 5 of table 2, since the former is based on the sector of the innovation, whilst the latter is based on the sector of principal activity of the innovating firm.

[17]   This percentage is not identical to the one in column 1 of table 2, for the reasons given in footnote 16.

Table 6
Characteristics of innovations produced and used by firms producing principally textiles, and leather & footwear

Principal sector of firm's activity (2-digit) (1)	Innovations used that are produced by firm		Innovations produced by firms that are used in other sectors (3)		Size distribution of innovating firm (rows add up to 100%) (4)			Innovations produced by firms in sector (No. produced)
	% (2)	Number used	%	Number produced	10,000+	1000–9999	1–999	
XIII Textiles	15.6	377	23.4	77	35.1	40.3	24.7	77
XIV&XV Leather and Footwear	48.9	45	56.0	50	44.0	18.0	38.0	50
Total: All sectors in sample	49.3	1401 [a]	64.0	2265	53.1	21.9	24.9	2265

Principal sector of firm's activity (2-digit)	% [b] firms' innovations outside principal sector of activity are (5)		Innovations produced by firms in sector (No.)	% of innovations in firms' sector of activity produced by (6)			Innovations produced in sector (No.)
	Concentric/ conglomerate	Vertical		Firms principally in the sector	Firms principally in other sectors that produce and use the innovation	Other	
XIII Textiles	3.9	20.8	77	63.8	2.2	34.0	91
XIV&XV Leather and Footwear	42.0	8.0	50	73.5	–	26.5	34
Total: All sectors in sample	20.3	11.2	2265	68.6	11.2	20.3	2265

[a] Includes only those innovations used in sectors specified in table 2.
[b] The sum of the two percentages is equal to that in column 1 in table 4.

In the case of textile firms, table 6 shows a high degree of dependence on external sources for process technology (column 2), a relatively small proportion of innovative activity devoted to product innovations (column 3), a relatively small average size of innovating firm (column 4), technological diversification mainly vertically into production technology with very little movement into other product markets (column 5), and a relatively big contribution to innovations in the sector by firms with their principal activities elsewhere, but not from sectors using textiles (column 6). More detailed data show the considerable importance to textile firms of machinery firms in supplying process technology, and of chemical firms in supplying process technology and in making innovations in the textile sector itself.

Table 6 also shows that innovating firms principally producing in leather and footwear do not fall so neatly into the category of supplier dominated firms. Certainly they are relatively small (column 4), and their users make a relatively small contribution to innovation in their principal sector of activity (column 6). However, they also produce a sizeable proportion of product innovations (column 3), as well as making a strong contribution to their own process technology (column 2), and they have a high degree of concentric/conglomerate technological diversification (column 5).

Close examination shows that all this technological diversification is into textile machinery innovations that find their main use in the textile sector. This pattern reflects the coding practice used by Townsend and his colleagues in their survey [60]. However, it does not reflect the fact that there is no separate SIC category for leather working machinery, that innovations in textile machinery have applications in the manufacture of leather goods, and that – although the main uses of the identified innovations in textile machinery were in the textile sector – they also found uses in the manufacture of leather goods. In other words, firms principally in leather goods were in fact making a major contribution to the development of their own process technology. In this case, they begin to join the production intensive category, which we shall now describe.

3.3. Production intensive firms

Adam Smith described some of the mechanisms associated with the emergence of production inten-

sive firms, namely, the increasing division of labour and simplification of production tasks, resulting from an increased size of market, and enabling a substitution of machines for labour and a consequent lowering of production costs. Improved transportation, increasing trade, higher living standards and greater industrial concentration have all contributed to this technological trajectory of increasing large-scale fabrication and assembly production. Similar opportunities for cost-cutting technical change exist in continuous processes producing standard materials, where the so-called two-thirds engineering law means that unit capacity costs can potentially be decreased by 1 percent by every 3 percent increase in plant capacity.

The technological skills to exploit these latent economies of scale have improved steadily over time. In fabrication and assembly, machines have been able to undertake progressively more complex and demanding tasks reliably, as a result of improvements in the quality of metals and the precision and complexity of metal forming and cutting, and in power sources and control systems. In continuous processes, increased scale and high temperatures and pressures have resulted from improvements in materials, control instrumentation and power sources. [18]

The economic pressure and incentives to exploit these scale economies are particularly strong in firms producing for two classes of price-sensitive users: first, those producing standard materials; second, those producing durable consumer goods and vehicles. In reality (if not in various models of technical change), it is difficult to make these scale-intensive processes work up to full capacity. Operating conditions are exacting, with regard to equipment performance, controlling physical interdependencies and flows, and the skills of operatives. In such complex and interdependent production systems, the external costs of failure in any one part are considerable. If only for purposes of "trouble-shooting," trained and specialist groups for "production engineering" and "process engineering" have been established. As Rosenberg [42] has shown, these groups develop the capacity to identify technical imbalances and bottlenecks which, once corrected, enable improvements in productivity. Eventually they are able either to specify or design new equipment that will improve

---

[18]   See Levin [22] for well documented examples.

productivity still further. Thus, one important source of process technology in production-intensive firms are production engineering departments.

Adam Smith also pointed out that process innovations are also made "... by the ingenuity of the makers of machines when to make them became the business of a peculiar trade" [54]. The other important source of process innovations in production-intensive firms are the relative small and specialised firms that supply them with equipment and instrumentation, and with whom they have a close and complementary relationship. Large users provide operating experience, testing facilities and even design and development resources for specialised equipment suppliers. Such suppliers in turn provide their large customers with specialised knowledge and experience as a result of designing and building equipment for a variety of users, often spread across a number of industries. Rosenberg [42] describes this pattern as "vertical disintegration" and "technological convergence". He draws his examples from metal-forming machinery; the same process can be seen at work today in the functions of production monitoring and control performed by instruments. These specialised firms have a different technological trajectory from their users. Given the scale and interdependence of the production systems to which they contribute, the costs of poor operating performance can be considerable. The technological trajectories are therefore more strongly oriented towards performance-increasing product innovation, and less towards cost-reducing process innovation.

The way in which innovating firms appropriate technological advantage varies considerably between the large-scale producers, and the small-scale equipment and instrument suppliers. For the large-scale producers, particular inventions are not in general of great significance. Technological leads are reflected in the capacity to design, build and operate large-scale continuous processes, or to design and integrate large-scale assembly systems in order to produce a final product. Technological leads are maintained through know-how and secrecy around process innovations, and through inevitable technical lags in imitation, as well as through patent protection. For specialised suppliers, secrecy, process know-how and lengthy technical lags are not available to the same extent as a means of appropriating technology. Competi-

tive success depends to a considerable degree on firm-specific skills reflected in continuous improvements in product design and in product reliability, and in the ability to respond sensitively and quickly to users' needs.

The characteristics of large-scale producers and of specialised suppliers in the production intensive category are reflected in tables 7 and 8. Table 7 shows that, in our sample of innovations, firms with their principal activities in five of the two digit sectors in our sample have the characteristics of scale-intensive producers in the production intensive category: food products, metal manufacturing, shipbuilding, motor vehicles, and glass and cement. In these categories, innovative firms produce a relatively high proportion of their own process technology (column 2), to which they devote a relatively high proportion of their own innovative resources (column 3). Innovating firms are also relatively big (column 4), they have a relatively high level of vertical technological diversification into equipment related to their own process technology (column 5), and they make a relatively big contribution to all the innovations produced in their principal sectors of activity (column 6).

Table 8 shows the very different pattern in mechanical and instrument engineering firms. They also produce a relatively high proportion of their own process technology (column 2), but the main focus of their innovative activities is the production of product innovations for use in other sectors (column 3). Innovating firms are relatively small (column 4); they diversify technologically relatively little, either vertically or otherwise (column 5); and they do not make a relatively big contribution to all the innovations produced in their principal sector of activity, where users and other firms outside the sectors make significant contributions (column 6).

A more detailed examinations of the data at the three digit level shows that, within mechanical engineering, firms in all the product groups in the sample have a high proportion of their innovative resources devoted to product innovation, are technologically relatively specialised, and (with the exception of firms principally producing industrial plant) are relatively small. However, about 20 percent of the innovations are made by general engineering firms that produce in a range of mechanical engineering products, and the size distribu-

Table 7
Characteristics of innovations produced by firms producing principally in scale-intensive sectors

Principal sector of firm's activity (2-digit) (1)	Innovations used that are produced by firm (2)		Innovations produced by firms that are used in other sectors (3)		Size distribution of innovating firm (rows add up to 100%) (4)			Innovations produced by firms in sector (No. produced)
	%	Number used	%	Number produced	10,000+	1000–9999	1–999	
III Food	58.8	68	48.8	78	79.5	7.7	12.8	78
VI Metal manufacturing	62.3	130	43.4	143	62.9	32.8	4.2	143
X Shipbuilding	64.5	90	34.8	89	61.8	34.8	3.3	89
XI Motor vehicles	45.7	221	36.9	158	72.2	20.3	7.6	158
XVI Glass and cement	68.3	63	50.6	87	74.7	16.1	9.1	87
Total: All sectors in sample	49.3	1401 a	64.0	2265	53.1	21.9	24.9	2265

Principal sector of firm's activity (2-digit)	% b firms' innovations outside principal sector of activity are (5)		Innovations produced by firms in sector (No.)	% of innovations in firms' sector of activity produced by (6)			Innovations produced in sector (No.)
	Concentric/ conglomerate	Vertical		Firms principally in the sector	Firms principally in other sectors that produce and use the innovation	Other	
III Food	16.7	14.1	78	83.1	3.1	13.9	65
VI Metal manufacturing	17.5	16.8	143	68.6	8.0	23.4	137
X Shipbuilding	21.3	37.1	89	71.2	13.5	15.4	52
XI Motor vehicles	12.6	20.9	158	82.0	1.6	16.4	128
XVI Glass and cement	13.8	18.4	87	81.9	5.6	12.5	72
Total: All sectors in sample	20.3	11.2	2265	68.6	11.2	20.3	2265

a Includes only those innovations used in sectors specified in table 2.
b The sum of the two percentages is equal to that in column 1 in table 4.

Table 8

Characteristics of innovations produced and used by firms producing production equipment

Principal sector of firm's activity (2-digit)	Innovations used that are produced by firm		Innovations produced by firms that are used in other sectors		Size distribution of innovating firm (rows add up to 100%)			Innovations produced by firms in sector (No. produced)
	%	Number used	%	Number produced	10,000+	1000–9999	1–999	
	(2)		(3)		(4)			
(1)								
VII Mechanical engineering	55.1	169	82.6	536	24.3	36.9	38.8	536
VIII Instrument engineering	58.4	60	81.4	187	24.6	21.4	54.0	187
Total: All sectors in sample	49.3	1401 [b]	64.0	2265	53.1	21.9	24.9	2265

Principal sector of firm's activity (2-digit)	Innovations produced by firms in sector (No.)	% [b] firms' innovations outside principal sector of activity are		% of innovations in firms' sector of activity produced by			Innovations produced in sector (No.)
		Concentric/ conglomerate	Vertical	Firms principally in the sector	Firms principally in other sectors that produce and use the innovation	Other	
		(5)		(6)			
VII Mechanical engineering	536	15.1	0.9	68.1	15.3	16.8	633
VIII Instrument engineering	187	9.7	10.2	45.2	19.3	35.5	332
Total: All sectors in sample	2265	20.3	11.2	68.6	11.2	20.3	2265

[a] Includes only those innovations used in sectors specified in table 2.
[b] The sum of the two percentages is equal to that in column 1 in table 4.

tion of which is bigger than other mechanical engineering, being close to the average for the sample of innovations as a whole. In instrument engineering, innovations are produced by firms in a wide range of user sectors, as well as by firms principally in mechanical engineering and in electronic capital goods.

3.4. Science-based firms

The third category, namely science-based firms, was also foreseen (if not observed) by Adam Smith who spoke of the contribution to technical of "... those who are called philosophers or men of speculation, whose trade it is not to do anything, but to observe everything; and who, upon that account, are often capable of combining together the powers of the most distant and dissimilar objects." From the data on innovations described above, science-based firms are to be found in the chemical and the electronic/electrical sectors. In both of them, the main sources of technology are the R&D activities of firms in the sectors, based on the rapid development of the underlying sciences in the universities and elsewhere.

As Freeman et al. [12] have shown, the development of successive waves of products has depended on *prior* development of the relevant basic science: in particular, of synthetic chemistry and biochemistry for the chemical industry; and of electromagnetism, radio waves and solid state physics for the electrical/electronic industry. Synthetic chemistry has enabled the development of a wide range of products, with useful structural, mechanical, electrical, chemical or biological characteristics, ranging from bulk materials replacing wood, steel and natural textiles, to specialised and expensive chemical and biological agents for medical or other uses. Post-war advances in the fundamentals of biochemistry are enabling the extension of these skills and techniques into biological products and processes.

Advances in electromagnetism, radio waves and solid state physics have enabled products and applications related to the availability of cheap, decentralised and reliable electricity, communications and (now) information processing, storage and retrieval. Applications in electricity vary from huge transformers to small motors within mechanical systems, in communications from expensive radar and satellite tracking systems to cheap tran-

sistor radios, and in information from huge computers to electronic wristwatches.

This pervasiveness has dictated the technological trajectories of firms in the science based sectors. The rich range of applications based on underlying science has meant that successful and innovative firms in them have grown rapidly, [19] and have had little incentive to look for innovative opportunities beyond their principal sector. Given the sophistication of the technologies and underlying sciences, it has been difficult for firms outside the sectors to enter them. The pervasive applications have also meant a wide variance in relative emphasis on production and process technology within each of the sectors, reflecting the different cost/performance trade-off for consumer goods, standard materials and specialised professional applications.

Firms appropriate their innovating leads through a mix of methods (i.e. patents, secrecy, natural technical lags, and firm-specific skills). Patent protection is particularly important in fine chemicals, with specific high grade applications, where the predominant product innovations can be quickly and cheaply imitated without it. [20] In addition, dynamic learning economies in production have been an important barrier to the entry of imitators in continuous process technology, large-scale assembly and – over the past 25 years – in the production of electronic components. According to Dosi [8], the particularly rapid rate and the form of technical change in electronic components involved a "paradigm shift." New firms have been able to enter the electronics industry, and to grow rapidly by aggressive product innovation coupled with the exploitation of steep dynamic economies of scale.

In the data on innovations in the UK collected by Townsend and his colleagues, characteristics of science-based firms emerge most clearly for those principally in chemicals, Table 9 shows that they produce a relatively high proportion of their own process technology (column 2), as well as a high proportion of product innovations that are used in other sectors (column 3). They are also relatively big (column 4), most of their technological diversification is concentric/conglomerate rather

---

[19] See, for example, the research of Rumelt [56] on the growth and diversification of US firms.

[20] See, in particular, the empirical studies of Taylor and Silberston [56].

Table 9
Characteristics of innovations produced and used by firms producing principally chemicals and electrical/electronic products

Principal sector of firm's activity (2-digit)	Innovations used that are produced by firm		Innovations produced by firms that are used in other sectors		Size distribution of innovating firm (rows add up to 100%)			Innovations produced by firms in sector (No. produced)
	%	Number used	%	Number produced	10,000+	1000–9999	1–999	
(1)	(2)		(3)		(4)			
V Chemicals	77.4	71	78.0	290	82.4	7.9	9.6	290
IX Electrical and electronic engineering	80.2	107	60.9	343	65.9	12.2	22.0	343
Total: All sectors in sample	49.3	1401 ᵃ	64.0	2265	53.1	21.9	24.9	2265

Principal sector of firm's activity (2-digit)	% ᵇ firms' innovations outside principal sector of activity are		Innovations produced by firms in sector (No.)	% of innovations in firms' sector of activity produced by			Innovations produced in sector (No.)
	Concentric/ conglomerate	Vertical		Firms principally in the sector	Firms principally in other sectors that produce and use the innovation	Other	
	(5)			(6)			
V Chemicals	21.7	4.8	290	84.8	2.4	12.8	251
IX Electrical and electronic engineering	21.5	2.3	343	77.0	11.5	11.5	339
Total: All sectors in sample	20.3	11.2	2265	68.6	11.2	20.3	2265

ᵃ Includes only those innovations used in sectors specified in table 2.
ᵇ The sum of the two percentages is equal to that in column 1 in table 4.

than vertical (column 5), and they produce a relatively high proportion of all the innovations made in their principal sector of activity (column 6). More detailed data also show that, within the two digit chemical sector, the detergent product group has a relatively high proportion of process innovations; and that the technological diversification of chemical firms outside their principal two digit sector is mainly into instruments, machinery and textiles. According to table 9, firms principally in electronic and electrical engineering also have most of the predicted characteristics of science-based firms: a relatively high contribution to own process technology (column 2), relatively big innovating firms (column 4), mainly concentric/conglomerate diversification [21] (column 5), and a relatively big contribution to all innovations in their principal sector of activity (column 6).

However, the proportion of product innovations, although absolutely large, is relatively small (column 3); more detailed data show that this cannot be explained simply by the preponderance of process innovations in broadcasting equipment, but also reflects a high proportion of innovations in electronic components that are produced and used by firms principally producing electronic capital goods. Furthermore, the relatively big contribution to the production of innovations made by firms with less than 1000 employees (table 9, column 4) reflects the increasing contribution made in the 1970s by such firms in the computer product group.

Finally, more detailed data suggest that large, diversified firms make a bigger contribution to innovations by science-based firms, than to those by specialised equipment supplies. As we saw in subsection 3.3, general engineering firms produced 20 percent of all the innovations in mechanical engineering. In chemicals, firms principally in general chemicals produced about 40 percent of the whole; and in electronics/electrical products, firms principally in electronics capital goods produced about 50 percent.

3.5. Technological linkages and changing trajectories

Linkages amongst the different categories of firm go beyond those described in the production

[21] More detailed data show that this is mainly into the mechanical engineering and scientific instruments sectors.

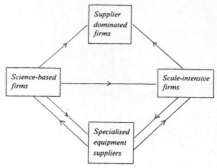

Fig. 1. The main technological linkages amongst different categories of firm.

intensive category (subsection 3.3. above). Figure 1 tries to represent the main technological flows emerging from our taxonomy and theory. Supplier dominated firms get most of their technology from production intensive and science-based firms (e.g power tools and transport equipment from the former; consumer electronics and plastics from the latter). Science-based firms also transfer technology to production intensive ones (e.g. the use of plastics, and of electronics, in the automobile industry). And, as we have seen, science-based and production intensive firms both receive and give technology to specialised suppliers of production equipment.

We have also argued that technological linkages amongst sectors can go beyond transactions involving the purchase and sale of goods embodying technology. They can include flows of information and skills, as well as technological diversification into the main product areas of suppliers and customers. Examples include the contribution of scale-intensive firms to the technology of their equipment suppliers and of chemical and electronics firms to innovations in textiles, scientific instruments and mechanical engineering.

Our data do not yet enable us to analyse if and how patterns of technical change in specific sectors change over time. We have hinted in subsection 3.3 that sectors can shift from the supplier dominated to the production-intensive pattern as a result of access to larger markets by individual firms, and of autonomous and induced improvements in capital goods: a contemporary example might be certain commercial and financial services, given producer concentration and rapid technical progress in information processing equipment. On the other hand, analysts like Utterback and Abernathy [70] would predict on the basis of their "product cycle" model that, where process technology has matured, sectors may shift from the production intensive to the supplier dominated pattern: for example, in bulk synthetic chemicals today, it is said that this focus has shifted from the chemical firms to their specialised suppliers of process equipment [6]. Whatever regularities in such change are eventually observed the above two examples suggest that trends in the *rate* of technological change will be an important explanatory variable. Of particular interest will be a systematic exploration of the effects of radical technological changes (e.g. solid state electronics) on prevailing paths and patterns of technical change. [22]

4. Some analytical implications

Our proposed taxonomy and theory have a number of implications for analysis of the nature, sources, determinants and economic effects of technical change. We shall now identify some of the most obvious ones, without pretending to be comprehensive in either breadth or depth of discussion.

4.1. Science and technology push versus demand pull

There is the continuing debate about the relative importance of "science and technology push" and "demand pull" in determining patterns of innovative activity, and in triggering innovative activity. As Mowery and Rosenberg [26] and others have pointed out, both technology push and demand pull are necessary for any successful innovation, and much of the debate about the relative importance of the two has been ill-conceived. Nonetheless, according to Schmookler [53], "demand-pull" has been a stronger influence than "science and technology push" on patterns of innovative activity, both across industry and over time. Across industry, he found a stronger statistical association between the volume of innovative

activity in capital goods (as measured by patents) and the volume of investment activity in user industries, than between the volume of innovative activity and of output in the supplier industries. Over time, he found that changes in the volume of innovative activity followed changes in the volume of investment activity. Using a more comprehensive data base, Scherer [52] has recently confirmed the former of Schmookler's findings, but could find no evidence of a lag between investment and innovative activities.

In our taxonomy, the close relationship between investment in user sectors and innovative activities in upstream capital goods comes as no surprise. Investment activities in supplier dominated and production intensive firms are likely to stimulate innovative activities in both the production engineering departments of user firms, and the upstream firms supplying capital goods. [23] To the extent that these investment activities are planned in advance, and co-ordinated with the activities of production engineering departments of investing firms and with firms supplying production equipment, we would also expect – as Scherer found – that the lag between investment and innovative activities would tend to disappear.

However, we would not expect in our science-based firms a similarly neat and lagged correspondence between the volume of investment in user sectors, and of innovative activities. Recent research by Walsh [68] has shown that the emergence of major new product families in the chemical industry in the twentieth century has been *preceded* by an upsurge of scientific and inventive activities. Furthermore, Scherer [52] found that in materials sectors, in contrasts to capital goods, the statistical relationship between the volume of innovative activities and of investment in user sectors is much weaker; given the role of the chemical industry in developing synthetic *substitute* materials, this should not surprise us. Finally, Scherer [52] found that the relationship between the volume of innovative activities and the output of the supply industry becomes much stronger when account is taken of difference amongst sectors in scientific and technological opportunity – the relationship between the two being particularly strong in the

---

[22] For further discussion on the automobile industry see Anderson et al. [1]. More generally see Ergas [9].

[23] User sectors covered in Schmookler's analysis included petroleum refining, synthetic fibres, glass, sugar, tobacco, railroads, textiles and apparel, and timber and paper.

organic chemicals and electronics sectors, where we would expect science-based technical opportunities to be particularly strong.

4.2. Product versus process innovation

Our proposed theory also offers an explanation of the balance in different sectors between product and process innovation. We would expect the relative importance of product innovation in a sector to be positively associated with its R & D and patent intensity; and negatively associated with proxy measures of the scale and complexity of its process technology, such as its capital/labour ratio, average size of production plant, or sales concentration ratios.

The reasoning behind such an expectation runs as follows. In product groups with a high proportion of science-based firms, we would expect a relatively high R & D intensity, and a high proportion of product/market opportunities generated outside the product groups. The relationship should be even stronger between patent intensity and product innovation, given that – in addition to R & D activities – patent statistics reflect the innovative activities in small firms, and the production engineering departments of large firms, both of which are particularly important sources of product innovation in mechanical and instrument engineering. On the other hand, in sectors with a relatively high proportion of production intensive firms, we would expect both a realtively high proportion of resources to be devoted to process innovations, on the one hand, and relatively high capital intensities, size of plant and industrial concentration on the other.

As can be seen in the Appendix to this paper, the regression based on our (very imperfect) statistics are consistent with our expectations (E1, E2, E3). [24] The signs are correctly predicted and, in some equations, explanatory variables are significant at the 1 percent and $2\frac{1}{2}$ percent level. Only the capital–labour ratio has a low explanatory power in all of the equations that we tried, which may say as much about the problems of measuring capital as about the predictive powers of our theory.

[24]   E1, E2 etc. refers to the relevant equations in the Appendix.

4.3. The locus of process innovation

Our taxonomy and theory also lead to expectations about the degree to which firms develop their own process innovations, or buy them from "upstream" suppliers of production equipment. In sectors with supplier-dominated firms, we would expect firms and production plant to be small in size, and innovations to come by definition from suppliers. In sectors with production intensive firms, we would expect firms and plant to be large in size, and a high proportion of process technology to be generated in-house. The same will be the case in science-based firms, especially in products involving continuous process and assembly technologies. In other words, we would expect a positive relationship between the proportion of a sector's process technology generated in-house, on the one hand, and the size of firms and of plant in the sector on the other.

Other writers have made related but somewhat different predictions, namely, that upstream equipment suppliers became relatively more important sources of process innovations as the absolute size of the market for the production process equipment grows. For Rosenberg [42], this reflects a greater division of labour in production resulting from a larger size of market. For Utterback and Abernathy [70], it reflects the large size and technological stability in firms at the later stages of the product cycle.

Von Hippel [67] and Buer [5] make predictions from a different basis, arguing that the balance between in-house development and recourse to upstream suppliers depends on the prospective benefits to be appropriated by the user of the production equipment. They argue that the benefits of appropriation by the user – compared to those of the supplier – increase with the degree of concentration in the user sector. The proportion of process technology developed in-house will therefore increase with the degree of user concentration. The data at present at our disposal does not enable an authoritative statistical test of these various hypotheses. Our measure of the proportion of process technology developed in-house is somewhat shaky, and we do not have comprehensive data on sources of process technology for sectors outside manufacturing. However, we can explore the relationship across sectors between the proportion of process technology developed in-house, on the one hand, and a range of variables reflecting the different hypotheses described above: average size of innovating firms, capital–labour ratio and average plant size (this writer's hypothesis); volume of investment in plant and equipment in equipment-using sectors (Rosenberg; Utterback and Abernathy); five firm concentration ratios in equipment using sectors (von Hippel; Buer).

This author's explanatory variables perform least well. Although the signs are all corectly predicted, none is statistically significant. However, the other hypotheses receive strong statistical confirmation (E4). The proportion of process technology developed by firms in the sector is negatively related to the absolute size of the market for process equipment, and positively to the degree of concentration of sales in the user sector.

4.4. Diversification

On the economic impact of technical change, our taxonomy and theory may also offer some insights into mechanisms of diversification, whether in terms of R&D and technology, or in terms of economic activity. Nelson [27] once suggested a positive relationship between the performance of basic research by firms and the diversity of their output, given that the uncertain results of basic research are more likely to find a use in a diversified firm than a specialised one. According to Scherer, however, the results of statistical analysis of the relationship between spending on basic research, and total R&D, on the one hand, and diversification, on the other "... have been mixed and to some extend contradictory" [49, p. 422].

According to our taxonomy, those related to total R&D are likely to be so, since we postulate a different causality, and predict an indeterminate and messy relationship between the variables. It is indeterminate (or, at least, non-linear), given that we predict relatively low levels of technology-based two digit diversification in sectors that are both R&D intensive (chemicals, instruments, and electrical/electronics), and low R&D spenders (supplier dominated). It is messy, given that the potential for technology-based diversification in science-based firms is much higher at the three digit than at the two digit level.

Furthermore, in both production intensive and supplier dominated firms the links between technology and production diversification may be

weak. This emerges from a comparison of Hassid's data on production diversification [17] in British firms with those for technology in table 4. Production intensive firms diversify less in production than in technology, possibly because they do not exploit themselves all the opportunities open to them for technology-based diversification upstream into equipment supply. Textile firms, on the other hand, diversify more in production than in technology, possibly because of non-technological complementarities with other sectors.

However, we can, on the basis of our taxonomy, make some predictions about the factors determining potential technological paths of diversification in innovating firms, as a function of their principal activity. The relative importance of upstream (i.e. vertical) technological diversification into sectors supplying equipment is likely to be negatively associated with R&D intensity (which tends to provide technological opportunities concentrically or downstream), and positively associated with the scale and complexity of production technology (which induces innovative activities on production techniques and upstream equipment). Using the capital–labour ratio, and average plant size as proxy measures for scale and complexity of production technology, we find none of the expected statistical relationships at the three digit level. However, at the two digit level, and using the 20 firm concentration ratio as a proxy for scale and complexity of process technology, the statistical relations are as expected, and significant at the 1 percent level (E5).

Our taxonomy and theory may also help us better understand the links at the level of the firm between firm strategy and R&D strategy. Although much study has been devoted to the "tactical" problems of the management of activities necessary for innovations, [25] relatively little attention has been devoted to the "strategic" question of the role of technology in determining the future activities of the firm, and in particular its future product lines.

We propose a model that identifies the "technological trajectories" of firms as a function of their principal activities, and that enables us to predict possible paths of technological diversification across product lines and sectors. Given the wealth and detail of statistical data now becoming availa-

[25] See the survey by Rothwell [45].

ble on individual firms' technological activities, it will be possible to put our predictions to the statistical test by answering two questions. First, do firms with the same principal activities have statistically similar distributions of technological activity across product groups and technical areas? Second, are the distributions those predicted from our taxonomy and theory? Whilst we should not claim to be able to predict the specific competitive strengths and weaknesses of particular firms, we would at least be able to identify and explain the technological opportunities and constraints that in part govern their behaviour and choice.

However, we can predict with greater certainty that, at the level of individual firms, the degree of technological diversification will be positively associated with its size. This will reflect three mechanisms in our taxonomy and theory: first, large-scale production intensive firms procuding innovations upstream, principally in mechanical engineering and instruments; second, the possibilities open to small and specialised firms producing production equipment to remain small, competitive and technologically dynamic; third, the possibilities open to science-based firms for technological diversification beyond their principal three digit (but within their principal two digit) sector. Given these patterns of technological diversification in science-based firms, we would expect this relationship to be stronger at the three digit than at the two digit level.

Our data on innovations confirm these predictions. The size distribution of firms producing innovations outside their principal three digit sector is more skewed than average innovating firms towards large size: 69.9 (53.2) percent with 10,000 and more employees; 14.0 (23.2) percent with between 1000 and 9999 employees; 16.1 (23.7) percent with fewer than 1000 employees. [26] Across three digit sectors, we find a positive and statistically significant relationship (at the 5 percent level) between the degree to which innovating firms diversify technologically outside their three digit sector, and their average size in each sector.

Finally, we would predict on the basis of our taxonomy that, amongst science-based firms, relatively high levels of basic research will allow more innovations, more diversification beyond three to

four digit sectors and more growth. In a recent study, Link and Long [24] found that the two most significant factors explaining differences amongst 250 US manufacturing firms in the proportion of sales spent on basic research were diversification at the four digit level, and having principal activities in science-based sectors. Although our proposed causality runs the other way, our results are consistent with those of Link and Long. Similarly, in a study of US firms in the petrochemicals industry, Mansfield [25] recently found a positive relationship between basic research as a percentage of value added, on the one hand, and the rate of growth of total factor productivity on the other hand. If one assumes further that growth of total factor productivitiy is positively associated with growth of output, then Mansfield's results are consistent with our taxonomy and theory.

4.5. Firm size and industrial structure

The causal links running from innovation to firm growth and to firm size are central to the recent research on the dynamics of Schumpeterian competition by Nelson and Winter [32]. They predict that, in industry with rapid rates of technical change, with uncertainty in the outcomes of investments in innovative activities, and with the strong possibilities for innovative firms to appropriate their innovative advantage, there are powerful tendencies over time towards the concentration of both production and innovative activities.

Our data and theory are consistent with these assumptions and outcomes for our science-based category of firms, but not for our supplier dominated or production intensive categories. In supplier dominated firms, any increase in firm size usually cannot be attributed to innovation, given that not much of it is generated in the sector, although increased size may enable (as described by Adam Smith) the introduction of more efficient process technology. In production intensive firms, innovation is associated with large and increasing size not, as Nelson and Winter [32] suggest, through the uneven exploitation amongst firms of a rich crop of new product/market opportunities, but through the search for increasing static scale economies in production. [27]

---

[26]   Numbers in brackets refer to the percentage for all innovations: see table 4.

[27]   See, for example, Levin [22].

The most important difference between Nelson and Winter's and our proposed model is the stable existence of small firms making innovations in production equipment and instrumentation. Rosenberg's description of textile machinery firms in the first half of 19th centry [42] is not very different – apart from the state of the technological art – from Rothwell's description of textile machinery firms in the second half of the 20th century [44]. As we have been in subsection 2.4, small, specialised and technologically dynamic equipment suppliers in mechanical and instrument engineering continue to live in symbiosis with even larger production intensive and science-based firms, and to confound trends towards Schumpeterian concentration. This is puzzling given that, as Rosenberg [42] has pointed out, common skills, techniques and know-how underlie all mechanical engineering products, just as they do in chemical-based and electrical/electronic-based firms. Why, then, have firms in these science-based sectors typically diversified and grown big on the basis of their accumulated skills, whereas those in mechanical and instrument engineering typically have not?

No definite answer can be given in this paper. Suffice here to suggest that explanations probably lie in sectoral differences in technology sources, users' requirements and appropriability. [28] Compared to chemical and electronic firms, those in mechanical and instrument engineering depend more on their customers for information and skills related to the operating performance, and to the design, development and testing of their products; they therefore can afford to remain small, but do not accumulate the same range and depth of technological skills. They also sell in markets that do not have such pronounced product cycle characteristics, and therefore have less market pressure to diversify. Finally, they find it more difficult to appropriate the benefits of their innovations, given the overwhelming importance of produce innovation, and relatively low barriers to entry, resulting from relatively small scale expenditures on product development, and the existence of many independent sources of skills and know-how in the production engineering departments of large firms.

Innovative small firms are now to be found not only in instruments and mechanical engineering, but also in electronics: according to Townsend et al. the share of firms with up to 1000 employees increased in electronics in the 1970s. There has been one essential difference between innovative firms in instruments and mechanical engineering innovations, and those in electronics. Whilst the former have on the whole remained relatively small and specialised, a few of the latter became very large through precisely the mechanism of innovation and growth described by Nelson and Winter.

According to Dosi [8], new small firms can become big in a sector when there is a "paradigm shift" in technology, which alters radically the rate, direction and skills associated with a technological trajectory. However, whilst this might serve to explain the entry of new firms in the US electronics industry from 1950 to 1970, based on advances in solid state technology, it cannot explain the relative stability of structure of the world chemical industry over the past 60 years, in spite of successive waves of radical innovations – or "paradigm shifts" – growing out of synthetic chemistry.

The reasons for this difference must probably be sought once again in the nature of the scale barriers facing new entrants. In electronics (especially solid-state components and related equipment), static scale barriers are low, but there are very steep dynamic economies in production. This means that a small and successful innovator can quickly become very big, since imitators are chasing the innovator down steeply declined cost curves. In chemicals, on the other hand, there are high static scale barriers to new entrants: in bulk chemicals, there are big static economies of scale; in fine chemicals, there are systems of public regulation and control for new products that require heavy expenditures on testing and screening.

This discussion suggests that formal models of the dynamics of Schumpeterian competition, like those developed by Nelson and Winter, would more accurately reflect a varied reality in technological trajectories, if they were to explore a range of assumptions about new entrants and static and dynamic economies of scale; about pressures for market diversification; and about complementary relations between producers and users of capital goods.

---

[28] For a more detailed exploration of this question, see Ergas [10].

5. Future perspectives

We began this paper with some dissatisfaction with existing conceptualisations of technical change. Based on systematic empirical data, we have tried to show why; and we have proposed another conceptualisation which, we hope, more accurately reflects the cumulative and varied nature of the technical change to be found in a modern economy. It is not necessary here to summarise the main conclusions of our analysis, since this is done at the beginning of the paper. Suffice to suggest some directions for the future.

First, our proposed taxonomy needs to be tested on the basis of complete sectoral coverage of the characteristics of innovations in Britain, of accumulated case studies, and of other data on innovative activities that become available. Our analysis suggests that R&D statistics do not measure two important sources of technical change: the production engineering departments of production intensive firms, and the design and development activities of small and specialised suppliers of production equipment. For reasons that are discussed elsewhere [37], it is probably that statistics on patenting activity capture innovative activity from these sources more effectively than do R&D statistics. The detailed information now becoming available on patenting activity by company should therefore enable a considerable step forward. As Rosenberg has observed [42], theoretical and practical advances have depended on good systems of measurement, and on accurate and comprehensive data. US patenting statistics could eventually enable the thorough econometric analysis that we considered and rejected at the beginning of this paper.

Second, our taxonomy itself needs to be modified and extended. Greater emphasis should be given to the exploitation of natural resources in the use of large-scale production equipment and instrumentation, [29] and therefore included in our production intensive category. And a fourth category should be added to cover purchases by government and utilities of expensive capital goods related to defence, energy, communications and transport.

Third, our taxonomy may have a variety of uses

for policy makers and analysts. At the very least it may help to avoid general and sterile debates about the relative contribution of large and small firms to innovation, and the relative importance of "science and technology push" compared to "demand pull." It may also increase the value and effectiveness of micro-studies and micro-policies for technical change, by suggesting questions to ask at the beginning, and by putting results in a broader perspective at the end.

Fourth, the taxonomy and the theory may turn out to have more powerful uses. As we have seen in section 4 of this paper, they cast a different and perhaps fresh light on a number of important aspects of technical change: for example, the sources and directions of innovative activities; their role in the diversification activities of industrial firms and in the evolution of industrial structures; and the accumulation of technological skills and advantages within industrial firms. They may also give us a firmer understanding of the determinants of the sectoral patterns of comparative technological advantage that have emerged in different countries. [30] Nelson and Winter [31] have rightly observed that analysis of technical change has been "balkanised"; perhaps the concepts in this paper will help towards re-unification.

Fifth, our taxonomy and theory contain one obvious and important warning for both practitioners of policies for technical change, and academic social scientists concerned with is conceptualisation. Given the variety in patterns of technical change that we have observed, most generalisations are likely to be wrong, if they are based on very specific practical experience, however deep, or on a simple analytical model, however elegant.

For policy makers – many of whom come from the hard sciences and engineering – this means accepting that personal experience and anecdotal evidence from colleagues are an insufficient basis for policies that cover a range of technical activities. It also implies a need for sympathy towards systematic data collection on scientific and technological activities. Such data may be flawed in precision, but they do have the advantage of being comprehensive.

For the academic social scientists, one implica-

---

[29]   See, for example, Townsend [61].

[30]   For further discussion, see [38;40].

Table 10
Definition and description of variables

Symbol	Description	Source
Prop 3	Proportion of innovations used outside their 3 digit sector of production	Data bank on innovations
Prop 2	Proportion of innovations used outside 2-digit sector of production	Column 1, table 2
Inhouse 3	Proportion of innovations used in sector that are produced by sector/firms in the sector (3 digit)	Data bank on innovations
vertical	Proportion of innovations by firms principally in sector that are vertical diversification (2 digit)	Table 6–9, column 5
R/Y	Total R&D in manufacturing firms as a percentage of net output in 1975 (2 and 3 digit)	Business monitor, M014, 1979, table 20, (HMSO)
PSU	Average plant size (3 digit)	Information supplied by Dept. of Industry: based on industrial census, 1977
C_5	Proportion of sales in first five firms in 1970 (3-digit)	Business monitor, PA1002, 1975, table 9 (HMSO)
T/Y	Patents granted in the UK as a percentage of net output in 1975 (2 digit)	Same as R/Y; Townsend et al., table 11.1
D_{20}	Proportion of sales in first 20 firms in (2 digit)	Same as PSU
I	Expenditure on plant and machinery, 1970 (3 digit)	Same as C_5

tion is that analytical models of technical change are likely to become more complex and more numerous [31] Salter's vintage model of technical change [48] may be an accurate reflection of what happens outside industry and in traditional manufacturing; but in mass assembly and continuous process industries, the emphasis placed on investment and production as sources of technical change by such writers as Schmookler [53], Gold [15], Sahal [47] and even Kaldor [18] and Verdoorn [63] may be more appropriate; whilst the Schumpeterian dynamics of innovation, growth and concentration in science-based sectors are better reflected in the models and analyses of writers like Freeman [41;42], Nelson and Winter [32] and Dosi [8]. As we have seen in this paper, the variety in sectoral patterns of technical change was recognised by Adam Smith. Perhaps his is a tradition to which we should return.

[31] This same point is made by Gold [15].

Appendix

Some exploratory statistical analysis

As we pointed out in section 3 of this paper, inadequacies in data are one set of reasons why this paper is not econometric in nature. Some of the main inadequacies are as follows:

● The data bank on UK innovations, together with the other available data on industrial characteristics, allow at the most 11 data points at the two-digit level, and 26 points at the three-digit level;
● Whilst the data bank on UK innovations covers the period from 1945 to 1980, other systematic and detailed data on UK industrial activity began to emerge only at the end of 1960s;
● Some industrial statistics are not readily available in the degree of detail that suit the purposes of our analysis: for example, the patent intensity measure (T/Y) is not readily available at the three-digit level.

Table 11
Results of selected regressions

Equation	Dependent variable	Independent variables: sign and significance					$\overline{R}^2$	d.f.	F statistic
E1	Prop 3	$+R/Y$ [b]	$-PSU$ [a]				0.22	22	4.432 [b]
E2	Prop 3	$+R/Y$		$-C_5$ [b]			0.23	15	3.475
E3	Prop 2			$-D_{20}$	$+T/Y$ [a]		0.54	8	6.872 [b]
E4	Inhouse 3			$+C_5$ [a]		$-I$ [a]	0.56	15	11.786 [a]
E5	Vertical	$-R/Y$ [a]		$+D_{20}$ [a]		$-K/L$	0.71	7	9.013 [a]

[a] Significant at 1% level.
[b] Significant at $2\frac{1}{2}$% level.

Thus a proper statistical exercise, using the UK data base on innovations, will probably have to await the completion of sectoral coverage, and will require considerable statistical efforts to compile matching data from other sources. In the meantime, our statistical analysis can be only exploratory. The results discussed in section 4 of the paper are described in more detail in tables 10 and 11.

References

[1] M. Anderson, D. Jones and J. Womack, Competition in the World Auto Industry: Implications for Production Location, in: The Future of the Automobile: A Trilateral View (forthcoming, 1984).

[2] H. Ansoff, Corporate Strategy (Penguin Books, Harmondsworth, 1968).

[3] K. Arrow, Economic Welfare and the Allocation of Resources for Invention, in: The Rate and Direction of Inventive Activity (Princeton University Press, 1962).

[4] C. Berry, Corporate Growth and Diversification (Princeton University Press, 1975).

[5] T. Buer, Investigation of Consistent Make or Buy Patterns of Selected Process Machinery in Selected US Manufacturing Industries, Ph.D. dissertation, Sloane School of Management, MIT, 1982.

[6] Bureau de d'Economie Theorique et Appliquée, Les Perspectives de la Chimie en Europe (Université Louis Pasteur, Strasbourg, 1982).

[7] D. De Melto et al., Preliminary Report: Innovation and Technological Change in Five Canadian Industries Economic Council of Canada, Discussion Paper No. 176 (1980).

[8] G. Dosi, Technological Paradigms and Technological Trajectories, Research Policy 11 (1982).

[9] H. Ergas, Corporate Strategies in Transition, in: A. Jacquemin (ed.), Industrial Policy and International Trade (Cambride University Press, 1983).

[10] H. Ergas, The Inter-Industry Flow of Technology: Some Explanatory Hypotheses (mimeo) (OECD, Paris, 1983).

[11] C. Freeman, The Economics of Industrial Innovation, 2nd edition (Francis Pinter, London, 1982).

[12] C. Freeman, J. Clark and L. Soete, Unemployment and Technical Innovation: A study of Long Waves and Economic Development (Francis Pinter, London, 1982).

[13] J. Gershuny, After Industrial Society? (Macmillan, London, 1978).

[14] M. Gibbons and R. Johnstone, The Roles of Science in Technological Innovation, Research Policy 3 (1974).

[15] B. Gold, Productivity, Technology and Capital (Lexington Books, Lexington, MA, 1979).

[16] M. Gort, Diversification and Integration in American Industry (Princeton University Press, 1962).

[17] J. Hassid, Recent Evidence on Conglomerate Diversification in UK Manufacturing Industry, The Manchester School 43 (1976).

[18] N. Kaldor, The Causes of the Slow Rate of Economic Growth of the United Kingdom (Cambridge University Press, 1966).

[19] J. Kamin et al., Some Determinants of Cost Distributions in the Process of Technological Innovation, Research Policy 11 (1982).

[20] H. Kleinman, Indicators of the Output of New Technological Products from Industry, Report to US Science Foundation (National Technical Information Service, US Department of Commerce, 1975).

[21] J. Langrish et al., Wealth from Knowledge (Macmillan, London, 1972).

[22] R. Levin, Technical Change and Optimal Scale: Some Evidence and Implications, Southern Economic Journal 44 (1977).

[23] R. Levin and P. Reiss, Tests of a Schumpeterian Model of R and D Market Structure, in: Z Grilliches (eds.), R and D, Patents and Productivity (University of Chicago Press, 1984).

[24] A. Link and J. Long, The Simple Economies of Basic Scientific Research: A Test of Nelson's Diversification Hypothesis, Journal of Industrial Economics 30 (1981).

[25] E. Mansfield, Basic Research and Productivity Increase in Manufacturing, American Economic Review 20 (1980).

[26] D. Mowery and N. Rosenberg, The Influence of Market Demand upon Innovation: A Critical Review of Some Recent Empirical Studies, Research Policy 8 (1979).

[27] R. Nelson, The Simple Economics of Basic Scientific Research, Journal of Political Economy (1959).

[28] R. Nelson, *The Moon and the Ghetto* (Norton, New York, 1977).

[29] R. Nelson, Research on Productivity Growth and Productivity Differences: Dead Ends and New Departures, *Journal of Economic Literature* 19 (1981).

[30] R. Nelson, The Role of Knowledge in R and D Efficiency, *Quarterly Journal of Economics* (1982).

[31] R. Nelson and S. Winter, In Search of a Useful Theory of Innovation, *Research Policy* 5 (1977).

[32] R. Nelson and S. Winter, *An Evolutionary Theory of Economic Change* (Harvard University Press, Cambridge, MA, 1982).

[33] K. Oshima, in: B. Williams, *Science and Technology in Economic Growth* (Macmillan, London, 1973).

[34] K. Pavitt (ed.), *Technical Innovation and British Economic Performance* (Macmillan, London, 1980).

[35] K. Pavitt, Technology in British Industry: A Suitable Case for Improvement, in: C. Carter (ed.), *Industrial Policy and Innovation* (Heinemann, London, 1981).

[36] K. Pavitt, Some Characteristics of Innovative Activities in British Industry, *Omega* 11 (1983).

[37] K. Pavitt, R and D, Patenting and Innovative Activities: A Statistical Exploration, *Research Policy* 11 (1982).

[38] K. Pavitt, *Patterns of Technical Change: Evidence, Theory and Policy Implications*, Papers in Science, Technology and Public Policy, No. 3 (Imperial College/Science Policy Research Unit, 1983).

[39] K. Pavitt and S. Wald, *The Conditions for Success in Technological Innovation* (OECD, Paris, 1971).

[40] K. Pavitt and L. Soete, International Differences in Economic Growth and the International Location of Innovation, in: H. Giersch (ed.), *Emerging Technologies: Consequences for Economic Growth, Structural Change, and Employment* (JCB Mohr, 1981).

[41] E. Penrose, *The Theory of the Growth of the Firm* (Blackwell, Oxford, 1959).

[42] N. Rosenberg, *Perspectives on Technology* (Cambridge University Press, 1976).

[43] N. Rosenberg, *Inside the Black Box: Technology and Economics* (Cambridge University Press, 1982).

[44] R. Rothwell, *Innovation in Textile Machinery: Some Significant Factors in Success and Failure*, SPRU Occasional Paper No. 2 (University of Sussex, 1976).

[45] R. Rothwell, The Characteristics of Successful Innovators and Technically Progressive Firms, *R and D Management* 7 (1977).

[46]. R. Rumelt, *Strategy, Structure and Economic Performance* (Graduate School of Business Administration, Harvard University, 1974).

[47] D. Sahel, *Patterns of Technological Innovation* (Addison-Wesley, New York, 1981).

[48] W. Salter, *Productivity and Technical Change*, 2nd Edition (Cambridge University Press, 1966).

[49] F. Scherer, *Industrial Market Structure and Economic Performance* (Rand McNally, 1981).

[50] F. Scherer, *The Economic Effects of Compulsory Patent Licensing* (New York University, 1977).

[51] F. Scherer, Inter-industry Technology Flows in the United States, *Research Policy* 11 (1982).

[52] F. Scherer, Demand Pull and Technological Invention: Schmookler Revisited, *Journal of Industrial Economics* XXX (1982).

[53] J. Schmookler, *Invention and Economic Growth* (Harvard University Press, Cambridge, MA, 1966).

[54] A. Smith, *An Inquiry into the Nature and Causes of the Wealth of Nations* (G. Routledge (1895 Edition)).

[55] L. Soete, A General Test of Technological Gap Trade Theory, *Review of World Economics* 117 (1981).

[56] C. Taylor and A. Silberston, *The Economic Impact of the Patent System* (Cambridge University Press, 1973).

[57] D. Teece, *The Multinational Corporation and the Resource Cost of International Technology Transfer* (Ballinger, New York, 1977).

[58] J. Toledano, A Propos des Filières Industrielles, *Revue d'Economie Industrielle* (1978).

[59] M. Teubal, *The Role of Technological Learning in the Exports of Manufactured Goods: The Case of Selected Capital Goods in Brazil and Argentina*, Discussion Paper No. 82.07 (Maurice Falk Institute for Economic Research in Israel, 1982).

[60] J. Townsend, F. Henwood, G. Thomas, K. Pavitt and S. Wyatt, *Innovations in Britain Since 1945*, Occasional Paper No. 16 (Science Policy Research Unit, University of Sussex, 1981).

[61] J. Townsend, Innovation in Coal-Mining Machinery, in K. Pavitt [34].

[62] M. Utton, *Diversification and Competition* (Cambridge University Press, 1979).

[63] P. Verdoorn, I Fattori che Regolaro lo Suiluppo della produttivita de lavoro, *L'Industrie* (1949).

[64] E. von Hippel, The Dominant Role of Users in the Scientific Instrument Innovation Process, *Research Policy* 5 (1976).

[65] E. von Hippel, A Customer-Active Paradigm for Industrial Product Idea Generation, *Research Policy* 7 (1978).

[66] E. von Hippel, The User's Role in Industrial Innovation, in: B. Dean and J. Goldhar (Eds.), *Management of Research and Innovation*, Studies in the Management Sciences, Vol. 15 (North-Holland, Amsterdam, 1980).

[67] E. von Hippel, Appropriability of Innovation Benefit as a Predictor of the Source of Innovation, *Research Policy* (1982).

[68] V. Walsh, Invention and Innovation in the Chemical Industry: Demand Pull or Discovery Push?, *Research Policy* (1984), forthcoming.

[69] J. Woodward, *Management and Technology* (HMSO, London, 1958).

[70] J. Utterback and W. Abernathy, A Dynamic Model of Process and Product Innovation, *Omega* 3 (1975).

[3]

The continuing, widespread (and neglected) importance of improvements in mechanical technologies [1]

Parimal Patel and Keith Pavitt

Science Policy Research Unit, University of Sussex, Falmer, Brighton BN1 9RF, UK

Rosenberg's historical analyses of the varying sources and directions of technological change are confirmed by contemporary bibliometric data, in particular: (1) the growth of science-based technologies developed mainly in the R&D laboratories of large firms; (2) more pervasive improvements in production methods based on mechanical technology.

The considerable importance of the latter has persisted well into the late twentieth century, but has been neglected in analysis and policy. Greater attention, in particular, should be paid to:
– more refined measures of technological activities than R&D classified by large firms' principal product groups;
– the cumulative and complementary (rather than displacing and competitive) nature of successive 'technological paradigms';
– the central role of mechanical, instrumentation and software technologies in the decentralised and continuous improvements in products and production methods.

Correspondence to: Parimal Patel, Science Policy Research Unit, University of Sussex, Falmer, Brighton BN1 9RF, UK. Tel., +44 273 686758; fax, +44 273 685865.
[1] An earlier version of this paper was prepared for 'The Role of Technology in Economics: a Conference in Honour of Nathan Rosenberg', held Stanford University on 9 November 1992. It draws heavily on the results of research, funded by the ESRC in the Centre for Science, Technology, Energy and Environment Policy (STEEP) at the Science Policy Research Unit at Sussex University. In addition to the conference itself, we have benefited from seminars at Sussex and Aalborg Universities, and from more detailed comments by Daniele Archibugi, Martin Bell, John Cantwell, Chris Freeman, Franco Malerba, and an anonymous referee. Errors and omissions are our own.

Research Policy 23 (1994) 533–545
North-Holland

1. Historical analysis

One powerful influence on Nathan Rosenberg's research has been his dissatisfaction with the way in which the supply of innovations is treated in economics (and other social sciences). This is perhaps best expressed towards the end of his influential critique of Schmookler's emphasis on the dominance of demand factors in technical change:

... inventions are not equally possible in all industries ... because (of) ... a crucial intervening variable: the differential development of the state of sub-disciplines of science and bodies of useful knowledge generally at any moment in time. ... it is very important that we cease to talk about "the state of science" and begin thinking in terms of "sciences". A central problem is to trace out carefully the manner in which differences in the state of development of individual sciences and technologies have influenced the composition of inventive activities (Rosenberg, 1974).

He himself has made major contributions to our understanding of the development of different technologies, particularly in his papers on the machine tool industry in the nineteenth century (1963) and on the early applications of routine chemistry in process and materials industries (1985), and – in his book with David Mowery (1989) – on the growth of science-based technology in the twentieth century in the R&D laboratories of the large firms in the chemical and electrical industries. For the purposes of our own analysis of contemporary patterns of technological change, the following insights are particularly important.

First, until well into the nineteenth century, mechanical technology and the underlying scientific principles advanced on the basis of "...unassisted human observations, with little or no reliance upon complex instruments or experimental apparatus" (Rosenberg, 1974). Similar understanding in the more complex fields of chemistry, physics and biology required more sophisticated research techniques and therefore took more time to develop.

Second, and as a consequence of the above, advances in mechanical technology depended on skills in design that owed little to contemporary advances in underlying science, but that were greatly stimulated by the advent of steel as cheap, reliable and high-performance material. Furthermore, the use of machinery in the cutting and shaping of metal resulted in a common core of techniques that were used in the production of a variety of apparently unrelated products such as firearms, sewing machines and bicycles (Rosenberg, 1963). This 'technological convergence' reinforced the conditions for 'vertical disintegration'; namely, the 'spin-off' of small firms, specialising in the design and production of machinery for a specific metal cutting and/or shaping operation, from (often large) machinery-using firms with similar metal cutting or using operations within otherwise complex and different production systems.

Third, and later in the nineteenth century, routine chemical analysis and instrumentation found profitable applications in the emerging large-scale process industries (such as steel, cement, food processing) as a means of controlling quality and consistency within the various stages of production, and of varying the attributes of the final output (Rosenberg, 1985). Chemical engineering also emerged as a distinct activity for the design and operation of machinery and processes for increasingly complex chemical transformations. Given pervasive use in a number of industries, technological convergence led to vertical disintegration in chemical process equipment and related instruments, just as had happened earlier

with metal cutting and metal shaping (Mowery and Rosenberg, 1989).

Fourth, the twentieth century has seen the growth of science-based technologies, particularly in the chemical and electrical industries. These emerged from the exploitation of fundamental scientific discoveries and techniques by the ever more numerous R & D laboratories established in large firms, often as part of broader organisational processes of functional specialisation, and mainly for the purpose of continuous diversification into new and technologically related product fields (Mowery and Rosenberg, 1989).

Using bibliometric methods, we shall show below that the analysis of Rosenberg and Mowery, in addition to being an authoritative historical record, also describes contemporary patterns of technical change [2]. Science-based technologies are those that in general are growing most rapidly and they are dominated by large firms, but this is only part (about a half) of the story. Nearly 40% is made up of improvements in mechanical, instrumentation and process technologies, the development of which is hidden in the published data on large companies' R & D activities and dispersed amongst small firms whose technological activities are only very imperfectly and incompletely recorded.

2. Contemporary evidence

2.1. Data sources

Our main data source is US patents granted, broken down since 1963 into 91 sub-classes, based in the US Patent Office's own Patent Classes; these are listed later in Tables 2 and 7. Like Schmookler (1966) and many other scholars, we treat patents as imperfect indicators of technological activities, but, as we hope this paper will show, with something substantial to add to con-

---

[2] Von Hippel's analyses (1988) of the role of users in the development of innovations in instruments and other fields has identified very effectively some of the important contemporary processes of technological convergence and vertical disintegration.

ventional measures like R & D activities [3]. In addition, we have combined data from the Fortune list of more than 500 of the world's largest, technologically active firms with data from the US Patent Office that specifies, since 1969, the name and country of origin of companies granted patents in each technical class and sub-class. We have consolidated subsidiaries with names different from their parent companies for 1984, on the basis of 'Who Owns Whom?'. We consequently have the following for each named company: country of headquarters, principal activity, sales, employment, R & D expenditure (when published), and numbers of US patents, broken down by country of origin and by technical sub-class. For the main part of our analysis, we shall use information on US patenting from 1981 to 1988, and on the large firms that were granted patents in this period.

2.2. Innovative activity in different technical fields: what the US patenting data show

Table 1 confirms the growing importance of science-based (i.e. electrical and chemical) technologies which from the 1960s to the 1980s increased their share from 40.0 to 48.3% of total patenting in the USA. This is broadly consistent with what the R & D statistics show; according to a recent publication by the National Science Board (1992), firms in the corresponding sectors (i.e. chemicals, including pharmaceuticals and electrical, including computers) accounted for 49.7% of all business-funded R & D in the USA in 1985 (see Table 3, last column). Elsewhere, Narin and Olivastro (1992) have shown that patents in chemical and electrical–electronic products groups make much more frequent reference to published scientific papers than those in

[3] The potential uses and abuses of patent statistics and other indicators of scientific and technological activities have been extensively discussed elsewhere (see, for example, Pavitt, 1988; Grilliches, 1990; Patel and Pavitt, 1994). It is noteworthy that business practitioners are growing users of patent data for technological intelligence, and that Business Week (1992) has recently begun publishing a 'Patent Scoreboard', comparing 200 named firms' performance in the quality and quantity of patenting activity, in addition to the increasingly comprehensive 'R&D Scoreboard' that it has been publishing since 1976.

Table 1
Trend in the relative importance of innovative activity in five broad technical fields: 1963–88

Technical fields [a]	Percentage share of US patents (%)			
	1963–68	1969–74	1975–80	1981–88
Chemical	18.3	20.3	23.6	21.8
Electrical	21.7	24.8	22.8	26.5
Non-electrical machinery	47.6	42.7	40.3	38.0
Transport	4.2	4.1	4.7	4.7
Other	8.2	8.1	8.6	9.0
Total	100.0	100.0	100.0	100.0

[a] See Table 2 for the technical sub-classes included in each broad field.

mechanical product groups, thereby reflecting their closer links to science.

However, this still leaves slightly more than half the total patenting activities in the 1980s outside these science-based sectors. In Table 1, we have divided this share into the following categories, the composition of which is given in Tables 2 and 7:
– non-electrical machinery: mechanically based capital goods that are inputs into manufacturing and primary production;
– transport equipment: mainly mechanically based machinery and equipment used principally in ships, railways, road vehicles and aircraft;
– other: technologies related mainly to traditional manufacturing (e.g. textiles), and to non-manufacturing (e.g. construction, medicine, agriculture).

The share of transport technology rises to just under 5% of the total, which is much less than the 21.2% share of US business-funded R & D for transport firms; on the other hand, in spite of its declining share of the total, non-electrical machinery still accounted for 38% of all patenting in the 1980s, which is much more than the 4% share of non-electrical machinery firms in US business-funded R & D (see Table 3, second column).

In order to begin to identify the causes of these discrepancies, we show in Table 2 the share of each of the 91 sub-fields of technology in total patents granted between 1981 and 1988. The top 20 account for about half of total patenting, and include well-known fields based on chemistry (organic and synthetic chemicals, drugs, processes) and on physics (electrical and electronic instruments, computers, semiconductors, image reproduction, telecommunications). They also include seven mechanical classes, each with more than 10 000 patents granted, and which together accounted for about 20% of all US patenting. These cover non-electrical instruments, and machinery and components for cutting and shaping

Table 2

Innovative activity in 91 technical sub-fields, US Patents granted: 1981–88

Broad technical field	Technical sub-field	Number of Patents	Percentage share (%)
M	Non-electrical instruments (optics, projectors, weighers, etc.)	23786	4.32
M	Metal working machinery, elements, mechanisms	21253	3.86
E	Electrical and electronic, instruments (inc. X-rays, etc.)	21197	3.85
C	Other carbon-based organic chemicals and other organic compounds	20383	3.70
C	Synthetic resins,	18941	3.44
C	Drugs	16832	3.06
E	Calculators, computers, data-processing systems	16386	2.98
O	Dentistry and surgery	15992	2.90
M	Pumps, valves, lubrication, fluid handling, fluid reaction surfaces	15027	2.73
M	Miscellaneous metal products	14240	2.59
E	Television, facsimile	13944	2.53
M	Apparatus for gas separation, vaporization, fluid handling, etc.	13630	2.47
C	Composite materials and compounds, other stock materials	12823	2.33
E	Photography and photocopy	12769	2.32
E	Electrical devices and systems	12574	2.28
M	Miscellaneous machinery (coopering, button making, presses, etc.)	10986	1.99
M	Heating, combustion, heat exchanges, non-electric furnaces	10332	1.88
E	Semiconductors	10042	1.82
C	Adhesive bonding and miscellaneous chemical processes	9934	1.80
E	Telecommunications equipment	9232	1.68
M	Material handling and assembling apparatus	8836	1.60
T	Internal combustion engines	8619	1.56
E	Electrical transmission	8402	1.53
E	Electric motors, dynamos, generators, motive power systems	8021	1.46
C	Miscellaneous chemical compositions (including detergents, perfumes, etc.)	7826	1.42
T	Bicycles and other land vehicles	6694	1.22
O	Amusement and toys	6651	1.21
O	Building materials and static structures	6468	1.17
C	Hydrocarbons, mineral oils, fuels and igniting devices	6348	1.15
E	Electric heating and furnaces	6217	1.13
C	Industrial chemical processes (gas separation, cleaning, etc.)	6186	1.12
M	Mining machinery, wells machinery and techniques	6185	1.12
C	Bioengineering	5953	1.08
E	Acoustics (including recording and sound reproduction, music, etc.)	5884	1.07
M	Metallurgical and metal treatment processes, metal founding	5883	1.07
M	Nuts, bolts, locks, miscellaneous hardware	5594	1.02
M	Fluid pressure brakes, clutches, power stop controls, etc.	5524	1.00
E	Electrolytic apparatus, batteries, photoelectric batteries etc.	5370	0.98
E	Lamps and discharge devices	5341	0.97
M	Joints, pipes, connections, sockets	5093	0.92
M	Handling/storing sheets, webs, cables, etc.	4984	0.90
O	Plant and animal husbandry	4976	0.90
M	Paper-making apparatus	4863	0.88
M	Plastic and other non-metallic articles shaping or heating products	4639	0.84
O	Miscellaneous Articles (jewellery, coffins, bottles, luggage, etc.)	4415	0.80
C	Inorganic chemicals	4400	0.80
M	Machine controls	4099	0.74
M	Textile manufacturing equipment	4087	0.74

Table 2 (continued)

Broad technical field	Technical sub-field	Number of Patents	Percentage share (%)
E	Electrical-current-producing chemical processes and compounds	3903	0.71
M	Apparatus for distillation, mixing, disinfecting, etc.	3863	0.70
C	Food, sugar, starch, carbohydrates	3848	0.70
M	Power plants	3478	0.63
O	Ammunition and weapons (firearms, torpedoes, ordnance, explosives)	3387	0.61
M	Agricultural machinery	3359	0.61
M	Apparatus for coating, bonding, etc.	3295	0.60
M	Road structures, bridges (processes and engineering)	3104	0.56
E	Radio communication systems (inc. Radar)	3026	0.55
O	Wood products (beds, chairs, receptacles)	2953	0.54
M	Plastic and non-metallic articles machinery	2827	0.51
M	Expansible chamber motors and devices, rotary expansible chamber	2803	0.51
T	Ships and aquatic devices and marine propulsion	2793	0.51
M	Printing, type-setting, book making, etc.	2740	0.50
C	Agricultural chemicals	2697	0.49
O	Boots, shoes, apparel	2663	0.48
T	Motor vehicles	2478	0.45
M	Refrigeration machinery	2266	0.41
M	Apparel equipment, sewing, leather working, boot and shoe making	2205	0.40
M	Closures and safes	1973	0.36
E	Typewriters	1825	0.33
M	Glass manufacturing	1811	0.33
T	Aeronautics	1765	0.32
O	Induced nuclear reactions: systems and elements	1761	0.32
M	Metallurgical apparatus	1754	0.32
T	Railways and railways equipment	1627	0.30
E	Laser technology	1589	0.29
C	Bleaching and dyeing of textiles	1526	0.28
M	Food and beverages apparatus	1331	0.24
M	Miscellaneous apparatus (ventilation, coin handling etc.)	1268	0.23
C	General chemical analytical physical products, Fisher–Tropsch reactions	1143	0.21
M	Tobacco	1063	0.19
O	Education and demonstration	995	0.18
M	Woodworking tools and machinery	892	0.16
T	Tyres	882	0.16
C	Distillation processes	725	0.13
T	Wheels and axles	495	0.09
C	Processes for disinfecting, deodorizing, preserving, sterilizing	288	0.05
T	Other transport equipment parts	263	0.05
E	Electricity transmission to vehicles	105	0.02
O	Fabric structures, knots	61	0.01
M	Mineral oil apparatus	43	0.01
E	Superconductors	18	0.00
Total		550762	100.00

C, chemical; E, electrical; M, non-electrical machinery; T, transport; O, other.

metal, specialised applications, treating fluids and gases, and heating; in other words, production-related technologies identified by Rosenberg and Mowery as important and potentially pervasive in the nineteenth and early parts of the twentieth century. As we shall now see, the contemporary importance of these technologies is severely underestimated, when we concentrate solely on aggregate measures of technological activities in large firms according to their principal product group, and when we neglect the processes of technological convergence and vertical disintegration described by Rosenberg.

2.3. The technological diversity of large firms

The two columns of Table 3 compare the sectoral distribution of technological activities performed in the USA, according to firms' principal product group, when measured by (column 1) the number of US patents granted to our large firms for their technological activities performed in the USA and (column 2) industry-funded R & D activities performed in the USA, based on a survey undertaken by the US National Science Foundation. The similarity between the two distributions is, given what we have seen above, remarkably high with a correlation coefficient of 0.8, significant at the 1% level [4]. For our purposes, it is particularly important to note that the technological importance of non-electrical machinery firms is very similar according to both the patenting and the R & D measure (4–5.5%), and much less than non-electrical machinery technologies, as revealed by the patent measure in Table 1 (38%).

One reason for this is revealed in Table 4 which shows that our large firms in all product groups are typically active in more than one of the broad technological fields. In particular, activities to improve non-electrical machinery technologies are pervasive across large firms in all sectors, ranging from around 9 to 22% of the total in science-based sectors and instruments, to more than 50% in six out of the 17 sectors. Improvements in chemical technology account for more than 20% of the total, except in the industries where assembly methods of production pre-

vail: electrical, motor vehicles, instruments, non-electrical machinery, aircraft, and other transport. Improvements in electrical and electronic technologies, on the other hand, are much less pervasive in their development, accounting for more than 25% of all technological activities in firms in only four sectors, and for less than 10% in eight.

Given the pervasiveness of their development by firms in all product groups, Table 4 shows that non-electrical machinery technologies account for 28.9% of technological activities in large firms – only slightly less than electrical or chemical technologies – whilst Table 3 shows that non-electrical machinery firms account for less than 6%. Hence, the first reason for underestimating the importance of improvements in non-electrical machinery technology: compared with improvements in chemical and electrical technologies, a

Table 3
A comparison of the sectoral distribution of large firms' US patenting [a] with industry-funded R&D

Principal product group	Percentage shares (%)	
	US patents 1981–88	US R&D 1985
Electrical	24.1	20.3
Chemicals	16.9	8.5
Mining and Petroleum	11.2	3.8
Instruments	7.9	8.1
Computers	7.6	14.8
Pharmaceuticals	6.8	6.1
Aircraft	6.7	9.9
Non-electrical machinery	5.5	4.0
Motor vehicles	4.7	10.8
Food, drink and tobacco	2.6	2.0
Non-metallic minerals	2.0	1.4
Paper and wood	1.7	1.3
Metals	1.5	1.3
Rubber and plastics	1.2	1.2
Textiles, etc.	0.4	0.4
Other transport	0.1	0.5
Total	100.0	94.4 [b]

[a] Includes patenting of all our large firms, regardless of nationality, originating from the technological activities performed in the United States; in other words it excludes patenting based on technological activities performed outside the United States. This improves the basis of the comparison with R&D expenditure data, all of which relate to activities performed within the USA.
[b] Total of the above product groups.
Sources: US Patenting from the SPRU Large Firms Database; Industry-funded R&D from National Science Board (1992).

---

[4] It should be noted that the two sectoral distribution are not calculated on the basis of the same population of firms: the NSF R&D survey almost certainly covered more firms than the 400 or more, in our large firm patenting data base, on which the sectoral patent shares in Table 3 were based. Nonetheless, the basis for comparison between R&D and patenting is two strongly overlapping populations of large firms, defined according to their principal product group. (For related earlier work, see Scherer, 1983.) This is a firmer basis for comparison than that often made between R&D activities in large firms classified according to their principal product group, and total patenting activities, classified according to their technology.

higher proportion of those in non-electrical machinery are ignored in the official R & D statistics, since they are more frequently a secondary activity in firms principally active in other product groups.

It is also worth noting that, since large firms can be defined as 'multi-technology' (Archibugi, 1988; Granstrand and Sjolander, 1990; Oskarsson, 1991; Cantwell, 1993), the links from their 'core products' to their 'core technologies' are neither obvious nor simple. For example, Table 4 shows that our large motor vehicle companies are granted 30.6% of their patents in transport technologies, compared with more than 46.1% in non-electrical machinery. More detailed patenting data show that large motor vehicle companies are also amongst the major sources of large company technological activity in such fields as combustion, pumps and valves, brakes and clutches, power plants, and miscellaneous metal products. Similarly, our large chemical and electrical firms are both granted more patents in non-metallic minerals technologies than the large firms in non-metallic minerals themselves. More broadly,

the third row of Table 5 shows that our large electrical and transport firms each generate more new technology in non-electrical machinery than our large non-electrical machinery firms themselves, and that our large chemical firms are not far behind.

2.4. The varying importance of large firms in different technologies

Table 5 shows the sources of all patenting in each broad technical field, according to type of firm; either large firms, according to their principal product groups, or the other (i.e. smaller) firms and individuals not included in our data base. It summarises the two reasons why improvements in non-electrical machinery technologies are underestimated in R & D statistics compared with those in the science-based technologies. First, the proportion of the total made by large firms in the principal product group is much lower; 4.8% in non-electrical machinery compared with 36.6% in chemicals and 38.5% in electrical products. Second, the proportion made

Table 4
The distribution of large firms' technological activities [a] in each product group according to broad technical fields 1981–88

Principal Product Group (PPG)	Percentage share of the PPG's patents in technology field (%)					
	Chemical	Non-electrical machinery	Electrical	Transport	Other	Total
Electrical (excl. computers)	8.1	21.7	66.4	1.1	2.7	100.0
Chemicals (excl. pharmaceut)	66.9	16.4	13.7	0.6	2.4	100.0
Motor vehicles	3.8	46.1	17.3	30.6	2.0	100.0
Mining and petroleum	55.5	35.1	7.3	0.9	1.2	100.0
Instruments	12.2	20.5	63.5	0.2	3.5	100.0
Non-electrical machinery	7.8	66.4	12.6	9.4	3.8	100.0
Pharmaceuticals	80.4	8.6	2.7	0.0	8.3	100.0
Computers	5.5	16.8	76.6	0.2	0.9	100.0
Aircraft	13.9	46.8	27.9	8.1	3.4	100.0
Metals	30.5	51.4	12.0	2.4	3.6	100.0
Non-metallic minerals	29.5	53.9	9.7	1.0	5.9	100.0
Food	65.8	23.1	6.1	0.5	4.5	100.0
Rubber and plastics	40.4	32.4	4.2	19.7	3.3	100.0
Paper and wood	23.8	51.6	9.2	0.6	14.8	100.0
Textiles etc.	47.1	34.8	10.2	0.7	7.3	100.0
Drink and Tobacco	26.0	63.4	4.6	0.5	5.5	100.0
Other transport	10.9	65.5	6.4	15.5	1.8	100.0
All large firms	29.5	28.9	34.0	4.5	3.1	100.0
All patenting	21.8	38.0	26.5	4.7	9.0	100.0

[a] Measured by total patenting of our large firms: in other words, reflecting technological activities performed both inside and outside the USA.

Table 5
Sources of all US patenting [a] in the five broad technical fields: 1981–88

Technical field	Large chemical firms [b]	Large electrical firms [c]	Large non-electrical machinery firms	Large transport firms	Other large firms	Smaller firms and individuals	Total
Chemical	36.6	5.3	1.0	2.0	18.6	36.5	100.0
Electical	4.9	38.5	1.3	4.7	10.7	39.9	100.0
Non-electrical machinery	4.4	8.3	4.8	7.5	10.7	64.3	100.0
Transport	1.2	3.1	5.6	31.0	4.4	54.8	100.0
Other	4.6	4.0	1.2	1.6	4.6	84.0	100.0
All technologies	11.4	15.0	2.8	6.1	11.6	53.2	100.0

[a] In other words, reflecting technological activities performed both inside and outside the USA.
[b] Includes pharmaceuticals.
[c] Includes computers.

by large firms in any product group (and therefore more likely to be recorded in aggregate R&D statistics) is much smaller; 35.7% (i.e. 100.00–64.3%) in non-electrical technologies, compared with more than 60% in the science-based technologies. Thus, the development of non-electrical machinery technology is much more dispersed across firms in different product groups, and in smaller size categories, than either of the science-based ones. In the next section, we shall use one of Rosenberg's major insights to show why.

2.5. Technological convergence and vertical disintegration in non-electrical machinery technologies

The processes of technological convergence and vertical disintegration prevalent in non-electrical machinery technologies can help explain its very different sources compared with the

Table 6
Correlation matrix for measures of technological convergence and vertical disintegration

	Herfin-dahl Index	Num-bers of firms	Share of large firms	Share of other firms
Number of firms	−0.44 *			
Share of large firms	0.52 *	0.16		
Share of other firms	−0.56 *	0.11	−0.60 *	
Share of unassigned	−0.30 *	−0.26 *	−0.87 *	0.13

* Denotes that the correlation coefficient is significantly different from zero at the 5% level.

science-based technologies. In terms of our statistical analysis, technological convergence in a technical field will be reflected in a strong dispersion of patenting activities across the principal product sectors of our large firms. Vertical disintegration will be reflected in a high proportion of patents granted to sources other than our large firms.

We can use the information on our 91 technological sub-classes to test for any systematic statistical relationship between a high degree of technological convergence, measured as a low Herfindahl Index [5], on the one hand, and a high degree of vertical disintegration, measured as a high share of total patenting outside our large firms, on the other. Table 6 shows that, in general terms, our explanation is confirmed, with significant correlations between the Herfindahl Index and the number of technically active large firms (negative), share of large firms in total patenting (positive), share of other firms (negative), and share of unassigned, mainly individuals and small firms, patents (negative) [6].

The empirical bases for these results are spelt out in greater detail in Table 7, which ranks the technical sub-fields from the lowest Herfindahl

[5] Defined, for each technological field, as the sum of the squares of the shares of total patenting in each of the 17 principal product groups. A low Index therefore reflects a high degree of dispersion of patenting activity across firms in many industries; in other words, a high degree of technological convergence.

[6] These results confirm those we obtained in an earlier, more explanatory, analysis (see Patel and Pavitt, 1991).

Table 7
Measures of technological convergence and vertical disintegration, 91 technical sub-fields: 1981–88 (sorted by Herfindahl Index)

Broad technical field	Technical sub-field	Herfindahl Index	Number of firms	Share of large firms	Share of other firms	Share of unassigned
M	Miscellaneous metal products	0.104	395	22.47	36.30	41.23
M	Plastic and non-metallic articles machinery	0.107	241	32.86	44.18	22.96
O	Miscellaneous articles	0.108	161	14.41	35.99	49.60
M	Material handling and assembling apparatus	0.111	354	26.00	46.27	27.74
M	Apparatus for coating, bonding, etc.	0.115	270	39.91	42.70	17.39
M	Miscellaneous machinery	0.118	393	23.21	40.27	36.52
T	Railways and railways equipment	0.130	78	25.26	55.69	19.05
O	Boots, shoes, apparel (products)	0.131	60	7.21	35.60	57.19
M	Joints, pipes, connections, sockets	0.143	273	33.10	42.04	24.86
M	Apparatus for distillation, mixing, disinfecting, etc.	0.144	279	39.27	39.30	21.43
M	Pumps, valves, lubrication, fluid handling, etc.	0.146	409	37.25	38.21	24.54
M	Handling, storing sheets, webs, cables, etc.	0.147	218	37.94	44.32	17.74
M	Heating, combustion, heat exchanges, non-elec. furnaces	0.150	321	32.01	36.56	31.44
M	Paper making apparatus	0.152	245	33.00	50.22	16.78
C	Adhesive bonding and miscellaneous chemical products	0.152	407	54.79	35.10	10.11
M	Textile manufacturing equipment	0.152	141	28.77	56.96	14.26
M	Metal working machinery, elements, mechanisms	0.160	447	34.75	40.40	24.85
O	Plant and animal husbandry	0.161	106	5.33	27.83	66.84
C	Composite materials and compounds, oth. stock materials	0.161	422	58.19	32.10	9.71
O	Amusement and toys	0.162	70	6.65	26.18	67.18
E	Electrical-current-producing chemical products and compound	0.167	267	57.39	34.44	8.17
M	App. for gas separation, vaporization, fluid handling, etc.	0.168	423	40.41	38.91	20.67
O	Building materials and static structures	0.172	191	17.16	35.19	47.65
M	Plastic and other non-metallic articles shaping or heating	0.173	342	51.46	36.00	12.55
M	Woodworking tools and machinery	0.174	31	9.98	36.77	53.25
M	Nuts, bolts, locks, miscellaneous hardware	0.179	210	20.08	45.42	34.50
O	Ammunitions and weapons	0.180	105	26.87	50.81	22.32
M	Non-electrical instruments (optics, projectors, weighers, etc.)	0.186	449	42.09	39.89	18.02
M	Miscellaneous apparatus (ventilation, coin handling, etc.)	0.186	91	22.71	42.74	34.54
M	Printing, type-setting, book making, etc.	0.187	116	26.98	51.24	21.78
E	Electrolytic apparatus, batteries, photoelectric batteries, etc.	0.191	265	52.25	36.93	10.82
C	Industrial chemical processes (gas separation, cleaning, etc.)	0.194	331	42.92	39.69	17.39
M	Metallurgical and metal treatment products, metal founding	0.199	315	60.43	31.26	8.31
M	Food and beverages apparatus	0.205	71	14.35	44.63	41.02
M	Expansible chamber motors and devices	0.208	165	48.23	34.68	17.09
M	Power plants	0.211	162	57.33	21.56	21.10
O	Dentistry and surgery	0.218	205	18.74	41.54	39.72
C	General chemical analytical physical products	0.230	138	56.78	33.42	9.80
T	Wheels and axles	0.231	55	41.82	30.30	27.88

Table 7 (continued)

Broad technical field	Technical sub-field	Herfindahl Index	Number of firms	Share of large firms	Share of other firms	Share of unassigned
E	Electricity transmission to vehicles	0.242	22	33.33	43.81	22.86
M	Closures and safes	0.244	94	15.91	44.96	39.13
M	Road structures, bridges (processes and engineering)	0.246	142	23.16	42.78	34.05
T	Ships and aquatic devices and marine propulsion	0.246	97	19.48	28.11	52.42
C	Inorganic chemicals	0.247	232	63.57	30.61	5.82
E	Electric heating and furnaces	0.267	305	46.87	38.15	14.98
C	Bioengineering	0.275	201	41.17	49.89	8.94
C	Miscellaneous chemical compositions	0.276	335	61.56	31.75	6.68
M	Refrigeration machinery	0.279	137	46.03	27.36	26.61
E	Typewriters	0.281	76	61.75	29.81	8.44
M	Apparel equip., leather working, boot and shoe making	0.287	46	25.35	55.06	19.59
C	Products for disinfecting, deodorizing, preserving, sterilizing	0.295	55	42.01	43.40	14.58
M	Agricultural machinery	0.295	79	25.72	35.99	38.29
C	Distillation processes	0.298	101	56.83	24.41	18.76
M	Metallurgical apparatus	0.298	141	45.44	38.37	16.19
O	Education and demonstration	0.315	56	13.87	29.05	57.09
C	Synthetic resins,	0.329	315	75.22	21.75	3.03
E	Calculators, computers, data-processing systems	0.338	295	63.96	29.42	6.62
M	Mining machinery, wells machinery and techniques	0.344	164	55.30	27.24	17.46
C	Carbon-based organic chemicals and other organic compounds	0.346	249	77.71	20.37	1.92
M	Fluid pressure brakes, clutches, power stop controls, etc.	0.346	170	53.84	36.69	9.47
C	Food, sugar, starch, carbohydrates	0.367	176	45.40	40.12	14.48
E	Photography and photocopy	0.369	166	74.59	19.82	5.59
M	Glass manufacturing	0.386	101	72.50	20.93	6.57
M	Machine controls	0.405	224	63.70	20.00	16.30
E	Electric motors, dynamos, generators, motive power syste	0.411	258	55.68	31.41	12.92
O	Wood products (beds, chairs, receptacles)	0.427	83	16.49	37.76	45.75
C	Drugs	0.434	179	59.58	32.83	7.59
E	Radio communication systems (inc. Radar)	0.441	120	49.34	39.00	11.67
O	Fabric structures, knots	0.455	5	18.03	31.15	50.82
E	Electrical devices and systems	0.466	270	56.38	32.13	11.49
E	Laser technology	0.467	82	54.88	39.52	5.60
E	Acoustics (inc. recording and sound reproduction, music, etc.)	0.468	163	46.97	27.65	25.37
E	Television, facsimile	0.485	200	72.63	22.26	5.11
E	Superconductors	0.514	7	66.67	27.78	5.56
C	Bleaching and dyeing of textiles	0.530	118	67.69	26.47	5.83
T	Motor vehicles	0.535	116	59.40	19.94	20.66
T	Bicycles and other land vehicles	0.537	145	29.29	27.85	42.86
E	Electrical transmission	0.548	256	55.62	33.81	10.57
C	Hydrocarbons, mineral oils, fuels and igniting devices	0.552	196	77.38	17.85	4.77
M	Mineral oil apparatus	0.556	11	60.47	23.26	16.28
E	Semiconductors	0.560	165	81.56	15.60	2.84
C	Agricultural chemicals	0.582	122	76.05	19.39	4.56
E	Lamps and discharge devices	0.587	169	48.85	29.73	21.42
T	Internal combustion engines	0.601	146	68.87	15.28	15.85
T	Aeronautics	0.634	72	42.66	33.82	23.51
E	Telecommunications equipment	0.663	155	66.84	24.73	8.43
T	Other transport equipment parts	0.679	33	40.30	16.73	42.97
O	Induced nuclear reactions: systems and elements	0.691	42	52.19	42.36	5.45
T	Tyres	0.733	49	66.10	16.78	17.12
O	Tobacco	0.745	34	43.18	36.69	20.13

Index (i.e. highest degree of cross-sectoral technological convergence). It confirms the primacy of non-electrical machinery technologies amongst those with the highest degree of technological convergence. Seven out of the first ten, and 20 out of the first 30 fields with the lowest Herfindahls are non-electrical machinery, whilst only two are chemical, and one each is electrical and transport. At the other end of the distribution, twelve of the 30 sectors with the highest Herfindahls are electrical (including semiconductors), six are transport, five are chemical and only three non-electrical machinery.

Finally, and perhaps most interestingly, we find amongst the 30 technologies with the highest degree of technological convergence, the seven fields of non-electrical machinery technology already identified in section 2.2 of this paper and Table 2, each with more than 10 000 patents and accounting in aggregate for 20% of total patenting activity. The share of large firms in total patenting in these fields varies from 22 to 42%, and that of unassigned patentees from 18 to 41%. From this we may conclude that at least 20% of all technological activities are concentrated fields of non-electrical machinery with high incidences of technological convergence and low visibility in conventional R&D statistics.

3. Conclusions (and one important unanswered question)

Our first major conclusion from this analysis is that history (or, at least, nineteenth-century technological history) is not bunk, but alive and well, and highly relevant to what is going on at the end of the twentieth century. We have shown that when we treat technical change as synonymous with R&D activities in science-based industries, we are in danger of neglecting up to nearly 40% of what is going on in technical change, especially in non-electrical machinery and in small firms. Nor can it reasonably be argued that our analysis has simply captured the last gasps of dying technologies. The large volume of patenting in some of the most dispersed technologies that we have identified suggests otherwise. So does the importance of technological convergence and vertical disintegration in fields of advanced manufacturing technology, such as Computer Aided Design

(Kaplinsky, 1983) and Robots (Fleck, 1983), and in the most rapidly expanding and revolutionary of contemporary technologies, software engineering (Brady and Quintas, 1991).

The second conclusion is about measurement. It is clearly dangerous to rely on just one aggregate measure (i.e. R&D expenditures) of technological activities in firms, and it is no longer necessary. This paper has shown a little of the potential of computerised patent data. Other analysts are exploring the potential of data on scientific papers and citations to improve our understanding of the role of science in technological change. One major challenge is to find a systematic and reliable measure of the development of software technology. In the meantime, it is worth repeating the inevitable limitations of aggregate R&D statistics on company R&D expenditures; they neglect both the range and the specific mix of technologies necessary to develop and make any particular class of product.

The third conclusion is that the persistence of improvements in mechanical technologies, above all in chemical and electrical firms, suggests that new 'paradigms' do not destroy old ones, but complement and extend them. Just as classical physics remains useful after relativity and quantum mechanics, so non-electrical machinery remains useful, indeed essential, even after organic synthesis, radio waves and the electronic chip. Furthermore, at the firm level, any single technological advance, however radical, is unlikely to destroy the economic value of the entire range of useful technological competencies typically found in the technical departments of the large modern corporation.

The fourth conclusion is that it is the very fact that the development and production of a given product at any given time requires a mix of technological skills (see, for example, Granstrand and Sjolander, 1990) that may help us explain how and why firms diversify their product range. If each separate product were linked to a separate and single technological competence, possibilities for competence-based diversification would not exist. On the other hand, mastery of a range of technologies linked to more than one product at least offers the potential for entering new product markets. Describing and explaining these linkages, and the way in which they evolve over time, is of obvious interest to the historian, and to corporate and national policy makers; it also should be (in our view) an input into the development of theories of production and change.

One final fundamental question remains unanswered: why does technological convergence (and vertical disintegration) happen more frequently in non-electrical machinery than in science-based technologies? After all, a variety of chemical and electrical technologies are also pervasive inputs into production in a wide variety sectors of economic activity (e.g. synthetic materials, electronic chips). In this sense, they are no different from non-electrical machinery. But why are they not like non-electrical machinery technologies, since they are not pervasive in their development and production, as well as in their use?

A first and obvious answer is that it is easier to make inventions and innovations in machinery than science-based technologies, given the lower skill and equipment requirements. Our data in Table 7 on the sources of these technologies do suggest that the 'barriers to entry' to non-electrical machinery technologies are lower than those to science-based technologies, in the sense that more firms, and smaller firms, are involved in their development. But, in a world of Schumpeterian competition and firm-specific competencies, why would large firms with high levels of technological skills devote more than a quarter of their technological activities (see Table 4) in order to maintain an innovative lead in fields where it is relatively easy to imitate?

A second answer is that it is the very ease of entry into non-electrical machinery technologies that make them an essential part of the competitive armoury of large firms in nearly all industries. These firms typically develop and operate complex production and/or product systems that combine and integrate inputs from science-based and mechanical technologies. Continued adjustments and improvements in these systems during design and development, as well as after experience in use, are necessary if they are to be viable and to improve. In most cases, it is easier (i.e. cheaper) to make these adjustments along the mechanical dimensions than the electrical and chemical ones. Furthermore, in the light of his own analysis (1969), Rosenberg might himself argue that spectacular improvements in science-based technologies create systemic bottlenecks, imbalances and constraints that can be remedied

only through steady improvements in mechanical technologies. Kodama (1986a,b) has attributed part of Japan's technological success to the capacity to fuse the achievements of science-based with those of mechanical technologies.

More generally, we would speculate that a continuous stream of machinery innovations by users is an essential feature of incremental and decentralised learning, made possible since the advent of cheap steel and electric power. Subsequent developments in chemical analysis, and related techniques and instrumentation, has enabled further decentralised improvements, especially in continuous flow production. Contemporary developments in electronic chip technology now enables decentralised and differentiated improvements in software systems. In all these technologies, growing 'ease of entry', often as a result of rapid technical change in highly concentrated supplier industries, has increased their importance as a means of improving the efficiency and flexibility of production. It is an explanation that promises a similarly bright future for machinery, instrument and software technologies. For the moment at least, we find it appealing.

4. References

Archibugi, D., 1988, In search of a useful measure of technological innovation, Technological Forecasting and Social Change 34, 253–277.

Brady, T. and P. Quintas, 1991, Computer software: the IT constraint, in C. Freeman, M. Sharp and W. Walker (editors), Technology and the Future of Europe (Pinter, London), 117–136.

Business Week, 1992, Global Innovation: Who's in the Lead?, 3 August, 68–73.

Cantwell, J., 1993, Corporate technological specialisation in international industries, in M. Casson and J. Creedy (editors), Industrial Concentration and Economic Inequality: Essays in Honour of Peter Hart (Elgar, Aldershot) forthcoming.

Fleck, J., 1983, Robots in manufacturing organisations, in G. Winch (editor), Information Technology in Manufacturing Processes (Rossendale, London).

Grilliches, Z., 1990, Patent statistics as economic indicators: a survey, Journal of Economic Literature 28, 1661–1707.

Granstrand, O. and S. Sjolander, 1990, Managing innovation in multi-technology corporations, Research Policy 19, 35–60.

Kaplinsky, R., 1983, Firm size and technical change in a dynamic context, Journal of Industrial Economics 22, 39–59.

Kodama, F., 1986a, Japanese innovation in mechatronics technology, Science and Public Policy 13, 44–51.

Kodama, F., 1986b, Technological diversification of Japanese industry, Science 223, 219–233.

Mowery, D. and N. Rosenberg, 1989, Technology and the pursuit of economic growth (Cambridge University Press, Cambridge).

Narin, F. and D. Olivastro, 1992, Status report: linkage between technology and science, Research Policy 21, 237–250.

National Science Board, 1992, The Competitive Strength of US Industrial Science and Technology: Strategic Issues (National Science Foundation, Washington).

Oskarsson, C., 1991, Technology diversification: the phenomenon, its causes and effects: a study of Swedish industry (Department of Industrial Management and Economics, Chalmers University of Technology, Gothenberg).

Patel, P. and K. Pavitt, 1991, Large firms in the production of the world's technology: an important case of 'non-globalisation', Journal of International Business Studies 22, 1–21.

Patel, P. and K. Pavitt, 1994, Patterns of technological activity: evidence and explanations, in P. Stoneman (editor), Handbook on the Economics of Innovation and Technical Change (Blackwell, Oxford), forthcoming.

Pavitt, K., 1988, Uses and abuses of patent statistics, in A. van Raan (editor), Handbook of Quantitative Studies of Science and Technology (Elsevier, Amsterdam).

Rosenberg, N., 1963, Technological change in the machine tool industry, Journal of Economic History 23, 414–443.

Rosenberg, N., 1969, Directions of technological change: inducement mechanisms and focusing devices, Economic Development and Cultural Change 18, 1–24.

Rosenberg, N., 1974, Science, innovation and economic growth, The Economic Journal 84, 90–108.

Rosenberg, N., 1985, The commercial exploitation of science by American industry, in K. Clark, R. Hayes and C. Lorenz (editors), The Uneasy Alliance: Managing the Productivity–Technology Dilemma (Harvard Business School Press, Boston).

Scherer, F., 1983, The propensity to patent, International Journal of Industrial Organisation 1, 107–128.

Schmookler, J., 1966, Invention and Economic Growth (Harvard University Press, Boston).

Von Hippel, E., 1988, The Sources of Innovation (Oxford University Press, New York).

PART TWO

MANAGEMENT

4

CALIFORNIA MANAGEMENT REVIEW Reprint Series
© 1990 by The Regents of the University of California
CMR, Volume 32, Number 3, Spring 1990

17

What We Know about the Strategic Management of Technology

Keith Pavitt

I*n high wage countries, both the competitiveness of firms and more general welfare depend critically on the ability to keep up in innovative products and processes and in the underlying technologies. Recent statistical studies show that the levels of companies' investments in technology explain international differences in* productivity and in shares of world markets.[1] In the increasingly turbulent, uncertain, and competitive world since 1973, the rate of growth of business funded R&D activities in the OECD area has actually increased.[2] In sectors like electronics, aircraft, and fine chemicals, companies' expenditures on R&D are greater than their investments in fixed equipment and plant.

In the UK, it has been recognized for some time that in spite of improvements in certain aspects of economic performance, national technological activities and international competitiveness remain unsatisfactory in many sectors. Similar concerns about technological competitiveness have spread to other countries as a consequence of the dynamism of Japanese firms. These concerns have been particularly marked in the United States, where sectors of earlier technological strength—steel, automobiles, and now electronics—are under threat from Japanese firms that spend about 30 per cent more of their output on R&D activities than do their U.S. counterparts.[3]

As a result, there has been increased interest in the 1980s among management scholars, consultants, and practitioners in the role of technology

This paper is based on a keynote address to the Annual Meeting of the British Academy of Management. An earlier version has been published in R. Mansfield, ed., *Frontiers of Management* (London: Routledge, 1989). It is based on the research program on Science, Technology, and Energy Policy, funded by the UK Economic and Social Research Council. I am grateful to Alan Cawson, Mark Dodgson, Mike Hobday, Ian Miles, and David Mowery for their suggestions and criticisms.

in such matters as corporate strategy, operations management, global competition, strategic alliances, and the like. However, it must be put in a proper historical perspective. Technology became an explicit element in management practice and strategy at the end of the 19th century with the growth of large chemical and electrical companies, particularly in Germany and the U.S.[4] Indeed, the industrial R&D laboratories central to this growth can be seen as part of the functional and professional specialization that defines much of modern management practice. Even before World War I, firms in these and other industries had extensive networks of external technological contacts, competed globally, and formed strategic alliances, often as part of world cartels.

Parallel to this accumulation of practical experience in the management of technology has been the scholarly research related to it. Although the importance of technological change had been acknowledged by earlier writers, it was Schumpeter who stressed the central importance of innovation in competition among firms, in the evolution of industrial structures, and in processes of economic development; and it was Schumpeter who gave us the most useful definition of innovation as consisting not just of new products and processes, but also of new forms of organization, new markets, and new sources of raw materials.[5] Schumpeter also made the distinction between "administrative management," which is management of what is well known, and "entrepreneurship," which is the creation and implementation of the new. However, Freeman has pointed out that Schumpeter never developed a theory of the innovating firm and had little to say on the sources of innovation and the importance of continuous incremental improvements.[6] More specifically, Schumpeter had little to say about the organizational characteristics of the major sources of technical change in established firms.

Characteristics of Technological Innovation

Innovation research has helped delineate four key characteristics of innovative activities in the firm:

- First, they involve continuous and intensive collaboration and interaction among *functionally and professionally specialized groups:* R&D, production, and marketing for implementation; organization and finance for strategic decisions to move into new areas.
- Second, they remain profoundly *uncertain* activities. Only about one in ten R&D projects turns out to be a commercial success with the other nine either not meeting technical objectives or (more often) commercial ones.
- Third, they are *cumulative*. Most technological knowledge is specific, involving development and testing of prototypes and pilot plants. Although firms can buy-in technology and skills from the outside, what they have

been able to do in the past strongly conditions what they can hope to do in the future.

- Fourth, they are highly *differentiated*. Specific technological skills in one field (e.g., developing pharmaceutical products) may be applicable in closely related fields (e.g., developing pesticides), but they are not much use in many others (e.g., designing and building automobiles).

These characteristics have major implications for theory and action related to the *content* of technological strategy, to the *processes* through which they are developed and implemented, and to *institutional continuity* in the face of *technological discontinuity*.

The Content of Technological Strategies

The cumulative and differentiated nature of technological developments in firms suggests that the choices about the *content* of technological strategy normally presented in the management literature—broad front versus specialized, product versus process, and the leader versus follower—do not take into account the enormous variety between firms in sources of technological opportunities and in the rate and direction of their development.[7] In particular, the innovative opportunities open to a firm are strongly conditioned by a firm's size and by its core business.[8]

This technological variety is summarized in Table 1. Innovating small firms are typically specialized in their technological strategies, concentrating on product innovation in specific producers goods such as machine tools, scientific instruments, specialized chemicals, and software. Their key strengths are in their ability to match technology with specific customer requirements. The strategic management tasks are to find and maintain a stable product niche and to benefit systematically from user experience.

Large innovating firms, on the other hand, are typically broad front in their technological activities and are divisionalized in their organization. Their key technological strengths can be based in R&D laboratories (typically in chemicals and electrical-electronic products), or in the design and operation of complex production technology (typically in mass production and continuous process industries), or (increasingly) in the design and operation of complex information-processing technology (typically in finance and retailing).

In R&D-based technologies, the key opportunities are for horizontal diversification into new product markets. The strategic management tasks are those of mobilizing complementary assets to enter new product markets (e.g., obtaining marketing knowledge when a pharmaceutical firm moves into pesticides) and continuous revision of divisional responsibilities to exploit emerging technological opportunities (e.g., personal computers

Table 1. Basic Technological Trajectories

Definition

	Science-Based	Scale Intensive	Information Intensive	Specialized Suppliers
Source of Technology	R&D Laboratory	Production Engineering and Specialized Suppliers	Software/ Systems Dept. Specialized Suppliers	Small-Firm Design and Large-Scale Users
Trajectory	Synergetic New Products Applications Engineering	Efficient and Complex Production and Related Products	Efficient (and Complex) Information Processing, and Related Products	Improved Specialized Producers Goods (Reliability and Performance)
Typical Product Groups	• Electronics • Chemicals	• Basic Materials • Durable Consumer Goods	• Financial Services • Retailing	• Machinery • Instruments • Specialty Chemicals • Software
Strategic Problems for Management	• Complementary Assets • Integration to Exploit Synergies • Patient Money	• Balance and Choice in Production Technology among *Appropriation* (Secrecy and Patents), *Vertical Disintegration* (Cooperation with Supplier), and Profit Center • "Fusion" with Fast-Moving Technologies • Diffusion of Production Technology among Divisions • Exploiting Product Opportunities • Patient Money		• Matching Technological Opportunity with User • Absorbing User Experience • Finding Stable or New Product 'Niches.'

cutting across previous responsibilities in computers, office machinery, and even consumer electronics).

In production-based and information-based technologies, the key opportunities are in the progressive integration of radical technological advances into products and production systems and in diversification vertically upstream into potentially pervasive production inputs (e.g., CAD-CAM, robots, and software). The strategic management tasks are to ensure diffusion of best practice technology within the firm and to make choices about the degree of appropriation (i.e., internalization) of production technology.

Firms do not have completely free choice about whether or not to be broad front or specialized, and product or process oriented. Similarly, they do not have a completely free hand about being a leader or a follower. In many areas, it is not clear before the event who is in the innovation race, where the starting and finishing lines are, and what the race is about. Even when it is, firms may start out wishing to be a leader and end up being a follower. Teece has shown that while there are some advantages in being first, particularly when there are strong regimes of property rights of cumulative learning, it is sometimes advantageous to be second, particularly when product configurations are not fully fixed, so that followers can learn from the mistakes of leaders who find themselves without the required range of complementary assets.[9]

Furthermore, given that firms develop their technological competences cumulatively, the uncritical application of conventional project appraisal techniques will result in myopic technology strategies. Such strategies neglect the benefits from the knowledge accumulated in a project that can be deployed subsequently to exploit technological opportunities in the future. Since these accumulated benefits are time-consuming, dynamic strategies that take them fully into account are more likely to emerge in companies and countries where performance is judged over the long term, and where managers are capable of making informed and reasoned judgement about the strategic implications of likely future developments in technology.[10]

The Implementation of Technology Strategy

A major criticism of the "content" view of technological strategy is that it neglects the context within which—and the process whereby—technological strategies are generated, chosen, and implemented. These processes are bound to involve more than the purely technical function. Production and marketing are inevitably involved with R&D in implementation, with finance in setting ground rules for evaluating and monitoring programs and projects, and with organization and the strategic function in decisions about entering new areas.

Company structure and company strategy thus play a major role in the formation of technological strategy. Hobday has pointed out that the ambitious technological strategies of Japanese electronic components firms depend in large part on their vertical integration with electronic equipment manufacture and on the relatively strong emphasis put on long-term growth compared to short-term profits. More generally, Japanese firms are apparently more likely than those in Europe and the U.S. to have a member of the main board responsible for technological policy.[11] Sharp argues that recent initiatives in European technological cooperation in ESPRIT have taken off rapidly precisely because they involve chief executives rather than R&D directors.[12]

Given that technology strategy involves many functions and professions, as well as major uncertainties, its formation and implementation are bound to be a choice territory for the advocacy, battles, and negotiations to which analysis in the process school of strategy give such great importance.[13] This was recognized some time ago by Freeman when, after reviewing the disappointing experience that firms had had with formal, quantitative methods of R&D project selection and technological forecasting, he concluded that

> empirical evidence confirms that decision making in relation to R&D projects or general strategy is usually a matter of controversy within the firm . . . uncertainty means that many different views may be held and the situation is typically one of the advocacy and political debate in which project estimates are used by interest groups to buttress a particular point of view. Evaluation techniques and technological forecasting, like tribal war-dances, play a very important part in mobilizing, energizing and organizing.[14]

However, technology strategy cannot be described solely in terms of political negotiation between hostile professional and functional tribes. In the market system, the ability to satisfy user's needs better than the competition is the ultimate measure of success and profitability within the firm. Innovation research has come to robust conclusions about the management factors associated with successful innovations. In addition to the quality of technical work, these include strong horizontal linkages among functional departments, with customers, and with outside sources of relevant technical expertise.

Either by conscious choice or by trial and error, successful innovating firms are more likely to develop "routines" (or rules of thumb) that reflect these ingredients. Given the high uncertainties involved, trial and error are inevitable in the development and implementation of innovation. In fact, the major importance of *development*—as opposed to *research*—activities in industrial laboratories can be considered as a systematic form of trial and error. Theory and computer simulations are not powerful enough to predict the performance of technological artifacts with a high enough degree of certainty to eliminate the costly development and testing of prototypes and pilot plants.

In addition, the ability to learn from experience—whether internally (learning by doing) or from suppliers, customers, and competitors (learning by using, learning by failing, reverse engineering)—is of major importance in the management of innovation. As Dodgson has pointed out, learning from experience actually dissolves sharp distinctions in the strategy debate between content, process, and context.[15] This is because *processes* of learning about the *context* help define the *content* of strategy, the implementation of which in turn helps define both the nature and directions of subsequent learning *processes* and changes in *context*. More simply put, content definition and implementation become indistinguishable, given the central importance of learning.

Comparative empirical research has demonstrated the importance of employee training for the effective exploitation of technology.[16] Particularly in the large firm, learning is also a collective activity requiring frequent communication among specialists and functions. Since knowledge accumulated through experience is also partly tacit, and the task to which such knowledge is applied are complex are loosely structured, personal contact and discussions are the most frequent and effective means of communication and learning. Policies for effective learning therefore go beyond training and organization to include those of geographical location. Allen and other scholars have shown the importance of physical location in influencing patterns of communication, both within the technical function and between the technical and other functions in the firm.[17] Howells has shown that decisions about the location of R&D laboratories by firms in the UK pharmaceutical industry have been strongly influenced by the requirements for effective internal communication with other functional areas.[18]

Technological Discontinuities and Institutional Continuities

With the present wave of radical technological change in micro-electronics and information technology, considerable emphasis is being placed in management theory and practice on the notion of "technological discontinuities," which imply a radical increase in the rate of technical change and a marked shift in its associated skills and required organizational forms.[19] It is often argued, on the basis of either Schumpeter's notion of creative destruction or the so-called product cycle theory, that technological discontinuities are associated with the emergence of new small firms that exploit them, given the conservatism, obsolescence, and bureaucracy in established large firms.

The evidence does not necessarily confirm this view. In electronics (the main sector of "discontinuity") in the UK since 1945, the proportions of significant innovations made by both large firms (with more than 10,000 employees) and by small firms (with fewer than 1,000 employees) have both been increasing at the expense of the medium-sized firms in between.[20] Mowery has shown that the growth of industrial R&D in the 20th Century has been associated in certain periods with greater stability among large firms.[21] Established chemical firms have successfully survived and indeed benefitted from successive waves of radical innovations in synthetic products. IBM was a world leader in the earlier, traditional electro-mechanical technologies of office machinery before it moved into computers.[22]

Some of the most revolutionary business applications of information technology today are to be found not in new technology-based firms, but among the oldest, largest, and most conservative of capitalists: banks, financial services, and large-scale retailing.[23] Two factors help explain why *technological discontinuities* can co-exist with *institutional continuities:*

- First, large established firms normally have specialized and professionalized R&D laboratories and other technical functions with accumulated skills and experience in orchestrating and integrating inputs from a wide variety of scientific and technical disciplines. They are therefore experienced in hiring and integrating professionals from promising new areas. Examples include the hiring of computer experts by IBM[24] and of aerodynamic and hydraulic engineers by Sulzer for the development of the shuttle-less loom.[25]

- The second reason was identified by Schumpeter in his later writings. Large firms have considerable oligopolistic power. In some countries, they are not subject to a strong, short-term profit constraint. They therefore have both the resources and the time to explore the implications of technological discontinuities for their business and to link them with core competences within the firm, through learning and incremental change, before deciding whether or not to move into commercialization. One observable feature of innovating firms is precisely that they develop technological capabilities beyond those strictly related to their current output.

Perez correctly pointed to the dangers of a mismatch between institutional routines and skills, on the one hand, and the effective exploitation of technological discontinuities, on the other.[26] Given its long-term importance, we need to know more about how many established business firms successfully overcome any mismatch, and how they assimilate and exploit technological discontinuities. Recent analyses of information technology in service firms by Barras[27] and by Thomas and Miles[28] suggest a process can be described either as a "reverse product cycle" or as the equivalent of technical change in production-centered firms (information technology is first used in such firms to improved processes and, after a sometimes long period of learning, becomes the basis of products sold outside). Further empirical studies are needed to see whether this model can be extended to other sectors, or to other technical areas like biotechnology.

Conclusions for Management: Beware the Conventional Wisdom

The major conclusion for management to emerge from this review of research into the innovation process is that some of the conventional wisdom from business schools and management consultants about technology strategy is irrelevant and even misleading.

First, it is not useful for a firm's management to begin by asking whether its technology strategy should be leader or follower, broad or narrow front, product or process. These characteristics will be determined largely by the firm's size and the nature of its accumulated technological competences, which will jointly determine the range of potential technological and market opportunities that it might exploit. There is no easy and generalizable recipe for success.

Second, the implementation of technology strategy is just as important as its definition, and an integral part of it. Given the cumulative nature of firm-specific competences and the inevitable uncertainties surrounding innovative activities, the capacity for in-house learning from experience will be fundamental for success. Being an essentially collective activity, such learning will depend on good systems of communication.

Third, convention methods of project appraisal and divisional organization will result in myopic technology strategies that neglect the effects of innovative choices today on the ability to exploit technological opportunities in the future. Such strategies also hinder the development of product opportunities that do not fit tidily into established divisional markets or missions. The continuing stream of new high-technology firms established by former employees of large firms confirms the importance of this problem.

Management's role is inevitably constrained by the accumulated organizational and technological characteristics of the firm. At the same time, coping with continuous change is not easy and requires more than a tribal chief organizing war dances or a charismatic prince playing Machiavellian politics. The successful management of technology requires:

- the capacity to orchestrate and integrate functional and specialist groups for the implementation of innovations;
- continuous questioning of the appropriateness of existing divisional markets, missions, and skills for the exploitation of technological opportunities; and
- a willingness to take the long view of technological accumulation within the firm.

References

1. J. Fagerberg, "A Technology Gap Approach to Why Growth Rates Differ," *Research Policy*, 16/2-4 (1987); J. Fagerberg, "International Competitiveness," *Economic Journal*, 98/391 (1988).
2. OECD, *Research and Development in the Business Enterprise Sector, 1963-1979*, Basic Statistical Indicators, Volume D, (Paris: OECD, 1983); OECD, *Recent Results: Selected Science and Technology Indicators, 1979-1984* (Paris: OECD, 1984).
3. K. Pavitt and P. Patel, "The International Distribution of Determinants of Technological Activities," *Oxford Review of Economic Policy*, 4/4 (1988):1-21.
4. D. Mowery, Industrial Research and Firm Size, Survival and Growth in American Manufacturing, 1921-1946: An Assessment," *Journal of Economic History*, 43 (1983).
5. J.A. Schumpeter, *Capitalism, Socialism and Democracy* (New York, NY: Harper and Row, 1950).
6. C. Freeman, "Schumpeter's 'Business Cycles' Revisited," paper prepared for Schumpeter Society Conference, Siena, 1988.
7. E. von Hippel, *The Sources of Innovation* (Oxford: Oxford University Press, 1988); F. Scherer, "Inter-Industry Technology Flows in the United States," *Research Policy*, 11 (1982).
8. K. Pavitt, M. Robson, and J. Townsend, "Technological Accumulation, Diversification and Organization in UK Companies, 1945-1983," *Management Science*, 35/1 (1989).

9. D. Teece, "Profiting from Technological Innovation: Implications for Integration, Collaboration, Licensing and Public Policy," *Research Policy,* 15 (1986).

10. Pavitt and Patel, op. cit.

11. P. Patel and K. Pavitt, "Is Western Europe Losing the Technological Race?" *Research Policy,* 16/2-4 (1987): 59/85.

12. M. Sharp, "Corporate Strategies and Collaboration—The Case of ESPRIT and European Electronics," in M. Dodgson, ed., *Technology Strategy and the Firm: Management and Public Policy* (New York, NY: Longman, 1989).

13. A. Pettigrew, *The Management of Strategic Change* (New York, NY: Blackwell, 1987).

14. C. Freeman, *The Economics of Industrial Innovation* (London: Penguin, 1974).

15. M. Dodgson, "Introduction: Technology in a Strategic Perspective," in M. Dodgson, ed., *Technology Strategy and the Firm: Management and Public Policy* (New York, NY: Longman, 1989).

16. C. Pratten, *A Comparison of the Performance of Swedish and UK Companies* (Cambridge: Cambridge University Press, 1976); S. Prais, "Educating for Productivity: Comparisons of Japanese and English Schooling and Vocational Preparation," *National Institute Economic Review,* 119 (1987).

17. T. Allen, *Managing the Flow of Technology* (Cambridge, MA: MIT Press, 1977).

18. J. Howells, "The Location and Organization of Research and Development: New Horizons," *Research Policy* (forthcoming).

19. M. Tushman and P. Anderson, "Technological Discontinuities and Organization Environments, in Pettigrew, op. cit.

20. K. Pavitt, M. Robson, and J. Townsend, "A Fresh Look at the Size Distribution of Innovating Firms," in F. Arcangeli et al., eds., *Frontiers of Innovation Diffusion* (Oxford: Oxford University Press, forthcoming).

21. Mowery, op. cit.

22. K. Pavitt, "'Chips' and 'Trajectories': How Does the Semiconductor Influence the Sources and Directions of Technical Change?" in R. MacLeod, ed., *Technology and the Human Prospect* (London: Pinter, 1986), pp. 31-54.

23. R. Barras, "Towards a Theory of Innovation in Services," *Research Policy,* 15 (1986).

24. B. Katz and A. Phillips, "Government, Technological Opportunities and the Emergence of the Computer Industry," in H. Giersch, ed., *Emerging Technologies: Consequences for Economic Growth, Structural Change and Employment* (Tubingen: Mohr, 1982).

25. R. Rothwell, "The Characteristics of Successful Innovators and Technically Progressive Firms," *R&D Management,* 7 (1977).

26. C. Perez, "Structural Change and Assimilation of New Technologies in the Economic and Social Systems," *Futures,* 15/4 (1983): 357-375.

27. Barras, op. cit.

28. G. Thomas and I. Miles, "Strategic Options for New Telecommunications Services," in M. Dodgson, ed., op. cit.

Keith Pavitt is a Professor of Science and Technology Policy at Sussex University in England and Deputy Director of the Science Policy Research Unit. He has been a staff member at the Organization for Economic Cooperation and Development in Paris and a visiting scholar at the universities at Princeton, Strasbourg, and Padua. He has published extensively on the public policy and management dimension of science and technology.

5

ELSEVIER

Research Policy 26 (1997) 141–156

research
policy

The technological competencies of the world's largest firms: complex and path-dependent, but not much variety [1]

Pari Patel *, Keith Pavitt

Science Policy Research Unit, University of Sussex, Brighton, BN1 9RF, UK

Abstract

Firm-specific technological competencies help explain why firms are different, how they change over time, and whether or not they are capable of remaining competitive. Data on more than 400 of the world's largest firms show that their technological competencies have the following characteristics:
1. They are typically *multi-field*, and becoming more so over time, with competencies ranging beyond their product range, in technical fields outside their 'distinctive core'.
2. They are *highly stable* and *differentiated*, with both the *technology profile* and the *directions* of localised search strongly influenced by firms' *principal products*.
3. The *rate* of search is influenced by both the firm's *principal products*, and the conditions in its *home country*. However, *considerable unexplained variance* suggests *scope for managerial choice*.

These findings confirm the importance of *complexity* and *path dependency* in the accumulation of firm-specific technological competencies, and show that *managers are heavily constrained* in the *directions* of their technological search. They also show the limits of the notion of competition through *variety*, given that the same specific field of technological competence is often essential to the development of a range of possible product configurations. Technological imperatives still exist. © 1997 Elsevier Science B.V.

1. Introduction

The purpose of this paper is to throw light on the nature and determinants of the technological competencies of the world's largest firms. These are of increasing interest to practitioners, and to theorists, who are seeking to explain why firms provide different ranges of goods and services, why they diversify

at different rates and in different directions over time, and what makes them competitive (Rumelt, 1974, Prahalad and Hamel, 1990, Carlsson and Eliasson, 1991, Dosi et al., 1992, Teece and Pisano, 1994).

Our main contribution is empirical, and is based on a data source consisting of systematic information on US patenting by more than 400 of the world's largest firms, sorted according to their volume of sales. For each firm, we have information on its headquarters country, principal product group, and the technical field and country address of the inventor of each patent. Similar data has been used by Narin et al. (1987) at Computer Horizons Inc. for corporate and competitor analysis, by Jaffe (1989) at

* Corresponding author. Tel.: 44 1273 68 67 58; fax: 44 1273 68 58 65; e-mail: parip@sussex.ac.uk.
[1] This paper is based on research at the Centre for Science, Technology and Energy and Environment Policy (STEEP), funded by the Economic and Social Research Council (ESRC) within the Science Policy Research Unit.

the National Bureau of Economic Research to identify and measure technological 'spillovers', and by Cantwell (1989) at Reading University to explain patterns of international production. And a similar approach has been taken by Barre (1996) at the *Observatoire des Sciences et des Techniques* to data on the patenting activities of multinational firms compiled at the European Patent Office. [2]

1.1. Data sources

The data set has been compiled from information, provided by the US Patent Office, on the name of the company, the technical field, and the country of origin, of each patent granted in the USA from 1969 to 1990. One weakness with this source is that many patents are granted to large firms under the names of subsidiaries and divisions that are different from those of their parent companies. Consolidating patenting under the names of parent companies can only be done manually, on the basis of publication like *Who Owns Whom*. Our consolidation shows that some firms have considerably more patents in our consolidated classification than in the original compilations of the US Patent Office (see Patel and Pavitt, 1991).

We have also included in our data set the following information on each firm: country of origin, sales, employment, and (where possible) R & D expenditures. Given the requirements of our statistical analysis, we have excluded firms with 50 or fewer patents in the period 1981–90. The distribution of our firms according to product groups is shown in Table 1. The distribution according to nationality shows that 47% are of US origin, 29% from Europe, and 25% from Japan. Our earlier analysis shows that these firms account for more than 40% of total patents granted in the USA, with considerably higher shares in the chemicals, electrical-electronic, and transport sectors (Patel and Pavitt, 1991).

1.2. Limitations

Our paper has three sets of limitations. *First*, we measure only technological competence, and thereby

neglect many others that are important. Dosi and Teece (1993) have distinguished organisational-economic competencies from technical competencies, and have argued that the latter derives from the former, and is therefore more fundamental to the firm. [3] Our empirical results suggest that this is only partly correct. A firm's organisational competence does influence its level of commitment to technological activities, and its rate of entry into fast-growing sub-fields. However, a firm's accumulated technological competence strongly constrains the directions in which it searches: even the brightest and the best organisational capabilities will find it difficult (impossible?) to convert a firm making Harris Tweed jackets, or Italian high-fashion shoes, into a world class firm in personal computers. The differentiated nature of technical competencies is one the most important factors explaining the coherence and the boundaries of the firm. And a recent survey of 100 Italian firms by Malerba and Marengo (1993), ranked technological competencies as of greater long term importance than competencies to respond to either market signals or competitors' strategic actions. The subject therefore deserves analytical and empirical attention, even if it does not cover, and cannot explain, everything. [4]

The *second* limitation is that we measure technological competencies only imperfectly through patent

---

[2] Similar research is also being undertaken by Malerba and Orsenigo at CESPRI, Bocconi University, Milan; Soete and Verspagen at MERIT, University of Limbourg; and Grupp at ISI, Fraunhofer Institute, Karlsruhe.

[3] "Organisational/economic competence involves: (1) allocative competence—deciding what to produce and how to price it; (2) transactional competence—deciding whether to make or buy, and whether to do so alone or in partnership; and (3) administrative competence—how to design organisational structures and policies to enable efficient performance. Technical competence, on the other hand, includes the ability to develop and design products and processes, and to operate facilities effectively · · · A firm becomes superior in a particular technological domain because it has certain organisational capabilities: it allocates resources to more promising projects, it harnesses experience from prior projects, it hires and upgrades human resources, it integrates new findings from external sources, and it manages a set of problem-solving activities associated with that technology." (Dosi and Teece, 1993, pp. 6–7).

[4] In a similar manner (and using the jargon of another academic discipline), we are fully aware that technological competencies in large firms are "socially constructed" (Hughes and Pinch, 1987). But we concentrate here on the important cognitive factors that shape the social construction of technology.

P. Patel, K. Pavitt / Research Policy 26 (1997) 141–156 143

data. [5] Nonetheless, patenting in the USA (together with patenting in Europe) is a better measure than most of the alternatives, given its relative homogeneity, detail, accuracy, and (after recent advances in information technology) accessibility and cost: hence its increasing use by both analysts and practitioners. [6] However, in relation to the subject of this paper, three potential limitations of the US patenting measure must be mentioned:

- Patents do not measure the extent of the firm's external technological linkages. However, many studies have shown (most notably, Cohen and Levinthal, 1989) that external technological linkages are in general complementary to internal competencies, and these we do measure.
- Patents measure codified knowledge, whereas a high proportion of firm-specific competencies is non-codified (i.e. tacit) knowledge. We argue that the two forms of knowledge are complementary, not substitutes. Other measures that embody tacit knowledge (such as R&D expenditure, judgements of technological peers) give results very similar to those using patenting (see Patel and Pavitt, 1987).
- Patenting does not fully measure competencies in software technology, since copyright law is often used instead as the main means of protection against imitation (see Barton, 1993, Samuelson, 1993). We readily admit this to be the major empirical shortcoming of our analysis, and plead only that no-one has yet found a satisfactory, accessible and systematic measure of competencies in software technology that we could use. [7] And as we shall see below, we have nonetheless been able to identify the growing importance of competencies in information technology.

The *third* limitation to our analysis is that we do not assess how differences in the rate and direction

of technological accumulation affect firms' economic and competitive performance. Suffice to say that an increasing number of studies confirm the competitive importance of technological competencies at the level of the firm, [8] which should in principle heighten interest in studies like ours that attempt to describe and explain how they are acquired.

1.3. The basic framework: coping with complexity and continuous change

Although empirically based, our analysis has a range of theoretical implications that we shall identify when presenting and analysing our results. Our basic framework is based on the pioneering work of Nelson and Winter (1982) that combines the insights of Schumpeter on the central importance of innovation in the dynamics of competition, and of Simon and his colleagues on the satisfying behaviour of business firms. Technological artefacts, and the organisational and economic worlds in which they are embedded, are *complex and everchanging*: they each comprise so many variables and interactions that it is impossible fully to model, predict and control their behaviour through explicit and codified theories and guidelines. Certainty about the future, probabilistic risk, and optimisation are therefore impossible. Management solves problems and makes improvements through step-by-step experimentation, in which changes are made in one feature or component at a time, and ends and means re-interpreted in the light of the subsequently observed changes. In addition to codified knowledge, experience and tacit knowledge improve the effectiveness of: (i) the choices of the feature or component to vary at each stage; (ii) the subsequent modifications in means and ends, made after observation of the effects of variations in features or components.

This method is called 'learning', or 'experimentation', or 'trial and error'. Essentially the same approach underlies Lindblom's prescriptions in public policy (Lindblom, 1959), Quinn's in corporate strategy (Quinn, 1980), and Kline's in engineering design and development (Kline, 1995). It helps explain the

---

[5] The uses and abuses of patent data have been extensively discussed elsewhere. See, for example, Basberg (1987, Pavitt (1988, Griliches (1990, Patel and Pavitt (1995b).

[6] In addition to Jaffe (1989) and Narin et al. (1987) see, for example, Griliches (1984) and Business Week (1993).

[7] Recent research by Jacobsson and Oskarsson (1995) uses very interesting data on the technical field of specialisation of Swedish engineers working in Swedish firms. Unfortunately, this method cannot easily be reproduced in other countries, because of lack of data.

[8] See, for example, Cantwell (1989, Franko (1989, Geroski et al. (1993, Oskarsson (1993).

144   *P. Patel. K. Pavitt / Research Policy 26 (1997) 141–156*

characteristics of the technological competencies that we observe in our large firms.

In Section 2, we show that the technological competencies of large firms are *spread over a large number of fields*. This reflects the complex and multivariate nature of specific products and methods of production, which require the combination and application of advances in many fields of specialised knowledge. Complexity constrains firms to search and experiment in and around what they already know. As a consequence, we find in Section 3 that each firm's technology profile is very *stable over time*, is *similar to that of most other firms in the same industry*, but is *strongly differentiated from those of most firms in other industries*. We then show in Section 4 that the future *directions* of corporate technological accumulation are strongly constrained by what firms already know, whilst the *rates* of such accumulation are strongly influenced also by both their home-country environment and by managerial discretion. As a consequence, we conclude in Section 5 that the central concepts in evolutionary theory of *complexity and path-dependency* are strongly supported by our findings, whilst the (often fuzzy) notion of *variety* is not.

2. The prevalence of the 'multi-technology' firm

2.1. Multi-field competencies

The most striking feature of the technological competencies of large firms is the *wide range* or *diversity* of technological fields in which they are active. Table 1 shows the distribution of US patenting of our large firms, in each of the 16 principal product groups, across five major technological families: chemical, electrical-electronic, non-electrical machinery, transport, and 'other'.[9] In broad terms.

---

[9] The method for distributing firms' technological activities amongst five technological families is described more fully in Patel and Pavitt (1994). Briefly stated we re-classified the US Patent classes and sub-classes into 34 technical fields (33 of which are listed in Table 3), and 91 sub-fields. On the basis of the 91 sub-fields, we re-combined patenting into the four technological families shown in Table 1. The 'other' category includes traditional manufacturing (e.g. textiles) and non-manufacturing (e.g. construction, medicine, agriculture).

they confirm what we would expect to be the core technological competencies of different industries— for example, 71% of chemical firms' patenting is in chemical technologies. At the same time, firms have substantial technological competencies outside what would appear to be their core fields. In particular, competencies in non-electrical machinery make up more than 10% of the total in all industries except pharmaceuticals.

A more refined measure of technological diversity is the number of technical fields—out of the total of 34 used in our analysis [10]—in which our firms have been granted a patent and are therefore technically competent. Only 4% of our firms were active sometime in the 1980s in ten or fewer of these technical fields, whilst 52% were active in between ten and 20, and 44% in more than 20—hence the term 'multi-technology' firm (see Archibugi, 1988; Granstrand and Sjolander, 1990).

2.2. Corporate technological reach is greater than product reach

Our large firms are, of course, diversified in their product portfolios, as well as their technological portfolios. However, our data suggest that they are more so in the latter than the former. Limitations in data on large firms' product mix do not allow a detailed, firm-by-firm comparison with their technology mix. But Table 2 confirms that our firms have a much broader range of technologies than products. It compares the number of firms with their principal activity in selected product groups with the number of firms active in their corresponding distinctive technologies. In all cases, the latter is considerably larger than the former. In part, this is because the former measures only firms' *principal* product activity, whilst the latter encompasses their total, rather than principal, technological activity. However, it is

---

[10] See Table 3 for the names of 33 of these technical fields. The last field is 'other' which includes non-manufacturing (e.g. construction, medicine, agriculture) as well as other fields not elsewhere specified.

P. Patel, K. Pavitt / Research Policy 26 (1997) 141–156

Table 1
The distribution of large firms' technological activities in five broad technological fields, according to their principal product group: 1981–90

Principal product group (PPG)	Percentage share of the PPG's patents in technology field					
	Chemical	Non-electrical machinery	Electrical	Transport	Other	Total
Chemicals (66)	71.0	16.9	8.9	0.6	2.6	100.0
Pharmaceuticals (25)	80.2	8.0	2.1	0.0	9.7	100.0
Mining and petroleum (31)	57.1	34.2	6.7	0.9	1.1	100.0
Textiles etc. (10)	52.9	31.7	9.5	0.6	5.3	100.0
Rubber and plastics (9)	43.2	29.3	4.7	20.1	2.7	100.0
Paper and wood (18)	25.4	47.1	12.4	0.4	14.6	100.0
Food (14)	70.6	21.9	3.0	0.1	4.3	100.0
Drink and tobacco (8)	40.8	50.3	4.6	0.3	3.9	100.0
Building materials (16)	30.5	51.3	10.0	0.9	7.3	100.0
Metals (38)	26.8	54.9	13.9	2.1	2.2	100.0
Machinery (58)	7.6	64.9	13.9	10.2	3.3	100.0
Electrical (56)	7.6	21.2	67.0	1.3	2.8	100.0
Computers (17)	5.2	16.3	77.3	0.2	1.0	100.0
Instruments (21)	14.3	18.3	64.2	0.1	3.0	100.0
Motor vehicles (35)	3.8	44.8	20.7	28.8	1.9	100.0
Aircraft (18)	8.1	48.5	31.2	8.3	3.9	100.0
All 440 large firms	28.8	27.9	35.7	4.4	3.1	100.0

Number of firms in each product group in parentheses.
Source: calculated from data supplied to SPRU by the US Patent and Trademark Office.

implausible that anywhere near 400 of our large firms are making and selling instruments or non-electrical machinery. It is also implausible that more than 94 of our firms make and sell computers (i.e. those with their principal products in computers, electrical and electronic products, and instruments). Yet a much larger number of firms are mobilising technological competencies in non-electrical machinery, instruments, and computing, in order to make other products.

Table 2
Number of active large firms in selected principal products, compared with number active in closely related technologies, 1985–90

Principal product (out of 16; see Tables 1 and 3)	No. of active firms (out of 440)	Technological field (out of 33 listed in Table 3)	No. of active firms (out of 440)
Computers	17	Computers	288
Electrical	56	Semiconductors	166
Instruments	21	Instruments	407
Chemicals	66	Organic chemicals	284
Pharmaceuticals	25	Drugs and bioengineering	204
Mining and petroleum	31	Chemical processes	413
		Chemical apparatus	393
Machinery	58	Non-elect. machinery	377
		Specialised machinery	394
		Metal working equipment	366
Motor vehicles	35	Vehicles and engines	142
Aircraft	18	Aircraft	73

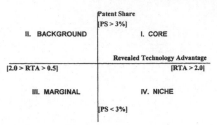

Fig. 1. A classification for firms' technological profiles.

2.3. Corporate technological profiles

Given these results, it is misleading to define a large firm's technological competencies simply in terms of a few fields of excellence. [11] It is more useful to think in terms of *profiles* of competencies, with varying levels of commitment and competitive advantage in a range of technological fields. With our data, these profiles can have two dimensions, as shown in Fig. 1:

- on the y-axis, the shares of a firm's total patenting in each of the 34 technological fields: in other words, the relative importance for the firm of competencies in each of these technological fields. We shall called this the *patent share (PS)* profile.
- on the x-axis, the shares of the firm in total patenting in each of the 34 technological fields, divided by the firm's aggregate share in all the fields: in other words, the relative importance of the firm to each field of technological competence, after taking account of the firm's total volume of competencies. We call this the *revealed technology advantage (RTA)* profile. [12]

On this basis:

- the 'distinctive' technical competencies of the firms are those in which the RTA is relatively

---

[11] For example, "Few companies are likely to build world leadership in more than five or six fundamental competencies. A company that compiles a list of 20 to 30 capabilities has probably not produced a list of core competencies." (Prahalad and Hamel, 1990).

[12] The firm's *RTA* in each of the 34 technological fields is similar to the *revealed comparative advantage (RCA)* measure used to assess the export performance of countries. The higher the *RTA*, the greater the relative strength of a firm in a technological field.

high: namely, in the *first* quadrant, which defines and describes the *core* of its competencies, and in the *fourth* quadrant, where it may have *niche* advantages in relatively small technological fields;

- in the *second* quadrant, the firm may have *background* competencies, in fields where it allocates a relatively high share of its technological resources, but where—given the relatively large size of the field—it does not achieve a relatively high technological advantage compared with its competitors;
- and in the *third* quadrant, it may have *marginal* competencies, where it neither allocates a large share of its own resources, nor achieves a distinctive advantage compared with other firms.

In Table 3, we classify the nature of contribution of competencies in each of our 34 technical fields to our firms in each of the 16 product groups. From this, it emerges that technical fields vary greatly in the nature and extent of their contributions to firm-specific competencies:

- organic chemistry and materials are *core* competencies in five product groups; drugs, non-electrical machinery, and image and sound each in three product groups; instruments—in spite of its overall importance—in only one; and five fields in none at all;
- *background* competencies are located mainly in machinery, instrumentation, and chemical processes;
- *niche* competencies are restricted to relatively few fields such as agricultural chemicals, bleaching and dyeing, power plant and nuclear energy;
- the most prevalent of *marginal* technologies are assembly and materials handling, plastic and rubber, and metallurgical processes;
- *computer-related* technological competence is so far identifiable beyond the usual 'high-tech' industries only in machinery and motor vehicles.

Many analysts would argue that most corporate technological resources should be concentrated in fields located in the first and fourth quadrants of Fig. 1. Table 3 shows that this is far from the typical case:

- in only four sectors (chemicals, pharmaceuticals, mining and petroleum, and electrical products) do the number of *core* technological fields outnumber the number of *background* fields;
- more detailed calculations show that, in at least six sectors, competencies in *background* fields

P. Patel, K. Pavitt / Research Policy 26 (1997) 141–156

Table 3

Firms' technical profile according to their principal product group: 1981–90

Technical field	Principal product group															
	Chem.	Phar.	Mini.	Text.	Rubb.	Pape.	Food	Drin.	Buil.	Meta.	Mach.	Elec.	Comp.	Inst.	Moto.	Airc.
Inorg. chemistry	***	****	****	****	****	**	**		*	****	*					*
Org. chemistry	****	****	****	****	****	**	**	**	**	**				**		**
Agr. chemistry	****	***	**	**	**	**	**	**	**	x						
Chem. process.	****	**	****	**	**	**	****	**	**	**	**	**	**			**
Hydrocarbons	*		****		*	*	**	**	**	*						**
Bleach	*	***	*	****		**	***									
Drugs and biotech.	****	****	*	**	****	***	***	*	**	*						
Plastic	**	**		***	****	***	**	*	***	**	**	**		*	*	**
Materials	**			**	**	***	*	**	**	**	***	**	**	*		**
Food and tobacco	*	*				*	*	****								*
Metallu. process.	**		***	*					**	****	***	*		*	**	**
Chem. apparatus			***						*	****	***	**				**
Non-el. machin.			**				**	**	**	***	***					*
El. equipment						**	**	**		**	**	**				**
Spec. machin.				**	**	***	**		***	***	****	**			*	**
Metal work. equ.					**	**	*	**	**	***	****	**	*	*	**	**
Ass. hand appar.				*	*	*	*	**	**	*	***	**	**	**	***	**
Nuclear											**					*
Power plant										*	**	*				**
Vehic. and engines					*	*					*	**			**	***
Other transport					**	**				*	***	**	*	*	**	**
Aircraft																***
Mining			****								****	**				
Telecoms												**	****			
Semiconductors												***	****	**		**
Electr. devices									**		**	***	****	***	**	**
Computers										**		**	****	****	***	**
Image and sound								**			**	**	**	***		
Photo. and copy	**			*	**	***	**	**	**	**	*			***	**	**
Instruments			**	**	**	***					**	**	*	****	***	**
Misc. met. prods.				*								*		**	*	*
Textil. wood etc.						**										
Medicine etc.	****													*		

*****, core; ***, niche; **, background; *, marginal.

account for more than 50% of firms' total patenting in all technological fields.

But why, then, do large firms spread their technological resources over a wider spectrum than their products, and particularly into fields where they do not have a distinctive advantage? Together with Granstrand (Granstrand et al., 1996), we have elsewhere identified two causes:

1. *Technical interdependence* between improvements and changes in the complex products and production systems developed and produced by our large firms, and the complementary improvements and changes required from suppliers of materials, components and production machinery: for example, a large automobile firm may not make either the window glass or the tyres that it uses, but it will need (at the very least) to have its own technical capacity in these fields to judge whether its suppliers can be expected to provide (say) more streamlined glass shapes and higher quality tyres, as complements to its own development of more powerful internal combustion engines. Such background competencies are essential for the effective identification, integration, and adaptation to firm-specific requirements of technological changes in the firm's supply chain

and related production technologies. In other words, they are essential for 'learning by doing' and 'learning by using' (Arrow, 1962, Rosenberg, 1982, von Hippel, 1988).

2. *Emerging technological opportunities* that the firm must master in order to identify potential contributions to future business opportunities: at least in the early stages, emerging fields will be *marginal* in the total technological portfolio of the firm, but this will change as a function of the richness of the stream of potential opportunities that are identified. For example, between the 1960s and the 1980s, computing technology became a *core* competence in electrical firms and a *background* competence in automobile firms (see Table 3). More generally, emerging technological opportunities (especially in computing, materials, and biotechnology) have increased over time the number of fields in which our large firms have been constrained to accumulate competencies.

3. Stability and differentiation

Coping with complexity in products and production systems, not only requires a wide variety of

Table 4
Stability of technological profiles across 34 technical fields, 1969–74 to 1985–90

	No. of firms	Analysis of correlation coefficients					
		Patent shares			Revealed technology advantage		
		Not sig. at 5%	Sig. at 5%	Sig. at 1%	Not sig. at 5%	Sig. at 5%	Sig. at 1%
Chemicals	66	1	1	64	5	7	54
Pharmaceuticals	25	2	3	20	0	0	25
Mining and petroleum	31	7	7	17	5	5	21
Textiles etc.	10	2	5	3	3	5	2
Rubber and plastics	9	0	0	9	1	1	9
Paper and wood	18	1	3	14	4	4	10
Food	14	0	1	13	1	1	12
Drink and tobacco	8	0	1	7	0	2	6
Building materials	16	0	0	16	0	0	16
Metals	38	2	4	32	4	6	28
Machinery	58	2	4	52	4	8	46
Electrical	56	4	5	47	5	8	43
Computers	17	0	0	17	0	1	16
Instruments	21	0	0	21	2	4	15
Motor vehicles	35	0	0	37	2	2	33
Aircraft	18	0	0	18	1	1	18
All sectors	440	21	34	387	37	55	354

technological inputs, it also imposes incremental and localised processes of search. As a consequence, we shall now see that large firms' technological profiles are both stable and differentiated.

3.1. Stability

For nearly all our firms, these technological profiles are remarkably stable over time. For each firm, we correlated both the patent shares (PS) and the RTAs for the periods 1969–74 and 1985–90. Table 4 shows that, according to both measures, more than 90% of firms have profiles of technological competence that are statistically similar between 1969–74 and 1985–90, at the 1% level of significance. No systematic differences in stability can be detected between firms in different sectors. Large firms clearly do not shift around rapidly in their fields of technological competence.

This remains true even after taking account of acquisitions and divestments. We have shown elsewhere that, for 41 of the largest firms in our population, only one had a technological profile statistically different (at the 5% level) in 1987–92 from 1979–84 (Patel and Pavitt, 1995a). And only in very few of the cases involving substantial technological activities were the technological profiles of the acquired firm different from the acquiring firm, either before or after acquisition. [13]

3.2. Differentiation

Large firms' technological profiles are highly differentiated, according to the products that they make. The profiles of technological competence of each of the sixteen industries (i.e. aggregated sectoral data based on our firms) are in general very different. We systematically correlated each industry's technological profile, in terms of *patent shares* in each of the 34 technical fields, against all others and found that

23% of the cross-industry correlations are positive and significant at the 5% level: in other words, there are no similarities amongst industries in their technological profiles in more than three-quarters of all cases. We then did the same correlations in terms of *RTAs*, and found that the share of industries that are technologically similar is reduced to 5% (see Table 5).

In both cases, there are three major clusters of industries with significant similarities: (i) the chemicals, pharmaceuticals, and textile sectors; (ii) machinery and vehicles; (iii) electrical and computers. There is also one negative correlation of considerable importance: between the *RTAs* of firms in chemicals and in electrical products. Although both are often lumped together as 'high technology' or 'science-based' firms, they are clearly based on significantly different profiles of technological competence.

One drawback in the above analysis is its neglect of possible differences in profiles of technological competencies of firms *within* each industrial sector. For this reason, we systematically correlated each firm's profile of *patent shares* against that of all other firms, and summarise the results in Table 6. The results confirm and extend what has been shown in Table 5.

- In aggregate, any firm is nearly five times more likely to find another firm with a statistically similar profile of patent shares within its own sector than outside it. In computers and pharmaceuticals the probabilities increase to nearly eight and more than six, respectively. Profiles of technological competence are mainly sector-specific.
- However, the frequency of technological proximity between firms in different sectors is not evenly spread or random, but reveals distinct groupings, many of which have been anticipated in Table 5: in particular, those with competencies in chemistry, in electronics, and in mechanical machinery.

These sectoral similarities and differences amongst firms in the sources and directions of technological accumulation are broadly consistent with a sectoral taxonomy of technical change proposed earlier by one of us (Pavitt, 1984):

- two distinct science-based sectors centred on organic chemistry (chemicals, pharmaceuticals,

---

[13] Intriguingly, all these cases involved US-owned firms (*Black and Decker's* purchase of *Emhart*, *General Motors* of *Hughes*, *General Electric* of *RCA*, and *Kodak* of *Sterling Drug*). In contrast, *ATT* and the European firms (*ABB*, *Alcatel*, *Philips*, *Thomson*, and *Olivetti*) re-inforced their existing profiles through their acquisitions, as did *Hitachi* and *Fujitsu*.

150   P. Patel, K. Pavitt / Research Policy 26 (1997) 141–156

Table 5
Correlations of average RTAs across 16 principal product groups: 1981–90

	Chem.	Phar.	Mini.	Text.	Rubb.	Pape.	Food	Drin.	Buil.	Meta.	Mach.	Elec.	Comp.	Inst.	Moto.
Pharmaceuticals	0.53*														
Mining and petroleum	0.16	−0.05													
Textiles etc.	0.51*	0.20	−0.06												
Rubber and plastics	0.14	−0.02	0.02	0.20											
Paper and wood	0.01	−0.03	0.12	0.30	0.30										
Food	0.09	0.15	−0.07	0.08	−0.05	0.23									
Drink and tobacco	−0.03	0.05	−0.07	0.00	−0.04	0.24	0.99*								
Building materials	0.20	−0.07	−0.01	0.34	0.27	0.55*	0.01	0.01							
Metals	0.04	−0.13	0.11	0.01	0.01	−0.08	−0.08	−0.07	0.20						
Machinery	0.39*	−0.35*	0.06	−0.24	−0.04	−0.10	−0.20	−0.15	0.01	0.16					
Electrical	−0.36*	−0.29	−0.25	−0.24	−0.13	−0.35*	−0.20	−0.16	−0.17	−0.08	−0.06				
Computers	−0.25	−0.18	−0.17	−0.17	−0.11	−0.20	−0.12	−0.10	−0.14	−0.13	−0.20	0.65*			
Instruments	−0.01	−0.08	−0.12	−0.03	−0.07	0.13	−0.09	−0.07	−0.03	−0.12	−0.21	0.06	0.14		
Motor vehicles	−0.27	−0.20	−0.17	−0.18	−0.07	−0.26	−0.13	−0.11	−0.14	−0.05	0.52*	−0.04	−0.06	−0.14	
Aircraft	−0.20	−0.18	0.00	−0.13	0.02	−0.25	−0.11	−0.09	−0.10	0.01	0.04	−0.03	−0.05	−0.14	0.10

* Correlation coefficient significantly different from zero at the 5% level.

P. Patel, K. Pavitt / Research Policy 26 (1997) 141–156

Table 6

Correlations of firms' shares across 34 technical fields, by principal product group: 1981–90

Principal product group (PPG)	Percentage of the total that are positive and significant at 5% level																
	Own PPG	All other PPGs	Phar.	Mini.	Text.	Rubb.	Pape.	Food	Drin.	Buil.	Meta.	Mach.	Elec.	Comp.	Inst.	Moto.	Airc.
Chemicals	78.6	19.1	60.3	61.8	53.6	49.0	17.8	39.0	9.8	24.9	19.1	4.5	1.7	0.0	8.7	2.4	1.9
Pharmaceuticals	94.3	15.8		28.3	30.8	18.7	6.4	38.3	12.5	4.5	4.1	1.1	0.1	0.0	8.2	0.1	0.0
Mining and petroleum	69.0	19.3			39.0	34.4	15.9	22.1	3.6	17.5	17.7	7.0	1.3	0.2	5.5	4.8	3.4
Textiles etc.	55.6	24.7				35.6	45.0	25.0	17.5	45.6	15.3	14.8	10.5	5.9	13.3	4.6	10.0
Rubber and plastics	72.2	16.9					14.8	14.3	4.2	18.1	8.5	7.3	5.4	1.3	6.3	2.2	5.6
Paper and wood	58.2	13.3						17.1	26.4	34.0	9.6	23.9	5.4	2.6	6.3	4.9	3.4
Food	67.0	14.1							54.5	14.3	4.3	5.2	0.8	0.4	3.4	1.0	1.2
Drink and tobacco	50.0	11.4								18.0	5.9	20.9	3.3	1.5	4.8	7.9	10.4
Building materials	48.3	15.5									25.7	17.1	5.7	0.7	6.5	6.4	15.3
Metals	72.8	12.7										23.0	11.1	4.2	5.9	7.2	18.4
Machinery	45.0	12.8											12.7	5.9	7.6	44.3	30.1
Electrical	61.6	10.4												59.3	30.9	13.7	41.4
Computers	100.0	12.7													34.2	5.9	38.9
Instruments	55.2	12.2														11.6	30.4
Motor vehicles	85.7	12.0															36.8
Aircraft	72.5	18.5															
All sectors	67.3	14.5															

152   *P. Patel, K. Pavitt / Research Policy 26 (1997) 141–156*

petro-chemicals), and on physics-based technology (electrical, computers, instruments);

- machinery suppliers with areas of specialisation influenced by major users;
- a range of scale intensive sectors with production technologies dependant on improvements in chemical processes, instrumentation and production machinery.

4. Variety and constraint in managerial choice

One implication of the high stability and differentiation of large firms' technological competencies is the severe constraint it puts on managerial choice. We shall now show more specifically that:

- each firm's *direction* of technological search (and accumulation of competence) is strongly constrained by its prior competencies;
- each firm's *rate* of search is significantly influenced by its home country and the products that it makes, but there remains considerable scope for managerial discretion.

4.1. Heavily constrained directions of search

The stability over time that we have observed in firms' technological profiles is defined by relatively broad technological fields, and does not reflect the more localised processes of search that firms undertake *within* these fields. For this reason, we have identified in US patenting activities the 1000 (out of a total of around 100 000) technological sub-classes of the highest technological opportunity, as measured by their absolute increase in patenting from the

1960s to the late 1980s: their share of total patenting increased steeply from 3% to 18% of total US patenting over this period. A relatively high proportion of these fast growing fields are to be found in electronics and chemical technologies, but cases can be identified in all technological fields. We assume that they reflect the fields of greatest technological opportunity.

We show in Table 7 that firms are in fact heavily constrained by their prior competencies in the *directions* in which they exploit the opportunities in these fast-growing fields. Each firm's distribution of its fast-growing patenting in 1985–90 across each of the five broad fields of technology used in Table 1— chemicals, mechanical, electrical-electronic, transport, and 'other'—is strongly and positively correlated with the prior distribution of its total patenting in the same fields over the period 1969–84: firms that were patenting mainly in chemical technology exploit fast-growing opportunities mainly within the chemical field, and the same for the other fields. In other words, firms' capacities to exploit specific fields of high technological opportunity are strongly constrained by the fields of their prior competencies.

4.2. Lightly constrained rates of search

Thus, a firm's existing product mix and associated competencies strongly constrain the *directions* in which it seeks to exploit technological opportunities and acquire competence. The rate of a firm's search is influenced by the available technological opportunities, and the incentives and the capacity that it has to respond to them. These depend in turn on its competitive environment and its accumulated

Table 7
Correlations of each firm's distribution of its total patenting across five broad technical fields from 1969 to 1984, with the distribution of its fast-growing patenting across same five fields from 1985 to 1990

	Shares of patenting in fast-growing areas in 85–90 in:				
	Chemicals	Mechanical	Electrical	Transport	Other
Share of total in chem. 1969–84	0.91 *	−0.41	−0.61	−0.26	0.00
Share of total in mech. 1969–84	−0.41	0.68 *	−0.10	0.14 *	0.09
Share of total in elec. 1969–84	−0.58	−0.12	0.87 *	−0.17	−0.17
Share of total in trans. 1969–84	−0.34	0.18 *	−0.13	0.85 *	−0.04
Share of total in other 1969–84	0.06	−0.12	−0.18	−0.07	0.55 *

* Coefficient that is significant and positive at the 5% level.

P. Patel, K. Pavitt / Research Policy 26 (1997) 141–156

Table 8
Factors influencing firms' rate of technological accumulation

	Dependent variable					
	Patent intensity: 1985–90		Change in patent share: 1969–74 to 1985–90		Share of patents in fast-growing fields: 1985–90	
	Coeff.	Std error	Coeff.	Std error	Coeff.	Std error
Constant	−32.08 *	10.90	0.83	0.52	6.71 *	1.69
Industry average	1.03 *	0.07	1.88 *	0.42	0.82 *	0.05
Country average	0.62 *	0.16	0.68 *	0.07	0.38 *	0.07
R^2 (adj.)	0.33		0.20		0.44	
F	113.3 *		57.1 *		172.7 *	
N	440		440		440	

* Coefficient significantly different from zero at the 5% level.

competencies. In this context, we shall now show that:

- the firm's *industry* influences its rate and direction of technological accumulation, given that the firm's competencies and directions of search are determined in large part by what it produces, and that technological opportunities are unequal across fields (Malerba, 1992);
- the firm's *home country* influences its rate of technological accumulation, thereby confirming the importance of the nationally based supply and demand-side inducement mechanisms described by Porter (1990). These are likely to remain strong since large firms continue to perform an overwhelming proportion of their R & D activities (~ 90%) in their home countries (Patel and Pavitt, 1991, 1995b);
- there remains considerable *unexplained* variance in firms' levels and rates of increase in technological activities, reflecting the different bets made by different managements in the face of complexity and uncertainty (Nelson and Winter, 1982).

In Table 8, we present the results of our analysis of the effects of home country conditions and of industry (both measured through the appropriate aggregate country and industry indicators from our large firm database) on three measures of the rate of accumulation of technological competencies in each firm. From this it emerges that:

- both home country and industry have a statistically significant influence on the rate of technological accumulation, whether measured in terms of patents per unit of sales, growth in patent

share, or share of total patenting in fast-growing fields; [14]

- the unexplained variance amongst firms nonetheless remains considerable—56–80% of the total, which suggests that company-specific factors, and particularly managerial choice, remain important in the volume of resources allocated to technological accumulation.

5. Conclusions

Our empirical results confirm two key characteristics of the large innovating firm that are identified in evolutionary theory: *technical complexity* and *path dependency*. They also cast doubt on the validity of a third characteristic: *variety*. These results have implications for both policy and theory.

5.1. Policy: constraints on managerial choice

Given complexity and path-dependency, there are strong constraints on managerial choice in shaping the technology strategies of large firms.

- Technological strategies in large firms can only rarely be 'focused', since the products they develop and make require the integration of knowledge from a wide range of technological fields.

----

[14] Since all three dependant variables are based on patenting, part of the unexplained variance may reflect inter-firm differences in the propensity to patent the results of R&D and related technological activities. However, this is less likely to operate in shares of total patenting in fast-growing fields.

- 'Distinctive core' competencies in technology are not enough. Large firms must also be competent to co-ordinate technological change and improvement in their supply chains, and to evaluate and exploit emerging technological opportunities.
- The capacity to modify their profiles of technological competence is limited, and takes a long time (see also Rosenbloom and Cusumano, 1987).
- In addition to these constraints on the *directions* of technological accumulation, both home-country and industry characteristics have a significant influence on the firm's *rate* of competence accumulation.
- However, there remains considerable scope for managerial discretion in fixing the *rate* of competence accumulation. Given uncertainties, different managements make different bets, based on different 'rules of thumb' or 'routines', which may be influenced by their professional backgrounds and associated loyalties. Scherer and Huh (1992) have shown in the USA, and Bosworth and Wilson (1992) in the UK, that the volume of resources allocated by firms to technological activities is positively associated with the presence of graduate scientists and engineers in top management.

5.2. Theory: what is 'variety'

'Variety' is an often used term in evolutionary economics. [15] Our analysis identifies several possible definitions of the term, each of which should be carefully distinguished.

- Variety *within* firms in the technological competencies that they embody—in other words, multifield or complex competencies.
- Variety *between* firms in their mix (or profile) of technological competencies, largely defined by the products they develop and make—in other words, sector-determined differentiation.
- Once given each firm's profile of competencies and product mix, *lack* of variety in managerial decisions about their *directions* of search, but considerable variety in decisions about their *rate*

of search. In other words, we find only very weak signs of any technological variety in competitive processes *within* sectors. There at least two possible explanations for this.

The first is that technological variety is a necessary characteristic of 'revolutionary' rather than 'normal' technological change: in the former, variety is essential for effective experimentation and choice under conditions of great uncertainty. We cannot test this explanation with our large firm database, but other evidence casts doubt on it. In particular, the most fruitful *directions* of technical change in major technological revolutions are often well known and clear. For example, for the past 30 years, Moore's Law has told us of the considerable advantages in costs and in widening applications of continuing miniaturisation in semiconductor circuitry, and has been a clear technological beacon to firms in the field: the problem has not been deciding *where* to go, but *how fast and effectively* to get there. [16]

The second explanation is that we must distinguish clearly between *technologies* and *products*. Variety in the latter is essential (and especially during revolutionary technological changes) given the well known and well established difficulties of predicting users' reactions to innovations, and the equally well known variation in consumer's tastes. Given that the same technologies underpin a range of competing and differentiated product configurations, [17] product variety in an industry is compatible with technological homogeneity.

We find the second explanation more plausible. It offers a clear and useful heuristic to practitioners, in highlighting the importance of both the pacing of the acquisition of key technological skills, and their matching with specific product configurations and users' needs.

But it also has uncomfortable implications for theorists. It is compatible in the sphere of product

---

[15] See, for example, Metcalfe and Gibbons (1989).

[16] The other major revolution has been the increasing generation and application of software technology in service activities like banking, distribution etc. Again, evidence and experience suggest a clear pattern, beginning with process innovations emerging from the automation of routine 'back office' activities (Barras, 1990).

[17] For example, in the automobile industry, *Volkswagen* and *BMW* have statistically similar technological profiles.

development with variety, experimentation, social shaping, and trade-offs at the margin. But in the sphere of technology, it is underpinned by quite rigid, one-to-one technological imperatives: if you want to design and make automobiles, you must know (amongst other things) about mechanics; if you want to design and make aeroplanes, you must know (amongst other things) about aeronautics; if you want to design and make all manner of complex products, you must know (amongst other things) about computer applications.

This will come as no surprise to practising engineers and managers. That it might disconcert certain evolutionary theorists illustrates the dangers in the adoption of biological metaphors that are intellectually fashionable, politically correct but empirically untested.

References

Archibugi, D., 1988. In search of a useful measure of technological innovation. Technological Forecasting and Social Change 34.

Arrow, K., 1962. The economic implications of learning by doing. Review of Economic Studies 29.

Burras, R., 1990. Interactive innovation in financial and business services: the vanguard of the service revolution. Research Policy 19, 215–238.

Barre, R., 1996. Relationship between multinational firms' strategies and national innovation systems. In: Innovations, Patents and Technological Strategies. OECD, Paris.

Barton, J., 1993. Adapting the intellectual property system to new technologies. In: Wallerstein, J., Mogee, M., Schoen, R. (Eds.), Global Dimensions of Intellectual Property Rights in Science and Technology. National Academy Press, Washington, DC.

Basberg, B., 1987. Patents and the measurement of technological change: a survey of the literature. Research Policy 16.

Bosworth, D., Wilson, R., 1992. Technological Change: The Role of Scientists and Engineers. Avebury, Aldershot.

Business Week, 1993. The global patent race picks up speed. 9 August 1993.

Cantwell, J., 1989. Technological Innovation and Multinational Corporations. Blackwell, Oxford.

Carlsson, B., Eliasson, G., 1991. The nature and importance of economic competence. Working Paper, Swedish Board of Technical Development, Stockholm.

Cohen, W.M. and Levinthal, D.A., 1989. Innovation and Learning: the two faces of R&D, Economic Journal, 99, 569–596.

Dosi, G., Teece, D., 1993. Competencies and the boundaries of the firm. Center for Research in Management, CCC Working Paper No. 93-11, University of California at Berkeley.

Dosi, G., Teece, D., Winter, S., 1992. Towards a theory of corporate coherence: preliminary remarks. In: Dosi, G., Gian-

netti, R., Toninelli, P.A. (Eds.), Technology and Enterprise in a Historical Perspective. Clarendon Press, Oxford.

Franko, L., 1989. Global corporate competition: who's winning, who's losing, and the R&D factor as one reason why. Strategic Management Journal 10.

Geroski, P., Machin, S., van Reenen, J., 1993. The profitability of innovating firms. RAND Journal of Economics 24.

Granstrand, O., Sjolander, S., 1990. Managing innovation in multi-technology corporations. Research Policy 19.

Granstrand, O., Patel, P., Pavitt, K., 1996. Multi-technology corporations: why they have 'distributed' rather than 'distinctive cose' competencies: forthcoming in California Management Review.

Griliches, Z. (Ed.), 1984. Patents, R&D and Productivity. University of Chicago Press, Chicago.

Griliches, Z., 1990. Patent statistics as economic indicators. Journal of Economic Literature 28.

Hughes, T., Pinch, T. (Eds.), 1987. The Social Construction of Technological Systems. MIT Press, Cambridge, MA.

Jacobsson, S., Oskarsson, C., 1995. Educational statistics as an indicator of technological activity. Research Policy 24, 127–136.

Jaffe, A., 1989. Characterizing the 'technological position' of firms, with application to quantifying technological opportunity and research spillovers. Research Policy 18, 87–97.

Kline, S., 1995. Conceptual Foundations for Multi-Disciplinary Thinking. Stanford, Stanford University Press.

Lindblom, C., 1959. The science of muddling through. Public Administration Review 19, 79–88.

Malerba, F., 1992. Learning by firms and incremental technical change. Economic Journal 102.

Malerba, F., Marengo, L., 1993. Competence, innovative activities and economic performance in Italian high technology firms. Bocconi University, Milan (mimeo.).

Metcalfe, S., Gibbons, M., 1989. Technology, variety and organisation: a systematic perspective on the competitive process. Research on Technological Innovation, Management and Policy 4.

Narin, F., Noma, E., Perry, R., 1987. Patents as indicators of corporate technological strength. Research Policy 16.

Nelson, R., Winter, S., 1982. An Evolutionary Theory of Economic Change. Belknap, Cambridge, MA.

Oskarsson, C., 1993. Diversification and growth in US, Japanese and European multi-technology corporations. Dept. of Industrial Management and Economics, Chalmers University of Technology, Gothenburg (mimeo.).

Patel, P., Pavitt, K., 1987. Is Western Europe losing the technological race? Research Policy 16, 59–85.

Patel, P., Pavitt, K., 1991. Large firms in the production of the world's technology: an important case of 'non-globalisation'. Journal of International Business Studies 22, 1–21.

Patel, P., Pavitt, K., 1994. The continuing, widespread (and neglected) importance of improvements in mechanical technologies. Research Policy 23, 533–546.

Patel, P., Pavitt, K., 1995a. The wide (and increasing) spread of technological competencies in the world's largest firms: a challenge to conventional wisdom. In: Chandler, A., Hagstrom,

P., Solvell, O. (Eds.), The Dynamic Firm. Oxford University Press (in press).

Patel, P., Pavitt, K., 1995b. Patterns of technological activity: their measurement and interpretation. In: Stoneman, P. (Ed.), Handbook of the Economics of Innovation and Technical Change. Blackwell, Oxford.

Pavitt, K., 1984. Sectoral patterns of technical change: towards a taxonomy and a theory. Research Policy 13, 343–373.

Pavitt, K., 1988. Uses and abuses of patent statistics. In: van Raan, A. (Ed.), Handbook of Quantitative Studies of Science and Technology. North Holland, Amsterdam.

Porter, M., 1990. The Competitive Advantage of Nations. Macmillan, London.

Prahalad, C., Hamel, G., 1990. The core competence of the corporation. Harvard Business Review (May–June), 79–91.

Quinn, J., 1980. Strategies for Change: Logical Incrementalism. Irwin, Homewood, IL.

Rosenberg, N., 1982. Learning by using. In: Inside the Black Box: Technology and Economics. Cambridge University Press, Cambridge.

Rosenbloom, R., Cusumano, M., 1987. Technological pioneering and competitive advantage: the birth of the VCR industry. California Management Review 29.

Rumelt, R., 1974. Strategy, Structure and Economic Performance. Graduate School of Business Administration, Harvard University, Cambridge, MA.

Samuelson, P., 1993. A case study on computer programs. In: Wallerstein, J., Mogee, M., Schoen, R. (Eds.), Global Dimensions of Intellectual Property Rights in Science and Technology. National Academy Press, Washington, DC.

Scherer, F., Huh, K., 1992. Top management education and R&D investment. Research Policy 21.

Teece, D., Pisano, G., 1994. The dynamic capabilities of firms: an introduction. Industrial and Corporate Change 3, 537–556.

von Hippel, E., 1988. The Sources of Innovation. Oxford University Press, New York.

6

Multi-Technology Corporations:

Why They Have "Distributed" Rather Than "Distinctive Core" Competencies

Ove Granstrand
Pari Patel
Keith Pavitt

F ew economists or managers today would question the pivotal role played by technological change in contemporary businesses. This is why the notion of the "firm-specific competencies" appeals across a wide spectrum. At one extreme, it helps some economic theorists explain why firms persistently develop, make, and sell different products.[1] At the other, it helps management scholars, consultants, and corporate practitioners understand how and why firm-specific knowledge contributes to corporate success. Prahalad and Hamel have rightly emphasized the long-run competitive importance of accumulating firm-specific competencies that defy evaluation by established financial techniques and that must often be re-combined across established functional and divisional boundaries.[2]

The Importance of Technological Diversity

Technological diversity in corporations is a driving force behind four major features of contemporary business: corporate growth; increasing R&D investment; increasing external linkage for new technologies by various means (such as acquisitions, alliances, licensing); and opportunities to engage in technology-related new businesses. Corporate technological competencies are dispersed over a wider range of sectors than their production activities, and this range is increasing. Technologies are not the same as products and must be dealt with differently.

Helpful comments on this paper have been received from Jorge Niosi, and from two anonymous referees.

There are, in turn, three major driving forces behind technology diversification in companies: opportunities to introduce new technologies into products and systems for improved performance and new functionalities; the continuing relevance of old technologies; and the co-ordination of innovation and change in core products with complementary changes in the production system and supply chain.

Increases in technological diversity in both companies and products challenge conventional wisdom and a number of widely accepted management concepts. This article challenges four of these notions: first, that for every company there exists a narrow set of core (or distinctive) technological competencies on which the company should focus; second, that major new innovations are often associated with major "competence destruction;"[3] third, that companies should not only downsize but disintegrate (i.e., become "virtual" or "hollow") and outsource technological competencies just like production; and fourth, that companies should focus or specialize on a narrow set of core businesses ("back to basics"). These notions do contain some truth, and they do apply in some companies in some periods of their life. However, as with any simple concept, they can be dangerous when carried to extremes.

Sources and Methods

There are two primary methods for measuring the technological competencies of firms. The first is as the capacity to achieve a certain level of functional performance in a generic product, component, or sub-system: for example, "the design, development, and manufacture of compact, high performance combustion engines." As a strategic technological target for a firm like Honda, this obviously makes sense.[4] But the achievement of target levels of functional performance requires the combination of technological competencies from a wide variety of underlying fields of knowledge, including mechanics, materials, heat transfer, combustion, fluid flow. Over time, competencies in other fields have become necessary: for example, ceramics, electronics, computer-aided design, simulation techniques, and software. Broad measures of functional performance thus tend to disguise the detailed structure of the underlying technological fields on which they are based.

This is why we prefer a second method, namely, measuring a combination of corporate competencies in different technological fields, defined by fairly stable, universal, and operational systems of classification. In this article, we report on research using the technological fields of U.S. patenting as the basic units of competence.[5] For the U.S. patent examiner, the granting of a patent reflects the judgement that the applicant has the competence to improve technology in a given field significantly, even though it is difficult to foresee its degree of usefulness at the time. This is one reason why patents differ greatly in their economic value.

We have made a systematic analysis of the patenting activities in the U.S. of the 440 most technologically active companies in the world. The data reflect corporate capacity to generate change and improvement, they are detailed and comprehensive, and they are used by practitioners themselves.[6] Their main drawback is that—until recently—they did not cover software inventions, and that firms sometimes use other methods than patenting to protect their technological lead. As shown in Tables 1 and 2, we assume companies are competent in a given technical field when they are granted five or more patents over a five-year period.

These comprehensive data have been complemented by several detailed case studies, of which two are referred to here. The British company Rolls-Royce has remained one of the world's leading producers of aircraft jet engines since World War II. Over the same period, the Swedish company Ericsson has shifted its main business from electro-mechanical switching and cables through computerized digital switching to cellular mobile communications. In both cases, patent analysis has been validated and complemented by interviews with corporate practitioners. These enabled a better understanding of the factors behind the observed diversity of technological competencies in the firms.

The Distributed Competencies of Large Corporations

Technological Diversity

We have classified each of our 440 large firms into one of sixteen principal product groups,[7] and its patenting activities into five broad technical fields and 34 more detailed fields.[8] This classification shows that large firms have significant competencies outside their intuitively obvious distinctive technologies.[9] For example: electrical/electronic firms = ~34% outside broad electrical/electronic field, of which ~20% in machinery; chemical firms = ~33% outside broad chemical field, of which ~16% in machinery; automobile firms = ~70% outside broad transport field, of which ~46% in machinery.

While most large firms are also heavily diversified in their product mix, they have a broader range of technological competencies than products.[10] Table 1 compares the number of firms with their principal activity in selected product groups with the number of firms active in their corresponding distinctive technologies (i.e., with five or more patents granted between 1985 and 1990). In all cases, the latter is considerably larger than the former, in part because the former measures only firms' principal product activity, while the latter encompasses all technological fields with five or more patents. However, nowhere near 300 of these large firms are making and selling instruments and controls or chemical processes; nor are more than 200 of the firms making and selling non-electrical machinery. This means that large numbers of firms are mobilizing technological competencies in instruments and controls, chemical processes, non-electrical machinery, computing, and so forth in order to make other products. The reasons for this technological diversity are systemic interdependence with the supply chain and widening technological opportunities.

TABLE 1. Number of Active Large Firms in Selected Principal Products and in Closely Related Technologies, 1985-90

Principal Product (out of 15, see reference 8)	No. of Firms (out of 440)	Technological Field (out of 34, see Table 2)	No. of Active* Firms (out of 440)
Computers	17	Calculators and Computers, etc.	151
Electrical & Electronic	56	Semiconductors	94
Instruments	21	Instruments and Controls	288
Chemicals	66	Organic Chemicals	190
Pharmaceuticals	25	Drugs and Bioengineering	114
Mining and Petroleum	31	Chemical Processes	304
		Apparatus for chemicals, food, glass	234
Non-Electrical Machinery	58	General Non-Electrical Industrial Equipment	246
		Non-Electrical Specialized Industrial Equipment	241
		Metallurgical & Metal Working Equipment	225
Automobiles	35	Road Vehicles & Engines	77
Aerospace	18	Aircraft	28

* With five or more patents granted 1985-90.

Co-ordinating Change in the Supply Chain

Close and complementary contributions to technical change are made by suppliers and users of producers' goods.[11] The cases of Rolls-Royce and Ericsson both show that, with complex products and production processes, there are strong technical interdependencies between what firms develop and make themselves and what they require from their suppliers of machinery, components, software, and materials. As a consequence, a decision to "buy not make" a production input does not automatically translate into a similar choice about the underlying technical knowledge. The effective use and improvement of outside components, sub-systems, and machinery requires a matching in-house capability to choose, integrate, and learn, as well as to co-ordinate and manage systemic change. The more complex the supply chain, the higher the proportion of technological resources large firms are likely to spend outside their distinctive technological competencies.

Thus, Rolls-Royce maintains *full design and manufacturing* capabilities for the inner core of the jet engine: fan, compressor, and combustion and turbine systems.[12] Although it may subcontract the production of the components related to the outer core of the engine, Rolls-Royce maintains a *full design* capability over them, together with a *systems integration* capability over the remaining components, and a capacity for being a knowledgeable purchaser. Similarly, Ericsson needed its own technological and design competence in semiconductors

Multi-Technology Corporations

TABLE 2. Number of World's Large Firms that Are Active* in Each of 34 Technical Fields (total number of firms = 440)

Technical field	1969-74	1985-90	Number of Net Entries (+) and Exits (–)
Materials (including glass and ceramics)	162	226	64
Instruments and Controls	246	288	42
Chemical Processes	268	304	36
Calculators & computers	117	151	34
Plastic and Rubber Products	86	118	32
Dentistry and Surgery	55	80	25
Miscellaneous Metal Products	210	234	24
Drugs and Bioengineering	92	114	22
General Electrical Industrial Apparatus	200	216	16
Image and Sound Equipment	103	118	15
General Non-Electrical Industrial Equipment	233	246	13
Hydrocarbons, Mineral Oils, Fuels	55	68	13
Road Vehicles and Engines	64	77	13
Metallurgical & Metal Processes	108	120	12
Semiconductors	85	94	9
Agricultural Chemicals	37	41	4
Inorganic Chemicals	87	90	3
Textile, Clothing, Leather, Wood Products	23	23	0
Other—(Ammunitions and Weapons)	152	150	–2
Telecommunications	140	138	–2
Photography and Photocopy	66	64	–2
Mining & Wells Machinery and Processes	51	49	–2
Apparatus for Chemicals, Food, Glass	238	234	–4
Electrical Devices and Systems	143	139	–4
Aircraft	32	28	–4
Organic Chemicals	195	190	–5
Induced Nuclear Reactions	17	11	–6
Power Plants	59	51	–8
Metallurgical and Metal Working Equipment	234	225	–9
Food & Tobacco (Processes and Products)	64	53	–11
Other Transport Equipment (excluding aircraft)	96	81	–15
Bleaching, Dyeing and Disinfecting	43	27	–16
Assembling and Material Handling Apparatus	169	147	–22
Non-Electrical Specialized Industrial Equipment	269	241	–28

* = five or more patents granted 1985-90.

for its mobile phones, even when it purchased them from independent suppliers.

Learning about Emerging Technological Opportunities

The technological competencies of our large firms depend heavily on their past and are fairly stable. Their patent mix by technological field depends on their principal product, and changes only slowly.[13] Nonetheless, the firms are on the whole becoming more technologically diversified over time as new opportunities emerge from general advances in science and technology. Table 2 shows the number of firms active in each of the 34 technological fields in the periods 1969-74 and 1985-90, with the fields ranked by the increased number of net entries by active firms over the period. The first two columns of figures confirm the continuing pervasiveness of instruments and controls, along with other production-related technologies. The final column confirms what practitioners have been saying over the past 20 years: competencies in materials and computing technologies have become increasingly widespread. Coupled with the corresponding increase in external alliances,[14] this reflects the complementarity in companies between external alliances and internal competence acquisition. The final column also shows a net increase in the number of fields in which the firms have acquired technological competencies: on average, every second firm has become active over the period in one additional technological field.

The evidence confirms that firms have distributed their competencies over a wider number of fields. However, commercial opportunities emerging from major scientific and technological breakthroughs were rarely clear immediately. Applications of new technologies spread out in unexpected directions, like the computer moving from scientific calculations through information processing and on to telecommunications switching. Large firms built up and maintained a broad technology base in order to explore and experiment with new technologies for possible deployment in the future. The creation of corporate competencies in new fields was a dynamic process of learning, often requiring a combination of external technology acquisition and in-house technological activities and usually resulting in an increase in R&D expenditures. While external technology sourcing was rarely a substitute for in-house R&D, it was an important complement to it. Furthermore, breakthroughs in new technological fields were often combined with or even "fused" with established technological competencies. This was a process of "creative accumulation" and "competence enhancement" rather than a process of "creative destruction" and "competence destruction."

Thus, although still making the same product and contracting out some of its production, Rolls-Royce has since the early 1970s substantially increased the range of technologies in which it is active, having exited only one field (piston engines). It has increasingly accumulated experience and knowledge in a variety of electronic-based technologies (e.g., sensors, displays, simulations). A materials research group that it had earlier established at a university was eventually integrated into the company when it proved important for core activities.[15]

The case of Ericsson has been even more spectacular. Until the 1950s, it had focused on a narrow range of core businesses (equipment for public telephone operators) and related core technologies (electro-mechanical switching and cable transmission technologies; radio transmission technology and associated businesses were considered peripheral). In the next four decades, Ericsson diversified into computerized, digital switching technologies in order to sustain existing businesses, and then it combined its radically improved switching technology with radio transmission technologies, thereby enabling successful diversification into the new business of cellular mobile communications. Although external sourcing of technologies was important (and especially the co-operation with the lead user—Telia, which provides telephone services), new technological developments were sourced mainly in-house.

In the process, R&D costs for Ericsson increased dramatically, approaching 20% of sales in the 1990s. During the period 1980-89, the total stock of engineers rose by 82%, and the diversity of competencies increased considerably. The traditional core competence in electrical engineering increased by only 32%, while mechanical engineering grew by 265%, physics by 124%, and chemistry by 44%. Additional engineering categories were added (e.g., computer science), and no broad category of engineering competence was scrapped.

Table 3 illustrates how technology diversification and external technology acquisition took place in Ericsson's development of successive generations of cellular phones and telecommunications cables. The products became more "multi-technology" and the company's technology base expanded. The new technological competencies that were required outnumbered the old ones that were made obsolete; and as a result of this process, "competence enhancement" dominated over "competence destruction" just as in the case of Rolls-Royce.[16]

Measuring and Classifying Firms' Technological Competencies

Based on this analysis, our firm-specific patenting data can be used to measure and classify the technological competencies of specific firms. In Figure 1, we develop a fourfold classification, based on two dimensions of any large firms' technological competencies.

- Along the Y-axis is the percentage share of each of our 34 technical fields in the total patenting of the firm (Patent Share), reflecting the relative importance of each field in the firm's total technological portfolio. Thus the average share per field is about 3%, which is where we have drawn the X-axis to distinguish above average from below average shares;

- Along the X-axis is an index of the firm's revealed technology advantage (RTA) in each of the 34 technical fields. The RTA index for a given firm in a given field is defined as the firm's shares in total patenting in each of the 34 technological fields, divided by the firm's share of total patenting in all the fields: in other words, the relative importance of the firm in each field of technological competence, after normalizing for the firm's size.[17] We have drawn the Y-axis at RTA = 2, to distinguish high from low.

Multi-Technology Corporations

TABLE 3. Increasing Technological Diversity in Ericsson's Product Generations

Product Generations	No. of Technologies (a)				R&D Costs (base = 100)	% Age of Technologies Acquired Externally	Main Technical Fields (e)	No. of Patent Classes (f)
	Old (b)	New (c)	Total	Obsoleted (d)				
Cellular Phones								
1. NMT-450	n.a.	n.a.	5	n.a.	100	12	E	17
2. NMT-900	5	5	10	0	200	28	EPM	25
3. GSM	9	5	14	1	500	29	EPMC	29
Telecoms Cables								
1. Coaxial	n.a.	n.a.	5	n.a.	100	30	EKM	14
2. Optical	4	6	10	1	500	47	EKMC	17

[n.a. = not applicable]
Notes:　(a) Technologies at roughly the same level of aggregation identified by experts.
(b) No. of technologies from the previous generation (i.e., sustained technologies).
(c) No. of new technologies, compared to previous generation (i.e., enhanced competencies).
(d) No. of technologies obsoleted from previous generation (i.e., destroyed competencies).
(e) "Main" = >15% of total engineering stock. Categories are: E = electrical; P = physics; K = chemistry; M = mechanical; C = computers.
(f) Number of International Patent Classes (IPC) at 4-digit level.
Source: O. Granstrand, E. Bohlin, C. Oskarsson, and N. Sjoberg, "External Technology Acquisition in Large Multi-technology Corporations," *R&D Management*, 22 (1992): 125.

Technical fields identified in quadrant I are the company's fields of *distinctive* or *core* competence, commanding both high shares of corporate technological resources and a strong revealed technology advantage compared to the competition. Technical fields in quadrant III are *marginal* fields taking only a small proportion of corporate technological resources and without a strong competitive position.

If each of our 34 technical fields contained the same number of patents, each company's competency profile would be a perfect straight line sloping down from quadrant I to quadrant III. According to Prahalad and Hamel, most corporate resources should then be concentrated in quadrant I. However, our 34 fields are of unequal size and importance. Those related to instrumentation and production technologies are relatively large. As a consequence, the allocation of a relatively large share of corporate resources to these fields does not necessarily result in *distinctive* competencies in our quadrant I. In our analysis, a technological field located in quadrant II typically procures a *background* competence, which enables the company to co-ordinate and benefit from technical change in its supply chain. At the other extreme, some technological fields are intrinsically small, so that firms can obtain a *niche* competitive position, with a relatively small share of corporate technological resources, in quadrant IV. Finally, quadrant III will include technological fields that are (at least at present) *marginal* to

FIGURE 1. A Classification For Corporate Technological Profiles

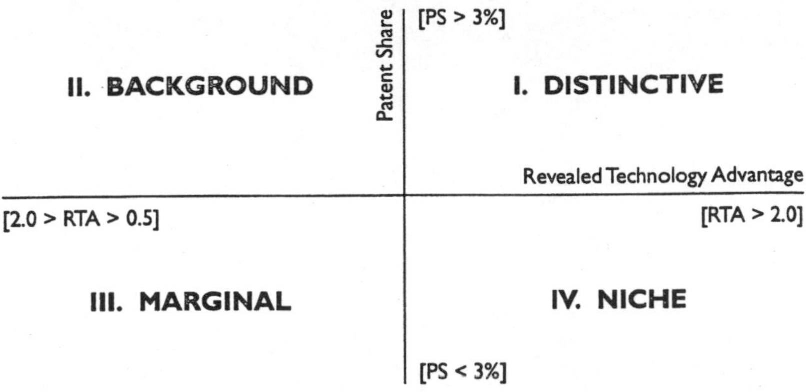

	Patent Share [PS > 3%]	
II. BACKGROUND		**I. DISTINCTIVE**
		Revealed Technology Advantage
[2.0 > RTA > 0.5]		[RTA > 2.0]
III. MARGINAL		**IV. NICHE**
	[PS < 3%]	

the corporation, but some of which may emerge as major opportunities for the future (e.g., computing in Ericsson in the 1960s).

In Figure 2, we use this framework to analyze the technological competencies of three well-known large companies: the chemical firm Bayer, the electrical and electronic firm Hitachi, and the automobile firm Ford. Our purpose is to illustrate differences in such competencies between sectors rather than to compare competitors in the same sector. We reproduce each company's competence profile for 1985-90, but we have also calculated the equivalent profiles for 1969-74. The comparison reflects major differences between sectors in: the fields of their distinctive competencies; the fields of growing technological opportunity that they exploit; and the balance between distinctive and background competencies.

Bayer had strong distinctive competencies in organic chemistry. The share of drugs in its patenting portfolio grew from 6 to 21% between the early 1970s and the late 1980s, while photography and bleaching and dyeing declined. 76% of all corporate patenting was in fields of *distinctive* competencies and only 7% in *background* competencies (chemical processes), reflecting the relatively simple supply chain in most chemical companies. At Bayer, 90% of the company's technological resources were concentrated in 9 out of our 34 technical fields.

By contrast, Hitachi's technological resources were distributed over a wider number of fields, with 90% of the total reached in 14 out of 34 fields. The *distinctive* competencies in computers, image and sound, and semiconductors accounted for only about 40% of all patenting. Computing increased from 6 to 17% over the period, while electrical devices and equipment declined from 15 to 10%. Nuclear technology remained a *niche* competence (perhaps against initial

Multi-Technology Corporations

FIGURE 2.   Technological Profiles of Three Large Companies

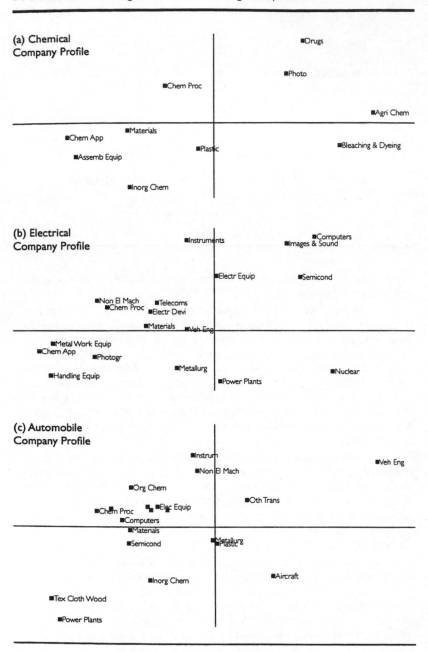

(a) Chemical
Company Profile

■Drugs

■Photo

■Chem Proc

■Agri Chem

■Materials

■Chem App

■Plastic

■Bleaching & Dyeing

■Assemb Equip

■Inorg Chem

(b) Electrical
Company Profile

■Instruments

■Computers
■Images & Sound

■Electr Equip

■Semicond

■Non El Mach
■Chem Proc

■Telecoms
■Electr Devi

■Materials

■Veh Eng

■Metal Work Equip
■Chem App

■Photogr

■Metallurg

■Nuclear

■Handling Equip

■Power Plants

(c) Automobile
Company Profile

■Instrum

■Veh Eng

■Non El Mach

■Org Chem

■Oth Trans

■Chem Proc
■Computers

■Elec Equip

■Materials

■Metallurg
■Plastics

■Semicond

■Inorg Chem

■Aircraft

■Tex Cloth Wood

■Power Plants

expectations). The 24% of all patenting in the *background* technologies of instruments and production equipment reflects a more complex supply chain than in chemicals.

The supply chain is even more complex at Ford. *Distinctive* competencies in vehicles and engines account for only 19% of all patenting, while *background* competencies in instrumentation and production technologies account for 43%. Interestingly, computing has also become a *background* competence, with its share increasing from 1 to 4.5% over the period. Reflecting an even greater spread of technological competencies than the firms in the other sectors, 90% of total patenting was spread among 16 of our 34 fields.

Implications for Management

"Distributed" rather than "Core" Technological Competencies

In addition to a focus on a number of the "distinctive" or "core" technological competencies as recommended by Prahalad and Hamel,[18] management in large firms needs to sustain a broader (if less deep) set of technological competencies in order to co-ordinate continuous improvement and innovation in the corporate production system and supply chain. Furthermore, they must do this in order to explore and exploit new opportunities emerging from scientific and technological breakthroughs. Thus, large firms typically must become multi-technology.

Rather than simply calling them "core" competencies, a more accurate description of large multi-technology firms' competencies is that they are "distributed." They are distributed in three senses:

- across a large and increasing number of technical fields—the precise mix of which is largely determined by the firm's principal product field;

- in different parts of the organization—for example, the corporate R&D units for exploring new and emerging opportunities; the divisional and subsidiary R&D units for sustaining and developing distinctive core competencies; and the production engineering (and even purchasing) departments for continuous change in production and the supply chain; and

- among different strategic objectives of the corporation—this has long been recognized by corporate practitioners, who use various concepts to classify technologies according to their commercial and economic impact in addition to their technical content (for example, key, base, emerging, pacing, generic, core).

These characteristics present two major challenges to management. First, given the distribution of technological competencies throughout the corporation, their identification, measurement, evaluation, and co-ordination is a non-trivial task. Their effective mobilization for reaching corporate objectives is also a demanding management task, since technological competencies cannot and should not be tidily located in one discrete part of the organization.

Second, given the differentiated nature of firms' mix of competencies, there is no universal recipe for deciding the appropriate mix of technological competencies. As noted above, advances in computing technology hardly affected the mix of competencies at Bayer, but emerged as a distinctive competence at Hitachi, and as a background competence at Ford. Identifying and integrating competencies essential for the corporation almost inevitably requires investment in in-house learning and patient experimentation, with the expectation of finding winning combinations of technologies through "competence enhancement" rather than through anticipating competence destruction.

Technology and the Boundaries of the Firm: Integration or Disintegration?

Chesbrough and Teece cast doubt on the assertion that large firms are becoming—or should become—technologically "virtual" by sub-contracting the generation and accumulation of their technological competencies to outside sources.[19] Our analysis confirms this view and also suggests that decisions about outsourcing technological competencies are in practice different to—and to some extent distinct from—decisions about out-sourcing production. Between full-scale integration (i.e., in-house R&D, design and production) and full-scale disintegration (i.e., simply purchasing inputs that are designed and produced externally) there are four intermediate positions.

- *Full Design Capability*—the competence to design and test all components that are produced externally.

- *Systems Integration Capability*—the competence to integrate changes and improvements in internally and externally designed and produced inputs into effective products and production systems.

- *Applied Research Capability*—the competence to specify, purchase, and control changes and improvements in externally produced equipment, components, and materials.

- *Exploratory Research Capability*—the competence to identify, evaluate, and integrate new opportunities emerging from general advances in science and technology.[20] These opportunities can over time have a major influence on all the three capabilities above.

Building on the earlier analysis by Chesbrough and Teece, we identify in Table 4 two sets of factors that influence corporate decisions on the degree to which technological competencies should be outsourced or internalized:

- The degree to which the innovation is *autonomous* (i.e., can be pursued independently from other innovations) or *systemic* (i.e., requires a cluster of related and complementary innovations)—the development of autonomous innovations can be more easily outsourced than systemic ones.

- The *number of independent sources* of technological competence outside the firm—the larger the number, the greater the need to monitor and absorb

TABLE 4. Matching the Nature of In-House Technological Competencies to the Nature of Innovation and to External Sources

		Type Of Innovation	
		Autonomous	**Systemic**
Number of External Technology and Related Product Competencies	High	• Exploratory and Applied Research	• Exploratory and Applied Research • Systems Integration
	Low	• Exploratory and Applied Research • Design (and Development)	• Exploratory and Applied Research • Systems Integration • Design and Development • (Production Engineering)

externally generated technological advances, and the smaller the danger of external monopoly power.

Whatever the type of innovation and the number of external sources, firms should always maintain capabilities in exploratory and applied research in order to have the competence to monitor and integrate external knowledge and production inputs (see Table 4). When the innovation is systemic and external knowledge sources are plentiful, the company should also have the systems competence to integrate in-house changes in products and processes with changes in the components and sub-systems purchased from outside (as shown in the first row of Table 4). When innovations are autonomous but outside sources are few, the company should also maintain a design (and perhaps a development) capability as a hedge against monopoly power in design by potential suppliers (second row, Table 4). When innovations are systemic, firms will need a further competence in product development and testing, as well as in production engineering when systemic effects spread to methods of production. These are related to the distinctive or core technological competencies of the firm.

Over time, the functions of corporate technological competencies can and do change. In some cases, the number of external sources of competence may increase. When the innovation is autonomous, it allows the firm to externalize its design competence. However, when the innovation is systemic, the growth of external sources becomes a major threat to the distinctiveness of its core competencies. In other cases, a rapidly evolving technological field may begin to have systemic effects that induce the firm to internalize, as Rolls-Royce has done in materials.

Technology Diversification—with or without Business Diversification

Just as products are becoming increasingly multi-technology, technologies are becoming increasingly multi-product and multi-firm. This explains why corporate technology diversification (competence enhancement) typically dominates over corporate technology substitution (competence destruction), and why technology-related business diversification is a feasible and often successful strategy provided that corporate technological and organizational competencies can be matched with major emerging opportunities.

Tables 1, 2, and 3 show that established large firms do diversify into new technologies. Our earlier studies of large firms in Europe, Japan, and the U.S. showed that increased technology diversification in the 1980s was a significant variable in explaining the growth of corporate sales and corporate R&D. [21] *Technology* diversification increased, even in firms where *product* diversification decreased—as has often been the case in Europe and the U.S. since the early 1980s, with the emphasis on "focus" and "back to basics." However, high growth firms often followed a sequential strategy, with technology diversification followed by product and/or market diversification (typically through internationalization). The success of this growth strategy was largely independent of the firm's main industry and country of origin. However, Japanese firms typically had the most developed managerial capability for concerted technology and business diversification into new product areas—but sometimes without success (e.g., steel firms entering bio-technology).

A more recent study by Gambardella and Torrisi of 32 of the largest U.S. and European electronics firms found that corporate performance is positively associated with technological diversification.[22] However, the authors also found that corporate performance is positively associated with greater focus in business operations. They attribute this result to the differentiated nature of downstream markets, which prevents extensive product diversification based on technology diversification. These conclusions are similar to those of Christensen and Rosenbloom, who found that the main difficulties in exploiting new technological opportunities lie in the challenges of corporate *organizational* competencies, rather than their technological competencies.[23] Our own case studies confirm this point. Different styles of management and organization in companies from Europe, Japan, and the U.S. influence the degree to which firms are successful in exploiting economies of scale and scope in the development and exploitation of technological competencies.

To begin with, R&D costs increase more than proportionately to the number of new technological competencies acquired, because these must not only be strengthened, but integrated with other competencies in the corporation. This involves not only the conventional costs of co-ordination, but also the sometimes heavy costs of integrating knowledge across well-established disciplinary frontiers. At Ericsson, for example, it took years to overcome barriers between the "elite" switching engineers and the "peripheral" radio engineers. In addition, scrapping or reducing established organizational competencies and related procedures is difficult, painful, and often heavily resisted, since they are often embodied in older people. These are generally well established in the corporate power structure, and their short-term interest is exploiting what they already know, rather than investing in costly, time-consuming, and risky change.

Organizational competencies also need "enhancement." For example, although Ericsson acquired personal computing technology in the 1980s, it failed organizationally. The marketing competence required to sell in this fast-moving mass market was very different from that the company's established competence for selling a smaller volume of expensive electronic switches and telecommunications systems in often highly political markets. Nonetheless, new technological competencies in electronic switching opened massive new business opportunities for Ericsson in mobile telephony, which the company then did exploit very successfully. In this case, the development of complementary organizational competencies was low-key and deliberately oriented towards learning from experience

Conclusions

Figure 3 summarizes the results of our analysis of the dynamic interactions between technology, products, firms, and markets. The challenge for management is to give more attention to the distribution of corporate technological competencies beyond the core, the enhancement and integration of new competencies, and the potential for related new product markets. Perhaps the most important single rule of thumb to follow is: *do not confuse technologies with products*. Although the two are obviously interrelated, corporate policies towards them differ in four important respects:

- *What may apply to outsourcing production does not apply to outsourcing technological competencies.* While the emphasis in production has been on increasing focus and out-sourcing, large firms have at the same time been spreading their technological competencies beyond their distinctive core. These competencies include the capacity to improve and to co-ordinate change in complex production systems and supply chains as well as to explore and exploit emerging new technologies.

- *"Creative destruction" in products—and even firms—is not associated with corporate "competence destruction" in technologies.* The careful, step-by-step accumulation and enhancement of corporate competencies in rapidly changing technologies like computing often opens opportunities to design, develop, produce, and sell radically changed or entirely new products. Many of the products will fail, especially in the early stages. Some of the firms will fail because they do not adapt their organization and management procedures to the requirements of new products, production methods, or markets. Failure to exploit radically new technologies has more to do with failure in product development, production, marketing,

FIGURE 3. The Dynamics of Corporate Technology Diversification

---

Technology-Product Interactions

- New Technologies in Products and Processes
- Sustained Old Technologies
- Changes in Supply Chain and Production

Corporate Competencies

- *Emergence of* Multi-Technology Corporations
- *with* Distributed Competencies
- *due to* Technology Diversification
- *and requiring* Organizational Change

Some Measurable Effects

- Growth of Sales
- Growth of R&D
- Growth of External Technology Sourcing
- Opportunities for Business Diversification

Challenges to Management

Current Concepts		A Revised Agenda
• Core Competencies	versus	• Distributed Competencies
• Competence Destruction	versus	• Competence Enhancement
• Outsourcing Production	versus	• Integrating Competencies
• Business Focus	versus	• New Product Markets

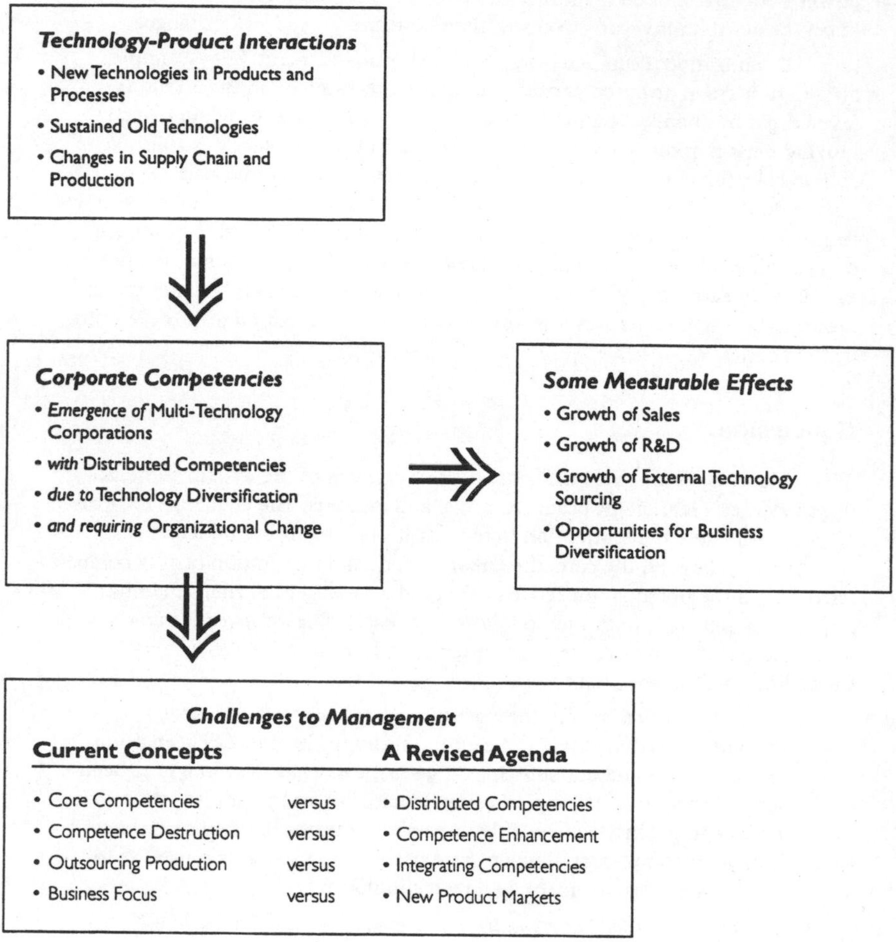

and organizational adaptation than with failure in technological competencies.

- *The rapid increase over the past fifteen years in technological alliances and exchanges among large firms is not the consequence of the progressive "outsourcing" of R&D activities in order to reduce their cost.* Our research results are consistent with those of Freeman, who has identified the rapid development and diffusion of new technologies, especially information technology, as the main reason for the growth of strategic alliances in technology.[24] This has been accompanied by an increase—rather than a decrease—in the cost of R&D, given the need to combine and integrate an increasing range of technological fields.

- *There is no clear match between technology diversification (or focus) and product diversification (or focus).* Technology diversification is associated with both increased product focus and increasing product diversity. Either way technology diversification is associated with better corporate performance.

Notes

1. G. Dosi, D. Teece, and S. Winter, "Towards a Theory of Corporate Coherence: Preliminary Remarks," in G. Dosi, R. Giannetti, and P. A. Toninelli, eds., *Technology and Enterprise in a Historical Perspective* (Oxford: Clarendon Press, 1992).
2. C. Prahalad and G. Hamel, "The Core Competencies of the Firm," *Harvard Business Review* (May/June 1990), pp. 79-91.
3. See M. Tushman and P. Anderson, "Technological Discontinuities and Organizational Environments," *Administrative Science Quarterly*, 31 (1986): 439-465.
4. See Prahalad and Hamel, op. cit.
5. One further possibility is using the technical field of qualification of corporate employees, especially engineers. This is used in some of our case studies. However, only a few countries apart from Sweden have collected comprehensive data on a regular basis. See S. Jacobsson and C. Oskarsson, "Educational Statistics as an Indicator of Technological Activity," *Research Policy*, 24 (1995): 127-136.
6. See F. Narin, E. Noma, and R. Perry, "Patents as Indicators of Corporate Technological Strength," *Research Policy*, 16 (1987): 143-155; "The Global Patent Race Picks Up Speed," *Business Week*, August 9, 1993, pp. 49-54.
7. These are pharmaceuticals, other chemicals, mining and petroleum, textiles, rubber and plastics, paper and wood, food, drink and tobacco, building materials, metals, machinery, electrical-electronic, computers, instruments, motor vehicles, and aircraft.
8. See Table 2 for the 34 technical fields, which are based on U.S. patent classes and sub-classes. The correspondence is available from P. Patel at SPRU, University of Sussex.
9. P. Patel and K. Pavitt, "The Wide (and Increasing) Spread of Technological Competencies in the World's Largest Firms: A Challenge to Conventional Wisdom," in A. Chandler, P. Hagstrom, and O. Solvell, eds., *The Dynamic Firm*, Oxford (Oxford: University Press, forthcoming).
10. Comparisons between technological and product diversity are particularly difficult given the problems of making one-to-one correspondences between the two sets of classification units. Suffice to say that, in spite of twice as many technological as product classes, firms in Table 1 are active in a higher number of the former than the latter.
11. N. Rosenberg, "Technological Change in the Machine Tool Industry," *Perspectives on Technology* (Cambridge: Cambridge University Press, 1976).
12. A. Prencipe, "Technological Competencies and Product's Evolutionary Dynamics: a Case Study from the Aero-Engine Industry," *Research Policy*, 25 (1997): 1261-1276.
13. Patel and Pavitt (forthcoming), op. cit.

14. J. Hagedoorn, "Trends and Patterns in Strategic Technology Partnering since the Early Seventies," *Review of Industrial Organization*, 11 (1992): 601-616.
15. Prencipe, op. cit.
16. A similar pattern can also be observed in Japanese firms in opto-electronics. In the late 1960s, they began to explore the potential opportunities offered by recent scientific advances in the field. Their initial experimentation was across a broad front and, after an extensive period of trial, error, and learning, they began to identify and target specific applications related to their core businesses. For example, Sharp was able to accumulate its competencies in liquid crystal displays through incremental learning and to find multiple applications: from simple matrix displays in watches, calculators, and word processors to active matrix displays in computers and television. See K. Miyazaki, *Building Competencies in the Firm: Lessons from Japanese and European Optoelectronics* (New York, NY: St. Martin's Press, 1995).
17. In other words, Patent Share $(PS) = Pij/\Sigma li\ Pij$, where Pij is the number of patents granted in technical field i to firm j.Similarly, Revealed Technology Advantage $(RTA) = PS \div \Sigma ljPij/\Sigma li\Sigma ljPij$.
18. "Few companies are likely to build world leadership in more than five or six fundamental competencies. A company that compiles a list of 20 to 30 capabilities has probably not produced a list of core competencies." Prahalad and Hamel, op. cit., p.84.
19. H. Chesbrough and D. Teece "When is Virtual Virtuous? Organizing for Innovation," *Harvard Business Review* (January/February 1996): 65-73.
20. This will include the competence to gain access to advances in university-based research. Hicks has recently shown that published research by large firms enables them to benefit from the much larger body of research performed in universities. D. Hicks, "Published Papers, Tacit Competencies and Corporate Management of the Public/Private Character of Knowledge," *Industrial and Corporate Change*, 4 (1995): 401-424.
21. The main interview and questionnaire study covered 14 large Japanese firms (including Hitachi, NEC, Toshiba, Canon, Toyota), 20 European (including Ericsson, Volvo, Siemens, Philips) and 16 U.S. (including IBM, General Electric, ATT, General Motors, Texas Instruments). A subsequent analysis of published data covered 57 large firms in OECD countries. Additional case studies of various companies and product areas have also been conducted. See O. Granstrand, *Technology, Management and Markets* (London: Pinter, 1982); O. Granstrand and S. Sjolander, "Managing Innovation in Multi-Technology Corporations," *Research Policy*, 19 (1990), 35-60. O. Granstrand and C. Oskarsson, "Technology Diversification in Multi-Tech Corporations," *IEEE Transactions on Engineering Management*, 41 (1994): 355-364. C. Oskarsson, *Technology Diversification—The Phenomenon, Its Causes and Effects*, Department of Industrial Management and Economics, Chalmers University of Technology, Gothenburg, 1993.
22. A. Gambardella and S. Torrisi, "Does Technological Convergence Imply Convergence in Markets? Evidence from the Electronics Industry," *Research Policy* (forthcoming).
23. C. Christensen and R. Rosenbloom, "Explaining the Attacker's Advantage: Technological Paradigms, Organizational Dynamics, and the Value Network," *Research Policy*, 24 (1995): 233-257.
24. C. Freeman, "Networks of Innovators: a Synthesis of Research Issues," *Research Policy*, 20 (1991): 499-514.

[7]

Technologies, Products and Organization in the Innovating Firm: What Adam Smith Tells Us and Joseph Schumpeter Doesn't

Keith Pavitt

(Science Policy Research Unit, Mantell Building, University of Sussex, Brighton BN1 9RF, UK)

Adam Smith's insights into the increasingly specialized nature of knowledge production are crucially important in understanding the contemporary problems of managing innovating firms. Products and firms are based on an increasing range of fields of specialized technological understanding. Competition is not based on technological diversity, but on diversity and experimentation in products, etc. Firms rarely fail because of an inability to master a new field of technology, but because they do not succeed in matching the firm's systems of coordination and control to the nature of the available technological opportunities.

All the improvements in machinery, however, have by no means been the inventions of those who had occasion to use the machines. Many . . . have been made by the makers of the machines, when to make them became the business of a peculiar trade: and some by . . . those who are called philosophers, or men of speculation, whose trade is not to do anything but to observe everything: and who, upon that account are often capable of combining together the powers of the most distant and dissimilar objects. . . . *Like every other employment . . . it is subdivided into a number of different branches, each of which affords occupation to a peculiar tribe or class of philosophers; and this subdivision of employment in philosophy, as well as in every other business, improves dexterity and saves time.* (Smith, 1776, p. 8, my italics)

Industrial and Corporate Change Volume 7 Number 3 1998

—————— *Technologies, Products and Organization in the Innovating Firm* ——————

1. Setting the Scene

Evolutionary Theory and the Innovating Firm

Attempts over the past twenty years to build an evolutionary theory of the firm have grown in part out of a dissatisfaction with the inability of mainstream theory to deal satisfactorily with two important, interrelated and empirically observable characteristics of contemporary society: continuous technical change, and the central role of the business firm in generating learning, improvement and innovation, through deliberate and purposive action. At the beginning, concepts like 'technological trajectories' and 'routines' were introduced by Nelson and Winter (1977, 1982) to reflect the cumulative and path-dependent nature of technical change, the often tacit nature of the knowledge underlying it, and the trial-and-error behaviour of business practitioners trying to cope with a complex and ever-changing world.[1]

Since then, two influential streams of analysis have helped deepen our knowledge of the innovating firm. First, numerous attempts have been made to apply the tools and techniques of evolutionary theory (and particularly biological evolution) directly to modelling and understanding technical change (e.g. Dosi and Marengo, 1993; Metcalfe and de Liso, 1995). Second, numerous empirical studies have attempted to formulate generally applicable laws that explain when and why established firms succeed in innovation, and when and why they fail (e.g. Iansiti and Clark, 1994; Teece, 1996). This paper will argue that, whilst both schools have made notable contributions to understanding of both the nature of the innovating firm and the conditions for successful innovation, much remains to be done.

Multi-technology Products and Firms

In particular, greater care and attention needs to be devoted to the distinctions between the *artefacts* (products, etc.) that the firm develops and produces, the firm-specific technological *knowledge* that underlies its ability to do so, and the *organizational* forms and procedures that it uses to transform one into the other. We shall argue that, in the late twentieth century, lack of technological knowledge is rarely the cause of innovation failure in large firms

---

[1] This paper concentrates on these aspects of firm behaviour, and not on the wider implications of evolutionary theory for the theory of the firm. For a concise evaluation of the latter, together with that of other recent theoretical developments, see Coriat and Weinstein (1995).

——————— *Technologies, Products and Organization in the Innovating Firm* ———————

based in OECD countries. The main problems arise in organization and, more specifically, in coordination and control.

This can best be understood if more attention is paid to what Adam Smith said about the division of labour, and less to what Schumpeter said about creative destruction. Smith's identification of the benefits of specialization in the production of knowledge has been amply confirmed by experience. Professional education, the establishment of laboratories, and improvements in techniques of measurement and experimentation have increased the efficiency of discovery, invention and innovation. Increasingly difficult problems can be tackled and solved.[2] Two complementary forms of specialization have happened in parallel.

First, new disciplines have emerged, with all the benefits of the division of labour highlighted by Smith himself at the beginning of this paper. These specialized bodies of knowledge have become useful over a growing range of applications, so that products incorporate a growing number of technologies: compare the eighteenth-century loom with today's equivalent, with its fluid flow, electrical, electronic and software elements improving the efficiency of its mechanical functions. In other words, products are becoming increasingly 'multi-technology', and so are the firms that produce them. Each specific body of technical knowledge cannot be associated uniquely with a single, specific class of product.[3] Products and related technologies co-evolve within firms, but their dynamics are different. For example, Gambardella and Torrisi (1998) found that the most successful electronics firms over the past ten years have been those that have simultaneously broadened their technological focus and narrowed their product focus. In other cases, firms have used their range of technological skills to create or enter new product markets (see Granstrand, 1982; Granstrand and Sjolander, 1990; Oskarsson, 1993; Granstrand and Oskarsson, 1994).

Second, in addition to the benefits of the cognitive division of labour into more specialized fields, the rate of technical change has been augmented by the functional division of labour within business firms, with the establishment of corporate R&D laboratories and similar groups devoted full-time to inventive and innovative activities. In addition to the Smithean benefits of specialization, professionalization and improved equipment, these laboratories enabled firms to monitor and benefit more systematically and effectively from

[2] The classic texts on this are Rosenberg (1974), de Solla Price (1984) and Mowery and Rosenberg (1989). See, for example, the reasons why problems in mechanics were solved more easily than those in medicine.

[3] Amongst other things, this is a source of frustration for economists who would like to match statistics on inventions from technology-based patent classes with product based trade and production statistics. See, for example, Scherer (1982).

——————— *Technologies, Products and Organization in the Innovating Firm* ———————

the outside advances in specialized academic disciplines. And with growing experience in the development and testing of prototypes, they have allowed systematic experimentation with a wider range of products and processes than had previously been possible through incremental improvements constrained by established products and production lines. In fields of rich technological opportunity, firms have in consequence become multi-product as well as multi-technology.

Two Bodies of Knowledge

Hence the importance, in analysing the innovating firm, of distinguishing clearly artefacts (products) from the knowledge sources on which they are based.[4] Nelson (1998) identifies two, complementary elements in firm-specific knowledge. First, there is a 'body of understanding', based on competencies in specific technological fields, and reflected in the qualifications of corporate technical personnel, and in the fields in which they patent and publish.[5] The second element is what Nelson (1998) calls a 'body of practice', related to the design, development, production, sale and use of a specific product model or a specific production line. This firm-specific practical technical knowledge is often obtained through the combination of experimentation, experience, and information and other exchanges amongst different parts of the organisation.[6] As such, it is an organizational task, so that 'a body of practice' consists largely of organizational knowledge that links 'a body of understanding' with commercially successful (or, more broadly, useful) artefacts.

----

[4] For an earlier discussion of this distinction, see Archibugi (1988).

[5] For measurement of corporate technological competencies through the fields of qualifications of technical personnel, see Jacobsson and Oskarsson (1995); through patenting, see Patel and Pavitt (1997); and through scientific papers, see Narin and Olivastro (1992), Godin (1996) and Hicks and Katz (1997).

[6] The difference between the two forms of knowledge is nicely illustrated in the following passage from Iansiti and Clark (1994), in relation to the firm-specific capabilities for the design and development of dies used in the production of body panels for automobiles: 'The knowledge that underlies that capability includes an understanding of metallurgy, the flow of metal under pressure and the relationship between the characteristics of the material, the forces and pressures applied to the material and the surface properties that result. These kinds of knowledge pertain to the fundamental properties of the die and its production system. But the firm must also have knowledge about how the fundamental concepts can be operationalised into effective actions. These include knowledge of techniques of die design, die modelling, die testing and finishing, for example. Additionally, knowledge can take the form of the skill of die designers in anticipating processing problems, customised software that allows for rapid and effective testing, patterns of communication and informal interaction between die designers and manufacturing engineers that allow for early identification of potential problems, and an attitude of co-operation that facilitates coordinated action between the die designers and the tool makers that will build the dies. These elements (and many others) define an organisational capability for die design and development' (p. 560).

—————— *Technologies, Products and Organization in the Innovating Firm* ——————

Method

The starting point for our analysis is the large, multi-divisional manufacturing firm, with established R&D activities and a product range that has grown out of a common, but evolving, technological competence. In part, this reflects the focus of this author's recent research (Patel and Pavitt, 1997; Tidd *et al.*, 1997). More important for the purpose of this paper, large multi-divisional firms are the largest single source of the new technological knowledge on which innovation depends. They perform most of the R&D activities, employ most of the qualified research scientists and engineers, perform and publish most of the corporate basic research, and maintain the closest links with academic research (Hicks, 1995). They also contribute to the development of knowledge and products for their suppliers of production equipment, components and software (Rosenberg, 1963; Patel and Pavitt, 1994). Finally, even when they fail in innovation themselves, they remain the major source of the technological and other competencies which enable new firms with different organizational approaches to succeed. Understanding the reasons for their success and failure therefore has the widest implications, not only for their managers, but also for the distribution of innovative activities amongst companies of different sizes and ages. It is not the main concern of this paper to argue in general for or against the large firm's ability to sustain radical innovation, but to understand better the reasons for its success and failure in trying to do so.[7]

We divide our analysis into four parts, reflecting four mechanisms identified by earlier analysts of the innovating firms: competition, cognition, coordination and control.[8] Table 1 sets out schematically how the division of labour, both in knowledge production and in corporate innovative activities, has influenced these four mechanisms. We argue in Section 2 that failure to distinguish between technologies and products has led to confusion in evolutionary

[7] Many analysts in the evolutionary tradition are pessimistic about the ability of large firms to sustain radical innovations, pointing to recent spectacular failures and to the emergence of new organization forms: Teece (1996) recently identified four types of firm: conglomerate, multi-product integrated hierarchy, virtual corporation and 'high flex' Silicon Valley type. Others argue that the obituary of the large innovative firm may well be premature, since there remain many examples of their success in developing and exploiting major innovations: for example, according to Methe *et al.* (1996): 'established firms, including industry incumbents and diversifying entrants, play vital and underemphasized roles as sources of major innovations in many industries" (p. 1181). And now even the oldest and best established of capitalists—grocers (supermarkets) and moneylenders (financial services)—have become major players in the development and exploitation of information technology.

[8] These dimensions of the innovating firm emerge from the original work of Nelson and Winter (1982), and from later work by Cohendet *et al.* (1994). In the language of Teece and Pisano (1994), in their analysis of the 'dynamic capabilities' of the firm, our competitive mechanisms relate their notions of corporate *position*, cognitive mechanisms to corporate *paths*, and coordination and control mechanisms to corporate *processes*.

——— Technologies, Products and Organization in the Innovating Firm ———

TABLE 1. Some Consequences of the Division of Labour in the Production of Technological Knowledge

Analytical implications	Technology ≠ products	Technological discontinuities ≠ product discontinuities	Technological diversity within firms, and within countries, but not within industries
Division of labour in knowledge production	Laboratories Disciplines Trained scientists and engineers → Increasing output, range and usefulness of knowledge →		
Division of labour in business functions	Specialized technical functions, inc. R&D labs → Increasing competence to understand and improve artefacts →		
	Multi-technology products →	Multi-technology firms →	Multi-product firms
Management implications	Co-ordination Organizational competence to experiment and learn across organizational boundaries	Competencies Technological competence enhancement > competence destruction	Control Organizational competence to reconfigure divisions and evaluate options in the light of technology characteristics

writings about what kind of 'diversity' is desirable in competitive processes. In Section 3, we further argue that, although there are clear cognitive limits on the range of technologies that a specific firm is capable of mastering, failure in innovation in established firms is not the result of the destruction of their technological competencies, but of their inability to match the technological opportunities with organizational forms and procedures appropriate for their development and exploitation. In Sections 4 and 5, we analyse in greater depth how the appropriate forms of two organizational elements that are central to corporate innovative activities—mechanisms of coordination and control—depend in part on the nature of the technology itself.

2. *Competitive Mechanisms and Technological Diversity*

It is around the notion of diversity[9] that the distinction between technologies (bodies of understanding) and products (bodies of practice) is most confused —and potentially most misleading, given the central importance accorded to diversity in the evolutionary theory of technical change.[10] In recent research undertaken at SPRU on the technological competencies of the world's largest firms, we have used the level and distribution of corporate patenting by technical field as a measure of the corporate body of technological understanding (Patel and Pavitt, 1997; O. Marsili, in preparation). This showed that technological diversity is prominent in some dimensions, but virtually absent in others.

- Large firms are active in a range of technologies *broader* than the products that they make. This reflects the multi-technology nature of their products, and the knowledge required to coordinate in-house product innovation with innovation in related production systems and supply chains. What is more, the range of technological competencies mastered by large firms is increasing over time, as new technological opportunities emerge.
- There is *high* diversity amongst large firms in the level and mix of their technological competencies, depending on the products that they produce. These largely sector-specific mixes of technological competence change only slowly over time, again in response to changing technological opportunities.

[9] A reading of the *Oxford Concise English Dictionary* suggests that the term 'diversity' is interchangeable with 'variety' and 'heterogeneity'.

[10] 'It is a basic proposition of evolutionary theory that a system's diversity affects its development' (Cohendet and Llerena, 1997, p. 227).

——————— *Technologies, Products and Organization in the Innovating Firm* ———————

- There is *low* diversity in the level and mix of technological competencies amongst large firms producing similar products. What is more, the degree of technological diversity is lowest in the product fields with the highest rate of technical change: computers and pharmaceuticals.

In other words, for the *individual firm*, technological diversity gives it the basis to make and improve its products. For the *economy as a whole*, more diversity amongst firms in their mixes of specialized technological knowledge enables them to explore and exploit a fuller range of product markets. But at the level of the *product market or industry*, there is similarity rather than diversity in the level and mix of technological activities in competing firms, and especially in those with high rates of technical change. Technological diversity is certainly not a characteristic of competition amongst innovating firms.[11] The diversity exists downstream in the body of practice, namely the product and process configurations that can be generated from the same or very similar base of technological knowledge. We know that some of these configurations do not work out technically, and many more do not work out commercially (Freeman and Soete, 1997). What emerges is a world where firms with broad and stable bundles of technological competencies have the capacity to generate and experiment with a range of product (and process and service) configurations, some of which succeed, but many of which fail.

At any given time, advances in some fields of technology open major opportunities for major performance improvements in materials, components and subsystems (e.g. economies of scale in continuous processes, economies of miniaturization in information processing). The directions of these improvements are easily recognized, even if they require the commitment of substantial resources for their achievement, e.g. Moore's Law in semi-conductors.[12] Thus, experimentation and diversity do not take place between different technologies. On the contrary, rich and well-known directions of

---

[11] A (frivolous) translation of these results into biological evolutionary terms might be (i) species need a range of genetic attributes for survival; (ii) since they live in different parts of the forest, the elephant and the mouse have different genetic mixes, which change only slowly; (iii) there is little room (or need) for genetic deviance when things are changing fast (and in predictable directions—see below).

[12] Recent comments by an IT expert make the point nicely: 'Precious little has happened in digital technology over the past five years. . . . Steady increases in processor speed and storage size have become as predictable as a child's growth. . . . Just as the computer industry is predicated on Moore's Law—that chips will double in speed every 18 months, which companies can literally plan on—the telecom industry can be predicated on the transparent network. . . . Change is routine and uneventful. . . . The fiber-optic backbone has joined the microprocessor on a steady predictable climb. Processing speeds will double every 18 months. Bandwidth will quadruple every two years. Corporate planners can rest easy' (Steinberg, 1998, pp. 80–84). In our framework, the conclusion of the last sentence does not follow from the preceding analysis. Such rapid if predictable change in underlying technology is bound to create a plethora of difficult-to-predict products and services.

——————— *Technologies, Products and Organization in the Innovating Firm* ———————

improvement in underlying technologies[13] create opportunities for diversity and experimentation in product configurations. Technological opportunities create product diversity. There is no convincing evidence that technological diversity creates product opportunities.[14]

3. *Cognitive Mechanisms and Creative Destruction*

Large firms may have competencies in a number of fields of technology but, in the contemporary world of highly specialized knowledge, the costs of mastering all of them clearly appear to outweigh the benefits. Firms develop their technological competencies incrementally, and constrain their search activities close to what that they already know. Thus, over the past 20 years, electronics firms have moved heavily into semiconductor technology (but not biotechnology), and drug firms into biotechnology (but not semiconductor technology). The firm's knowledge base both determines what it makes, and the directions in which it searches (Patel and Pavitt, 1997). In this sense, there are clear cognitive limits on what firms can and cannot do.

The central importance of firm-specific technological competencies has led some analysts to place technological discontinuities (i.e. major technological improvements) at the heart of the theory of the innovating firm. In particular, they argue that such discontinuities may either enhance established competencies and strengthen incumbent firms, or destroy established competencies and undermine them. Again, it must be stressed that technological discontinuities are not the same as product discontinuities, even if they are often treated as such. For example, perhaps the most influential paper on discontinuities by Tushman and Anderson (1986) talks of *technological* discontinuities in its title, whilst the basis for its empirical analysis are new products and processes (e.g. jet engines, oxygen steel-making).

Although they may have revolutionary effects, technological discontinuities rarely encompass all—or even most of—the fields of knowledge that feed into a product. Typically they may affect the performance of a key component (e.g. transistors vs. valves) or provide a major new technique (e.g. gene splicing). But they do not destroy the whole range of related and complementary

---

[13] Nelson and Winter (1977) originally called these 'natural trajectories' and were roundly criticized by those arguing that technologies are socially constructed. But perhaps Nelson and Winter were right. The range of opportunities in different technological fields depends heavily on what nature allows us to do. Compare rates of increase of information storage capacity over the last 20 years with rates of increase in energy storage capacity. In the former, newsprint, punch cards and analogue recording have been overwhelmed by digital methods. In the latter, petroleum remains supreme, in spite of considerable technological efforts to develop better alternatives.

[14] In this context, a recent paper, Stankiewicz (1998) proposes the notion of interrelated 'design space' for artefacts, and 'operands' for the underlying knowledge base, techniques, etc.

———— *Technologies, Products and Organization in the Innovating Firm* ————

technologies (e.g. sound reproduction in radios, memories in computers, molecular design in pharmaceuticals) that are necessary for a complete product.[15] Indeed, they create opportunities for product 'discontinuities' that often can be achieved only through improvements in complementary but long-established technologies (e.g. metal tolerances and reliability for robots).

Furthermore, as Gambardella and Torrisi (1998) have shown, corporate technological dynamics can have different dynamics from corporate product dynamics, with technological diversification going hand in hand with increasing product focus: for example (i) when a technological discontinuity is incorporated in a product family at the mature stage of its product cycle; or (ii) when a technological discontinuity provokes the emergence of radically new but technology-related product markets with different—but as yet ill-defined—characteristics; this is probably the case in the electronics industry studied by Gambardella and Torrisi (1998).

Finally, it should be noted that the predominance given to revolutionary technologies in the destruction of corporate competence has often been associated with the notion of paradigm shifts in technology (Dosi, 1982), similar to those in science (Kuhn, 1962). But this is a misinterpretation of the notion of paradigm. A new paradigm does not discredit and displace all the knowledge generated in the earlier paradigms, but instead adds to them. Newtonian physics still has major theoretical and practical uses, and at least a quarter of all the new technology created today is still in mechanical engineering. The development and commercial exploitation of technological discontinuities turns out to be a more cumulative process than is often supposed.

Certainly, there are many historical examples of firms that have failed because they did not master major emerging fields of technology (Cooper and Schendel, 1976). But competence-destroying technologies are the exception rather than the rule today, especially amongst large firms, who have demonstrated a strong capacity through their R&D departments to acquire and develop competencies in 'discontinuity-creating' technologies like computing and biotechnology (Patel and Pavitt, 1997). The key factors behind the success and failure of innovating firms must be sought elsewhere, in the organizational processes linking technologies, products, their production and their markets.

Cognitive mechanisms also underlie the taxonomy of innovation proposed by Abernathy and Clark (1985), which distinguishes four types: incremental, component, architectural and revolutionary. Based on an analysis of innova-

---

[15] On biotechnology, see McKelvey (1996).

tion in photolithographic aligners, Henderson and Clark (1990) argued that innovations in product architecture[16] destroy the usefulness of the architectural knowledge in established firms, and this is difficult to recognize and remedy. More recently, and based on analysis of innovations in computer disk drives, Christensen and Rosenbloom (1995) concluded that architectural innovations do not necessarily destroy established competencies. What does is a change in the 'value network' (i.e. user market) of the innovation.[17]

Whilst these studies throw interesting and important light on innovation processes within firms, they must be interpreted with care. An alternative reading of the Henderson and Clark story is that failure has less to do with cognitive failure by design engineers to recognize the value of alternative product architectures, than with organizational factors such as the inability of design engineers to recognize signals from users or the marketing department, or the unwillingness or inability of corporate management to establish a new design team or product division. Furthermore, it may be a mistake to generalize from the experience of US firms specialized in the IT sector to firms in other sectors and countries. In contrast, for example, to Christensen and Rosenbloom's emphasis on the difficulties of US firms making computer disk drives in switching end-user markets, most of the world's leading chemical firms have been very successful in the twentieth century in deploying their techniques and products deriving from organic synthesis in markets as diverse as textiles, building, health and agriculture (Hounshell and Smith, 1988; Plumpe, 1995).

4. Co-ordination Mechanisms and Learning across Organisational Boundaries

One of the most robust conclusions emerging from empirical research on the factors affecting success in innovation is the importance of coordinating learning and other change-related activities across functional boundaries (Burns and Stalker, 1961; Rothwell, 1977; Cooper, 1988; Wang, 1997).[18] Here we see the second major feature of the division of labour that is central to contemporary corporate innovative activities, namely coping with functional specialization, with the emergence of specialized departments for

[16] '. . . reconfiguration of an established system to link together existing components in a new way' (Henderson and Clark, 1990, p.12)

[17] The authors liken a change in the 'value network' (i.e. disk configuration and user market) to a paradigm shift, which implies a much broader definition of the notion of paradigm than that probably envisaged by Dosi (1982).

[18] Problems of such coordination have also figured largely in the works of Coase (1937), Penrose (1959), Aoki (1986) and Loasby (1998).

——————— *Technologies, Products and Organization in the Innovating Firm* ———————

R&D, production, marketing, logistics, strategy, finance, etc. Such coordination cannot realistically be reduced to designing flows of codified information across functional boundaries. It also involves coordinated experimentation (e.g. new product launches), and the interpretation of ambiguous or incomplete data, where tacit knowledge is essential. As the observations of Iansiti and Clark in footnote 6 (p. 436) show, personal contacts, mobility and interfunctional teams are therefore of more central importance than pure information flows.

In our present state of knowledge, effective coordination belongs in the field of practice rather than the field of understanding. Unlike purely technological processes, organizational processes are difficult to measure and evaluate, and do not lend themselves readily to rigorous modelling and controlled experiments. In addition, the coordination processes in which we are interested are complex. Experimentation and learning across critical organizational interfaces are particularly difficult when combining knowledge from different functions, professions and disciplines, each with their distinct and different analytical frameworks and decision rules—which is another reason why firms may try to compensate for greater technological complexity by greater market focus.[19]

In addition, the identification of the location of critical interfaces is not easy, for three reasons. First, there are potentially several such interfaces, involving a multitude of possible linkages between R&D, production, marketing and logistics within the firm, and a variety of sources of outside knowledge in universities, other firms (suppliers, customers, competitors, etc.) and other countries. Second, the interfaces that merit analysis and managerial attention vary considerably amongst technologies and products. Compare firms in pharmaceuticals and automobiles. In the former, strong interfaces between in-house R&D and the direct output of academic research (in medicine, biology and chemistry) are essential. In the latter, they are not, but strong interfaces between in-house R&D and production are of central importance. These differing characteristics have important implications for both the appropriate organizational forms, and geographical location of corporate innovative activities.

Finally, the key interfaces for organizational learning change over time, very often as a result of changes in technology-related factors themselves. Witness the growing importance for the pharmaceutical industry of the interface with

---

[19] Models of intra-corporate coordination have not got very far in grappling with these essential features of innovative activities. For example, in Aoki's models (1986), problems of coordination are in production and dealt with through information flows, rather than in learning and innovation mediated through tacit knowledge. Furthermore, sources of instability and change are in an exogenous environment, rather than created by the firms themselves.

─────── *Technologies, Products and Organization in the Innovating Firm* ───────

research in academic biology, the consumer market interface for producers of telephones and computers, and of interfaces with software and materials technologies for firms in virtually all sectors. One major source of failure in innovation is likely to be inadequate recognition of the importance of these new interfaces, or the inability of management to take effective action to establish them. We should look for what Leonard-Barton (1995) calls 'core rigidities', when individuals and groups with the established competencies for today's products are either ignorant of, or feel threatened by, the growing importance of new competencies.

5. *Control Mechanisms: Matching Strategic Styles with Technologies*

The lack of a one-to-one link between each product and each technology has at least two major implications for organizational practice in business firms.[20] The first is that firms which master fields of rich technological opportunity are often able to develop and produce several products based on the same body of knowledge. In other words, they compete and grow through technology-related diversification.[21] Second, the very existence of this broadly useful knowledge means that the classic M-form organization is unable to match tidily each field of its technology to one product or to one division.

As a consequence, systems of corporate control in the multi-product firm have a major influence on the rate and direction of its innovative activities. Chandler (1991) distinguishes two essential functions of corporate control: the *entrepreneurial* function of planning for the future health and growth of the enterprise, and the *administrative* function of controlling the operation of its many divisions.[22] The administrative function is normally exercised through systems of financial reporting and controls. The entrepreneurial function for technology is the capacity to recognize and exploit technology-based opportunities. This requires an ability to evaluate projects and programmes where the normal financial accounting techniques are often inoperable and inappropriate, since exploratory research programmes should be treated as options, rather than full-scale investments (Myers, 1984; Hamilton, 1986; Mitchell, 1986; Mitchell and Hamilton, 1988). It may also require the establishment of a central corporate research programme or laboratory, funded in part independently from the established product divisions

[20] For more extended discussions, see Kay (1979), Prahalad and Hamel (1990), von Tunzelmann (1995) and Marengo (1995)

[21] See Rumelt (1974). Numerous examples can be found in the electrical and chemical industries.

[22] See also the earlier pioneering work of Goold and Campbell (1987).

─────── *Technologies, Products and Organization in the Innovating Firm* ───────

(Graham, 1986). And it will certainly require the capacity to reconfigure the composition and objectives of established divisions in the light of changing opportunities (Prahaled and Hamel, 1990).

Different balances between the administrative and the entrepreneurial functions are likely to be appropriate to different levels of technological opportunity. In addition, the appropriate degree of decentralization of the entrepreneurial function within the corporation will depend in part on the nature of the firm's core technology.[23] The higher the costs of product development, the greater the need for central control of the entrepreneurial function. In other words, the appropriate system of corporate control will depend in part on the nature of the technology.

Thus, Table 2 suggests that firms with low technological opportunity are likely to be compatible with an emphasis on the administrative rather than the entrepreneurial function, and with more centralization with increasing capital intensity. Firms with high technological opportunities and high costs of product and process development—such as those in drugs and automobiles—are likely to be best suited to a strong entrepreneurial function at the corporate level. Those with high technological opportunities, but low costs of product development—like those in consumer electronics and the 3M Corporation—will be best served by more decentralized entrepreneurial initiative.

Table 2 also shows that there can be mismatches between strategic style and the nature of technological opportunities. For example, tight financial control and emphasis on short-term profitability do not allow investments in exploring longer-term options emerging from new technological opportunities: this is one reason why GEC in the UK and ITT in the USA have progressively excluded themselves from many high-technology markets (*Economist*, 1995, 1996). Similarly, the characteristics of technology, and the corresponding organizational requirements, change over time. Thus, one reason for the recent deliberate demerger of ICI was the reduced technological opportunities in the previously fast-moving field of bulk chemicals (Owen and Harrison, 1995). Similarly, the high costs of mainframe computers in the 1960s and 1970s, and their specificity to the corporate office market, imposed centralized entrepreneurship. With the advent of the microprocessor and packaged software, the costs of experimentation tumbled and new markets emerged. Mainframe firms had great difficulty in adjusting in time to the requirements of greater decentralization.

───

[23] Marengo (1995) models learning, and comes to some intuitively appealing conclusions about the balance between organizational centralization and decentralization. But his learning is also about changes in the environment, rather than about internally generated changes.

───────────────── ─────────────────

———— *Technologies, Products and Organization in the Innovating Firm* ————

TABLE 2. Technology and Corporate Control

		Strategic style	
		Entrepreneurial	Administrative
Levels of decision-making	HQ	*High-tech opportunity + high costs of product development* • drugs • automobiles • bulk chemicals in 1960s → • mainframes in 1970s ↓	*Low-tech opportunity + high cost of investments* → GEC (UK) → ITT (USA) • aluminium • steel
	Division	*High-tech opportunity + low costs of product development* • consumer electronics • 3M	*Low-tech opportunity + low cost of investments* • conglomerates

6. Conclusions

The main argument of this paper is that—as foreseen by Adam Smith—specialization in knowledge production is a central feature of the innovating firm. It is therefore of great importance to distinguish products (and other artefacts) from the underlying bodies of technological understanding on which they are based. Although the two evolve together, they do not have the same dynamics. Inadequate care in distinguishing the two can result in mistaken policy prescriptions (e.g. Granstrand *et al.*, 1997). And it can lead to too much emphasis in evolutionary theorizing on the economic benefits of technological diversity, on the frequency and causes of creative destruction, and on the nature and implications of changes in technological paradigms.

The main challenge is to improve understanding of the organizational processes of coordination and control that make for a successful matching between the development and deployment of bodies of technological knowledge, on the one hand, and commercially successful (or useful) working artefacts, on the other. We have stressed that our practical and theoretical knowledge of these organizational processes are less well grounded than our knowledge of the processes of technological advance *per se*. This why companies with outstanding technological competencies—Xerox and IBM in the early days of personal computers, for example—failed to develop organizational forms to exploit them. Nonetheless, large firms are capable of restructuring their activities to benefit from the new technological oppor-

——————— *Technologies, Products and Organization in the Innovating Firm* ———————

tunities that they have mastered. 'Routines' can and do change. 'Creative destruction' is not inevitable.

The appropriate organizational processes will depend on the characteristics of the technologies, such as their sources, the rate and direction of their change, and the costs of developing and building artefacts based on them. And since technologies vary greatly in these characteristics, and they change over time, any improved knowledge that we acquire will be highly contingent. Nonetheless, the research of Woodward (1965) and Chandler (1977) on the organizational dimensions of changes in process technologies shows that such research can make a major difference to our understanding of innovation in firms. The following avenues of research appear to be particularly fruitful:

1. Bibliometric studies and surveys that map linkages between knowledge, products and organization in business firms over a range of sectors. This is essential given intersectoral variety. The great challenge is to develop measures of organization that are conceptually clear and empirically robust.

2. Detailed studies by historians and sociologists of the interactions between the development of the technological knowledge base, and the associated artefacts that emerge from them (e.g. Constant, 1998; Stankiewicz, 1998).

3. Case studies of the effects on firms, their organization and their products of the introduction of technological discontinuities, whether in the form of new sources of useful knowledge, or of order-of-magnitude improvements in the performance in one field. If then our analysis is correct, large firms in advanced countries will have few difficulties in mastering the new technology, resultant product discontinuities will happen only after a extended period of learning,[24] and failure is likely to result from 'core rigidities', namely resistance from established groups within the organization.

Finally, our analysis suggests that truths about the real innovating firm will never be elegant, simple or easy to replicate. It is nonetheless to be hoped that formal theorizing will try to incorporate more real-world features of the innovating firm. In particular, evolutionary economics grew out of dissatis-

---

[24] See Miyazaki (1995) for an account of the extended period that Japanese firms spent learning about opto-electronics. It might be argued that the personal computer began with a component innovation (the microprocessor) which, after a number of complementary component innovations (e.g. memories), and architectural innovations (internalizing the disk drive) and incremental improvements, created the conditions for the emergence of the revolutionary innovation that was the PC.

———— *Technologies, Products and Organization in the Innovating Firm* ————

faction with mainstream formalizations of technical change. It would be a pity if it ended up going down the same path.

Acknowledgements

The author has benefited greatly in the preparation of this paper from the comments of Stefano Brusoni, Mike Hobday, Patrick Llerena, Richard Nelson, Ed Steinmueller and two anonymous referees. The usual disclaimers apply.

References

Abernathy, W. and K. Clark (1985), 'Innovation: Mapping the Winds of Creative Destruction,' *Research Policy*, 14, 3–22.

Aoki, M. (1986), 'Horizontal vs. Vertical Information Structure of the Firm,' *American Economic Review*, 76, 971–983.

Archibugi, D. (1988), 'In Search of a Useful Measure of Technological Innovation,' *Technological Forecasting and Social Change*, 34, 253–277.

Burns, T. and G. Stalker (1961), *The Management of Innovation*. Tavistock: London.

Chandler, A. D. (1977), *The Visible Hand: The Managerial Revolution in American Business*. Belknap Press: Cambridge, MA.

Chandler, A. (1991), 'The Functions of the HQ Unit in the Multibusiness Firm,' *Strategic Management Journal*, 12, 31–50.

Christensen, C. and R. Rosenbloom (1995), 'Explaining the Attacker's Advantage: Technological Paradigms, Organisational Dynamics, and the Value Network,' *Research Policy*, 24, 233–257.

Coase, R. (1937), 'The Nature of the Firm,' *Economica*, 4, 386–405.

Cohendet, P. and P. Llerena (1997), 'Learning, Technical Change and Public Policy: How to Create and Exploit Diversity,' in C. Edquist (ed.), *Systems of Innovation: Technologies, Institutions and Organisations*. Pinter: London.

Cohendet, P., P. Llerena and L. Marengo (1994),: 'Learning and Organizational Structure in Evolutionary Models of the Firm', EUNETIC Conference 'Evolutionary Economics of Technological Change: Assessment of Results and New Frontiers', Strasbourg.

Constant, E. (1998), 'Recursive Practice and the Evolution of Technological Knowledge,' in J. Ziman (ed.), *Technological Innovation as an Evolutionary Process*. Cambridge University Press: Cambridge (in press).

Cooper, A. and D. Schendel (1976), 'Strategic Responses to Technological Threats,' *Business Horizons*, February, 61–69.

Cooper, R. (1988),'The New Product Process: A Decision Guide for Management,' *Journal of Marketing Management*, 3, 238–255.

Coriat, B. and O. Weinstein (1995), *Les Nouvelles Theories de l'Entreprise*. Livre de Poche, Librarie Generale Française: Paris.

de Solla Price, D. (1984), 'The Science/Technology Relationship, the Craft of Experimental Science, and Policy for the Improvement of High Technology Innovation,' *Research Policy*, 13, 3–20.

Dosi, G. (1982), 'Technological Paradigms and Technological Trajectories: A Suggested Interpretation of the Determinants and Directions of Technical Change,' *Research Policy*, 11, 147–162.

Dosi, G. and L. Marengo (1993), 'Some Elements of and Evolutionary Theory of Organisational

——————— *Technologies, Products and Organization in the Innovating Firm* ———————

Competencies', in R. W. England (ed.), *Evolutionary Concepts in Contemporary Economics*. University of Michigan Press: Ann Arbor, MI.

Economist (1995), 'The Death of the Geneen Machine', June 17, 86–92.

Economist (1996), 'Changing of the Guard', September 7, 72.

Freeman, C. and L. Soete (1997), *The Economics of Industrial Innovation*. Cassel-Pinter: London.

Gambardella, A. and S. Torrisi (1998), 'Does Technological Convergence Imply Convergence in Markets? Evidence from the Electronics Industry,' *Research Policy* (in press).

Godin, B. (1996), 'Research and the Practice of Publication in Industry,' *Research Policy*, 25, 587–606.

Goold, M. and A. Campbell (1987), *Strategies and Styles: The Role of the Centre in Managing Diversified Corporations*. Blackwell: Oxford.

Graham, M. (1986), 'Corporate Research and Development: The Latest Transformation,' in M. Horwitch (ed.), *Technology in the Modern Corporation: a Strategic Perspective*. Pergamon Press: New York.

Granstrand, O. (1982), *Technology, Management and Markets*. Pinter: London.

Granstrand, O. and C. Oskarsson (1994), 'Technology Diversification in "Multi-tech" Corporations,' *IEEE Transactions on Engineering Management*, 41, 355–364.

Granstrand, O. and S. Sjolander (1990), 'Managing Innovation in Multi-technology Corporations,' *Research Policy*, 19, 35–60.

Granstrand, O., P. Patel and K. Pavitt (1997), 'Multi-technology Corporations: Why they Have 'Distributed' rather than 'Distinctive Core' Competencies', *California Management Review*, 39, 8–25

Hamilton, W. (1986), 'Corporate Strategies for Managing Emerging Technologies', in M. Horwitch (ed.), *Technology in the Modern Corporation: A Strategic Perspective*. Pergamon Press: New York.

Henderson, R. and K. Clark (1990), 'Architectural Innovation: The Reconfiguration of Existing Product Technologies and the Failure of Established Firms,' *Administrative Sciences Quarterly*, 35, 9–30.

Hicks, D. (1995), 'Published Papers (1995) Tacit Competencies and Corporate Management of the Public/Private Character of Knowledge,' *Industrial and Corporate Change*, 4, 401–424.

Hicks, D. and S. Katz (1997), *The Changing Shape of British Industrial Research*, STEEP Special Report no. 6, Science Policy Research Unit, University of Sussex.

Horwitch, M. (ed.) (1986) *Technology in the Modern Corporation: A Strategic Perspective*. Pergamon Press: New York.

Hounshell, D. and J. Smith (1988), *Science and Corporate Strategy: Du Pont R&D, 1902–1980*. Cambridge University Press: New York.

Iansiti, M. and K. Clark (1994), 'Integration and Dynamic Capability: Evidence from Product Development in Automobiles and Mainframe Computers,' *Industrial and Corporate Change*, 4, 557–605.

Jacobsson, S. and C. Oskarsson (1995), 'Educational Statistics as an Indicator of Technological Activity,' *Research Policy*, 24, 127–136.

Kay, N. (1979), *The Innovating Firm*. Macmillan: London.

Kuhn, T. (1962), *The Structure of Scientific Revolutions*. Chicago: University of Chicago Press.

Leonard-Barton, D. (1995), *Wellsprings of Knowledge*. Harvard Business School Press: Boston, MA.

Loasby, B. (1998), 'The Organisation of Capabilities,' *Journal of Economic Behaviour and Organisation* (in press).

Marengo, L. (1995), 'Structure, Competence and Learning in Organisations,' *Wirtschaftspolische Blatter*, 42, 454–464.

McKelvey, M. (1996), *Evolutionary Innovations: The Business of Biotechnology*. Oxford: Oxford University Press.

Metcalfe, S. and N. de Liso (1995), 'Innovation, Capabilities and Knowledge: The Epistemic Connection', in

─────── *Technologies, Products and Organization in the Innovating Firm* ───────

J. de la Mottre and G. Paquet (eds), *Evolutionary Economics and the New International Political Economy*. Pinter: London.

Methe, D., A. Swaminathan and W. Mitchell (1996), 'The Underemphasized Role of Established Firms as Sources of Major Innovations,' *Industrial and Corporate Change*, 5, 1181–1203.

Mitchell, G. (1986), 'New Approaches for the Strategic Management of Technology,' in M. Horwitch (ed.), *Technology in the Modern Corporation: a Strategic Perspective*. Pergamon Press: New York.

Mitchell, G. and W. Hamilton (1988), 'Managing R&D as a Strategic Option,' *Research- Technology Management*, 31, 15–22.

Miyazaki, K. (1995), *Building Competencies in the Firm: Lessons from Japanese and European Optoelectronics*. Macmillan/St Martin's Press: New York.

Mowery, D. and N. Rosenberg (1989), *Technology and the Pursuit of Economic Growth*. Cambridge: Cambridge University Press.

Myers, S. (1984), 'Finance Theory and Financial Strategy,' *Interfaces*, 14, 126–137.

Narin, F. and D. Olivastro (1992), 'Status Report: Linkage between Technology and Science,' *Research Policy*, 21, 237–249.

Nelson, R. (1998), 'Different Perspectives on Technological Evolution,' in J. Ziman (ed.), *Technological Innovation as an Evolutionary Process*. Cambridge: Cambridge University Press (in press).

Nelson, R. R. and S. G. Winter (1977), 'In Search of a Useful Theory of Innovation,' *Research Policy*, 6, 36–76

Nelson, R. R. and S. G. Winter (1982), *An Evolutionary Theory of Economic Change*. Belknap Press: Cambridge, MA.

Oskarsson, C. (1993), *Technology Diversification—The Phenomenon, its Causes and Effects*. Department of Industrial Management and Economics, Chalmers University of Technology, Gothenburg.

Owen, G. and T. Harrison (1995), 'Why ICI Chose to Demerge,' *Harvard Business Review*, March–April, 133–142.

Patel, P. and K. Pavitt (1994), 'The Continuing, Widespread (and Neglected) Importance of Improvements in Mechanical Technologies,' *Research Policy*, 23, 533–546.

Patel, P. and K. Pavitt, (1997), 'The Technological Competencies of the World's Largest Firms: Complex and Path-dependent, but not Much Variety,' *Research Policy*, 26, 141–156

Penrose, E. (1959), *The Theory of the Growth of the Firm*. Blackwell: Oxford.

Plumpe, G. (1995), 'Innovation and the Structure of the IG Farben,' in F. Caron, P. Erker and W. Fischer (eds), *Innovations in the European Economy between the Wars*. De Gruyter, Berlin.

Prahalad, C. K. and G. Hamel (1990), 'The Core Competencies of the Corporation,' *Harvard Business Review*, May–June, 79–91.

Rosenberg, N. (1963), 'Technological Change in the Machine Tool Industry, 1840–1910,' *Journal of Economic History*, 23, 414–446.

Rosenberg, N. (1974), 'Science, Invention and Economic Growth,' *Economic Journal*, 84, 333.

Rothwell, R. (1977), 'The Characteristics of Successful Innovators and Technically Progressive Firms,' *R&D Management*, 7, 191–206.

Rumelt, R. (1974), *Strategy, Structure and Economic Performance*. Graduate School of Business Administration, Harvard University.

Scherer. F. (1982), 'Inter-industry Technology Flows in the US,' *Research Policy*, 11, 227–45

Smith, A. (1776), *The Wealth of Nations*. Dent (1910): London.

Stankiewicz, R. (1998), 'Technological Change as an Evolution of Design Spaces', in J. Ziman (ed.), *Technological Innovation as an Evolutionary Process*. Cambridge: Cambridge University Press (in press).

——————— *Technologies, Products and Organization in the Innovating Firm* ———————

Steinberg, S. (1998), 'Schumpeter's Lesson: What Really Happened in Digital Technology in the Past Five Years,' *Wired*, January.

Teece, D. (1996), 'Firm Organisation, Industrial Structure and Technological Innovation,' *Journal of Economic Behaviour and Organisation*, 31, 193–224.

Teece, D. and G. Pisano (1994), 'The Dynamic Capabilities of Firms: An Introduction,' *Industrial and Corporate Change*, 3, 537–556.

Tidd, J., J. Bessant and K. Pavitt (1997), *Managing Innovation: Integrating Technological, Market and Organisational Change*. Wiley: Chichester.

von Tunzelmann, N. (1995), *Technology and Industrial Progress*. Elgar: Aldershot.

Tushman, M. and P. Anderson (1986), 'Technological Discontinuities and Organisational Environments,' *Administrative Science Quarterly*, 31, 439–465.

Wang, Q. (1997), 'R&D/Marketing Interface in a Firm's Capability-building Process: Evidence from Pharmaceutical Firms,' *International Journal of Innovation Management*, 1, 23–52.

Woodward, J. (1965), *Industrial Organisation: Theory and Practice*. Oxford: Oxford University Press.

Ziman, J. (1998), *Technological Innovation as an Evolutionary Process*. Cambridge: Cambridge University Press (in press).

[8]

LARGE FIRMS IN THE PRODUCTION
OF THE WORLD'S TECHNOLOGY: AN IMPORTANT CASE
OF "NON-GLOBALISATION"

Pari Patel* and Keith Pavitt**
University of Sussex

Abstract. US patenting by 686 of the world's largest manufacturing firms shows that their share of the world's production of technology is less than their share of R&D activities, and varies greatly amongst sectors. In most cases, the technological activities of these large firms are concentrated in their home country, the characteristics of which influence the volume and trends in their technological activities much more strongly than the international component of these activities. At the same time, these large firms are major elements in the volume and the pattern of sectoral specialisations in their home countries' technological activities.

In this paper, we shall use recently developed data, based on US patenting, to evaluate the importance of the technological activities of the world's largest firms in different sectors and countries. There are at least two interrelated reasons for doing this.

The first is that technological change is a central feature in economic development, structural change and improvements in efficiency in all countries.

---

*Pari Patel is an economist with previous experience in macroeconomic modelling at City University Business School, London. His earlier work at the Science Policy Research Unit (SPRU) at the University of Sussex was on technology and employment. His current work is on science, technology and energy policy in British economic development.

**Keith Pavitt read engineering and industrial management at Cambridge, and economics and public policy at Harvard. He was a staff member in the Directorate for Scientific Affairs at the OECD (Organisation for Economic Co-operation and Development), and a Visiting Lecturer at Princeton, before becoming a Senior Fellow at SPRU, and Professor in the University of Sussex. He has published numerous papers and books on the implications for management and policy of technological innovation. He is now Deputy Director of SPRU.

This paper is based on research funded by the Economic and Social Research Council in the Centre for Science, Technology and Energy Policy, at the Science Policy Research Unit. We are grateful to Sandra Wilson and her fellow students for research assistance, to Richard Dickins for computing assistance, and to John Cantwell, Mike Hobday, Margaret Sharp, Nick von Tunzelmann and three anonymous referees for comments on an earlier version.

Received: June 1989; Revised: March & June 1990; Accepted: June 1990.

1

Recent studies have shown that technological activities—as measured through R&D and international patenting—are statistically significant determinants of differences in export and productivity performance amongst the major OECD countries [Soete 1981; Fagerberg 1987, 1988]. At the same time, major international differences have emerged over the past twenty-five years in trends—and subsequent levels—of these activities, both in these countries and in the large firms based on them [Pavitt and Patel 1988; Franko 1989]. Briefly stated, the Japanese have had the strongest upward trend, the Anglo-Saxons (UK and USA) and the Dutch the weakest, and the other continental Western Europeans have grown at rates between the two. By the mid-1980s, the countries spending the highest proportion of national resources on business-funded technological activities were Sweden, Switzerland, FR Germany and Japan, followed at some distance by the USA, and then by Belgium, Canada, France, the Netherlands and the UK.

The second is that the debate continues about the degree to which these technological activities are localised in large firms. The heavy concentration of R&D activities in large firms [Freeman 1982] has led some analysts to conclude that they have a dominant position in countries' development of new technology. On the other hand, a number of studies using other measures show that R&D activities considerably underestimate the volume of technological activities in firms that are too small to have functionally specialised R&D departments [Acs and Audresch 1989; Kleinecht 1987; Pavitt et al. 1987].

A parallel debate is taking place about the extent and the implications of the international concentration of large firms' technological activities, most often when technology has been made a central explanatory variable in the internationalisation of business.[1] In Vernon's early formulation [1966] and in subsequent analyses by Dunning [1980] and Cantwell [1989], home markets are important determinants of large firms' technological advantage, through the nature and extent of inducement mechanisms that stimulate technical change, and of positive externalities that influence the effectiveness of firms' response to these stimuli.

However, in later formulations, Vernon [1979] and Dunning [1989] suggest that large firms are increasingly footloose in their R&D activities, thereby weakening the links between the development of their technology and their home country. Such trends towards "techno-globalism" are an essential component of currently fashionable predictions of the emergence of "The Stateless Corporation" [*Business Week* 1990]. In the context of these debates, we shall try to answer three questions.

First, how important are large firms in the production of the world's technology? Using US patenting statistics, we show in the third section that the aggregate level of importance is less than that shown by R&D expenditures, with considerable variations amongst technological sectors.

Second, how important are the technological activities of large firms in those of their home countries and elsewhere? We show in the fourth section the considerable variation both in the relative importance of large firms in their home countries' technological activities, and in the degree of internationalisation of their activities. But in most of the countries at the world's technological frontier, the foreign technological activities of large firms are still not the major feature.

Third, how do the volume, trends and sectoral patterns of large firms' technological activities relate to those of their home countries? In the fifth section we show that the two are closely correlated, and that country-specific factors dominate over firm-specific factors.

We begin with a description of the nature, strengths and weaknesses of the database.

THE DATA SET, ITS ADVANTAGES AND LIMITATIONS

The Nature of the Data Set

The data set has been compiled from information, provided by the US Patent Office, on the name of the company, the technical sector, and the country of origin, of each patent granted in the USA from 1969 to 1986. One difficulty with this source is that many patents are granted under the names of subsidiaries and divisions that are different from those of their parent companies, and therefore are listed separately.

Consolidating patenting under the names of parent companies can only be done manually, on the basis of publications like "Who Owns Whom." We have now extended our earlier consolidations for the UK and FR Germany [Patel and Pavitt 1989] to cover 686 of the world's largest firms. With the help of the Economics Department at the University of Reading, we have also included in our data set the following information on each firm: country of origin and sales, employment and R&D expenditures for years 1972, 1977, 1982 and 1984. Not all these three variables are available for all the firms for each of the years.

Table 1 lists the top twenty firms patenting in the USA in the period 1981-86, according to our own consolidated classification, and to the original classification by the US Patent Office. It shows that some firms have very similar numbers of patents in both classifications; in particular, US General Electric, Hitachi, IBM, Toshiba, RCA, Canon, Westinghouse, Dow, Nissan and Mobil. However, other firms have considerably more patents in our consolidated classification, and consequently higher rankings: in particular, Bayer, Siemens, Philips, AT&T, Du Pont, Hoechst, Allied, Matsushita and United Technologies. At the bottom of the sample, firms' annual sales in 1984 were about $900 million.

Table 2 shows the numbers of large firms in our database, according to their home country and to their principal sector of activity. Just under half

TABLE 1
Top 20 Patenting Firms in the USA (1981-86):
Patel and Pavitt list versus the US Patent Office List

Company Name	Patel and Pavitt	US Patent Office
General Electric Company (US)	4587	4527
Hitachi	3710	3416
Bayer	3352	2304
IBM	3207	3207
Siemens	3151	2480
Toshiba	3094	2855
Philips Corporation	2968	2464
AT&T	2732	1980
RCA	2716	2716
E.I. Du Pont	2401	1971
Hoechst	2270	1327
Canon	2266	2266
Westinghouse	2145	2090
Ciba-Geigy	1992	1709
Allied Corporation	1989	1085
Dow Chemical Company	1961	1816
Nissan	1960	1887
Mobil Oil	1907	1749
Matsushita	1895	1276
United Technologies	1889	1028

Note: Firms ranked by number of patents in the Patel and Pavitt classification.

the firms are US owned, about one-fifth are Japanese and just under one-third are European. The UK is the largest European contributor, followed by FR Germany and France. In terms of the industrial distribution, firms with their principal activity in mechanical engineering and metal goods account for 21% of the sample, those in chemicals and pharmaceuticals for 16%, and those in electrical, electronic and computing machinery for 12%.

Advantages and Disadvantages

Patent statistics have been used frequently by economists and other analysts as a proxy measure of technological activities.[3] Their general advantages compared to other measures, such as R&D expenditures, are that—with the advent of modern information technology—they are readily available over long time periods; they can be broken down in great statistical detail, according to firm, technical field and geographical location; and they capture technological activities undertaken outside R&D departments, such as design activities in small firms, and production engineering in large firms. Their main general disadvantage is that, like other routine measures of technological activities, they do not measure satisfactorily one of the major fields of technological growth, namely, software.

The advantages and disadvantages specific to our database are along three dimensions: the nature of the technological activities measured, variations in the propensity to patent, and the interpretation of trends over time.

TABLE 2
The Distribution of the 686 Large Firms in the Sample by Principal Activity and Country

	US	JP	CA	UK	GE	FR	SE	CH	NL	IT	BE	NO	FI	OT	Total
Chemicals	35	25	–	2	5	5	–	1	2	2	1	1	–	1 (AU)	80
Pharmaceuticals	18	4	–	3	2	–	–	2	–	–	–	–	–	–	29
Mining (Coal & Oil etc)	29	10	3	5	4	2	–	–	1	1	1	1	1	–	58
Textiles, Cloth. & Leather	12	5	–	2	1	1	–	–	–	–	–	–	–	–	21
Rubber and Plastics	6	3	1	1	1	1	–	–	–	1	–	–	–	–	14
Paper & Wood products	21	6	4	1	1	–	4	–	–	–	–	–	2	1 (IE)	40
Food	33	15	2	14	–	4	1	2	1	–	–	–	–	–	72
Drink and Tobacco	8	1	4	8	–	–	–	–	1	–	–	–	–	1 (AU)	23
Non-metallic Minerals	11	6	1	6	–	2	–	1	–	–	–	–	1	–	28
Metal Manufacture	22	13	6	2	13	4	1	1	1	1	2	1	–	1 (AU)	68
Mechanical Engineering	37	12	2	9	6	1	4	2	2	–	–	–	2	–	77
Electrical/Electronics	31	18	1	4	4	2	3	1	1	1	–	–	–	–	66
Computing Machinery	12	2	–	1	1	1	–	–	–	1	–	–	–	–	18
Instruments	10	6	–	–	1	–	–	–	1	–	–	–	–	–	18
Motor Vehicles	12	19	–	3	6	3	2	–	–	1	–	–	–	1 (ES)	47
Aircraft	14	–	–	2	1	4	–	–	–	–	–	–	–	–	21
Other Transport	3	1	–	1	–	–	–	–	–	–	–	–	1	–	6
Total	314	146	24	64	46	30	15	10	10	8	4	3	7	5	686

Notes:

(1) Country Definitions: US=United States, JP=Japan, CA=Canada, UK=United Kingdom, GE=FR Germany, FR=France, SE=Sweden, CH=Switzerland, NL=Netherlands, IT=Italy, BE=Belgium, NO=Norway, FI=Finland, OT=Others: AU=Austria; IE=Ireland; ES=Spain.

(2) There are two companies where the home country is not easily identifiable: Shell, which we regard as Dutch, and Unilever which we regard as British.

The Nature of the Technological Activities Measured

Since a patent is granted normally in recognition of technical novelty, our data is better able to capture technology creation than technology diffusion-transfer-imitation. For those who assume that technology is information (i.e., costly to create, but virtually costless to transfer and reproduce), this distinction is a rigid one. However, in the real world of technology that is complex, partially tacit and specific,[4] the diffusion-transfer imitation of technology generally requires technological activities by the imitator, which sometimes result in improvements over the original.[5]

Patenting activities do reflect this type of imitation, which is typical of advanced country companies competing close to the world's technological frontier. However, they do not reflect many other types of imitation and related technological activities not involving originality, such as trade in capital goods and know-how, on-the-job training, assimilative R&D and production engineering, and the foreign education of scientists and engineers. These are particularly important forms of imitation for developing countries (see Rosenberg and Frischtak [1985]).

Variations in the Propensity to Patent

Patenting is also an imperfect reflection of novel technological activity. Its primary function is to act as a legal barrier against imitation. Three kinds of variation in the propensity to patent the results of technological activities must therefore be borne in mind.

First, there are variations amongst countries, reflecting differences in the costs (e.g., patenting fees) and benefits (e.g. degree of protection, prospective size of market) of patenting. Patenting in the USA is a reliable metric, since screening procedures are homogeneous and rigorous, and success provides relatively strong protection in a large market. Thus, a recent survey of patenting behaviour of multinational firms shows that the USA is the first foreign country in which they normally seek patent protection [Bertin and Wyatt 1988]. For this reason, the international distribution of the sources of US patenting show statistically highly significant similarities to the international distribution of business enterprise R&D expenditures, both in aggregate and in specific sectors [Soete and Wyatt 1983; Soete 1987; Patel and Pavitt 1987].[6]

Second, there are variations in the propensity to patent amongst technical fields, reflecting differences in the relative importance of patenting as a protection against imitation, compared to other factors, such as secrecy, know-how, first-comer advantages on learning curves.[7] For this reason, it is advisable to normalise numbers of patents as a proportion of their respective technical fields.

Third, there are variations amongst firms in the propensity to patent, reflecting ex ante uncertainties and differing patenting practices over the wide range of patents with relatively low value.[8] Nonetheless, statistically significant correlations have been found in the USA between inter-firm differences in R&D, and in US patenting [Soete 1978; Pakes and Grilliches 1983].

Interpretation of Time Trends

Given lack of time and other resources, our consolidated classification of the 686 firms has been compiled for only for one year—1984. Our time-trend analyses of patenting by companies between 1969 and 1986 therefore reflect the firms as constituted in 1984, and none of the changes resulting from purchases or sales of divisions before or since then. Thus, measured changes over time are composed of those of the parts of the firm retained up to 1984, together with those of acquisitions made up to 1984: in other words, what the firm kept and what it bought, up to 1984.

**LARGE FIRMS IN THE PRODUCTION
OF THE WORLD'S TECHNOLOGY**

Table 3 shows, for thirty-three technical fields and in aggregate, the shares of US patents granted in 1981-86 to the large firms in our sample, to other firms, to government agencies, and to individuals.[9]

Aggregate

In aggregate, our set of large firms account for just under half the world's technological activities, as measured by US patenting, and for about 60%

TABLE 3
Sources of US Patenting in 33 Technical Sectors:
Percentage Shares in 1981-86

	Large Firms	Govt. Agen.	Priv. Ind.	Other Firms
Semiconductors	80.28(138)	3.94	2.69	13.08
Hydrocarbons, mineral oils, etc.	79.45(158)	0.82	5.77	13.96
Agricultural Chemicals	78.98(92)	0.96	4.29	15.76
Organic Chemicals	77.04(348)	1.73	2.71	18.52
Photography and photocopy	73.40(147)	0.39	5.84	20.36
Calculators, computers, etc.	69.23(281)	1.61	7.14	22.03
Inorganic Chemicals	67.37(218)	2.81	5.57	24.24
Bleaching Dyeing and Disinfecting	65.20(125)	1.94	7.75	25.11
Road vehicles and engines	62.45(179)	0.34	20.49	16.72
Electrical devices and systems	59.62(327)	3.26	11.38	25.74
Drugs and Bio-affecting agents	59.48(215)	3.35	8.08	29.09
Power Plants	58.17(153)	2.48	20.79	18.56
Telecommunications	57.41(289)	6.54	13.69	22.36
Image and sound equipment	57.42(207)	1.80	17.61	23.17
Chemical Processes	56.36(503)	2.36	10.91	30.36
Plastic and rubber products	55.58(327)	1.56	14.01	28.84
Metallurgical and other mineral proc.	53.30(372)	1.75	13.94	31.02
Gen. Electrical Industrial Apparatus	50.30(407)	2.17	15.73	31.80
Food & Tobacco (proc. and products)	48.96(175)	1.61	15.50	33.92
Non-metallic minerals, glass, etc.	48.50(431)	1.24	20.22	30.04
Mining and wells mach. and processes	47.68(178)	0.89	22.47	28.95
Nuclear Reactors and systems	47.45(38)	6.83	7.60	38.11
Aircraft	43.05(62)	14.44	23.47	19.04
Instruments and controls	40.93(491)	3.55	22.06	33.46
Gen. Non-electrical Industrial Equip.	39.86(433)	0.97	25.33	33.84
Appar. for chemicals, food, glass, etc.	39.76(516)	0.97	21.42	37.85
Metallurgical and metal working equip.	34.99(379)	0.68	27.18	37.16
Assembling & material handling appar.	29.97(377)	0.87	28.85	40.30
Other transport equip. (exc. aircraft)	28.46(197)	1.39	42.01	28.14
Non-electrical specialized machinery	27.63(481)	0.76	30.39	41.22
Miscellaneous metal products	23.35(444)	0.67	40.28	35.70
Other n.e.c.	13.49(241)	5.25	65.71	15.55
Textile, clothing, leather, wood prod.	13.08(117)	0.71	52.06	34.15
All Sectors	49.10(660)	2.11	19.68	29.10

Notes:
(1) Table is sorted by the share of large firms.
(2) Each row adds up to 100.
(3) The number of large firms active in technical sector is in parenthesis.

of that undertaken by firms. This distribution confirms what we found in an earlier study of the UK and FR Germany [Patel and Pavitt 1989], namely, a lower concentration of technological activities amongst large firms when measured by US patenting than by R&D expenditures. Although strict comparisons at the world level are not possible, national surveys in OECD countries show that typically about 80% of firms' R&D activities are concentrated in firms with 10,000 or more employees. Given that the cut-off level of employment at the lower end of our sample is about 8,000 employees, the proportion of total patenting accounted for by our large firms would have to be more than 80% to reach the same level of concentration as R&D expenditures.

Differences amongst Sectors

Table 3 also shows major differences amongst sectors in the relative importance of large firms and of the other sources of the world's technological activities. Government agencies are relatively unimportant in aggregate but account for more than 5% in nuclear reactors, aircraft and telecommunications—all technologies heavily influenced by military programmes. As in our earlier analyses, large firms are relatively important in chemicals (eight sectors with shares between 56% and 79%), motor vehicles (62%), and electrical and electronic products (five sectors between 57% and 80%), but unimportant in capital goods (seven sectors between 23% and 40%).

Table 4 confirms a significant positive correlation across sectors between our large firms' patenting shares, and the shares of the top twenty technically active firms ranked according to sales; and it confirms a significant negative relationship with shares of patenting of "Private Individuals." It also shows that the sectoral shares of "Other Firms," encompassing the very small up to 8,000 employees, are more similar to those of private individuals than to those of our large firms.

A Possible Explanation of Intersectoral Differences

Recent analysis has shown that intersectoral differences in the concentration of technological activities can be best understood in the context of dynamic interactions between technological opportunities and their appropriability, on the one hand, and the competitive growth of innovative firms, on the other. Briefly stated, higher technological opportunity and appropriability will result in higher concentration [Dasgupta and Stiglitz 1980; Nelson and Winter 1982; Levin et al. 1985]. Both R&D-intensive sectors (particularly chemical and electronic products) and capital goods sectors have abundant technological opportunities. One of us has shown elsewhere that the low appropriability and concentration in capital goods is positively related to a greater spread of technological activities in capital goods amongst UK firms with different principal sectors of activity [Pavitt et al. 1987]. See, also, Malerba and Orsenigo [1988].

Our data tend to confirm this pattern. Table 3 shows relatively low concentration of capital goods technology activities in our large firms, together with a relatively high proportion of these firms producing some capital goods technology, albeit at a relatively low level. This is reflected in the significant and positive correlations shown in Table 4 between sectoral levels of concentration of technological activity, on the one hand, and the

TABLE 4
Correlation Matrix of Various Measures of Concentration
of Technological Activities: 33 sectors, 1981-86

	Lfirms	Govt.	PInd	OthF	CRSale20
Govt.	−0.040				
PInd	−0.909*	−0.008			
OthF	−0.625*	−0.230	0.273		
CRSale20	0.661*	0.266	−0.564*	−0.576*	
HIPPG	0.606*	0.417	−0.524*	−0.573*	0.806*

Notes:
For each sector:
LFirms   = The share of large firms.
Govt   = The share of government agencies.
PInd   = The share of private individuals firms.
OthF   = The share of firms other than the large firms in our sample.
CRSale20 = The share of top 20 technologically active firms sorted according to sales.
HIPPG   = Hirfindahl Index calculated as the sum of squared shares of the firms active
                 in each technical sector aggregated according to their Principal Activity.

*Correlation Coefficient significantly different from zero at the 5% level.

Herfindahl index of concentration, aggregated according to the sectors of our large firms' principal activity, on the other.

This is because capital goods technology remains largely mechanical. Important mechanical inventions and innovations can still be made without the specialised equipment and range of formal skills required in chemical and electronic technologies [Freeman 1982]. The spatial and design skills of individuals and small groups remain important sources of technology, as do users with experience in operating capital goods. These competences are spread widely across industries and firms, which provide multiple possibilities of entry into promising areas of capital goods technology, thereby reducing the possibilities of appropriation by first-comers. We shall be considering this explanation in greater econometric depth in a future paper.

LARGE FIRMS IN HOME COUNTRIES' TECHNOLOGICAL ACTIVITIES

The previous section shows considerable variations amongst sectors in the technological activities of the world's largest firms. In this section, we also show variation in their contribution to the world's leading technology-producing countries.

This emerges from Table 5, which uses our data on patenting in the USA in the first half of the 1980s to compare the composition of the technological activities of the eleven countries that account for more than 95% of total OECD R&D expenditures funded by business enterprises, and of total US patenting. The first two columns show the shares of total national patenting in the USA granted to the nationally controlled large firms, and to the foreign-controlled large firms, in our database, whilst the third

TABLE 5
Large Firms in National Technological Activities, 1981-86

Country	National Sources of Patenting in US (3 columns add up to 100%)			Patenting in US by Nationally Controlled Firms from Outside Home Country (% of National Total)
	Large Firms			
	Nationally Controlled	Foreign Controlled	Other	
Belgium	8.8	39.7	51.5	14.7
France	36.8	10.0	53.2	3.4
FR Germany	44.8	10.5	44.2	6.9
Italy	24.1	11.6	64.3	2.2
Netherlands	51.9	8.7	39.4	82.0
Sweden	27.5	3.9	68.6	11.3
Switzerland	40.1	6.0	53.9	28.0
UK	32.0	19.1	49.0	16.7
W. Europe	44.1	6.2	49.7	8.1
Canada	11.0	16.9	72.1	8.0
Japan	62.5	1.2	36.3	0.6
USA	42.8	3.1	54.1	3.2

Note: All columns as percentage of total national patenting in US, 1981-86.

column gives the combined share for the other national sources (i.e., government agencies, other firms and individuals). Thus, assuming that US patenting reflects national technological activities, Table 5 shows that 8.8% of technological activity in Belgium came from Belgian large firms, 39.7% from non-Belgian large firms, and the remaining 51.5% from other sources in Belgium (firms, government agencies, individuals).

The fourth column shows US patenting by nationally controlled firms from outside their home country, expressed—like the other three columns—as a percentage of total national patenting in the USA. Thus, again by way of illustration, the technological activities of Belgian-controlled large firms undertaken outside Belgium amount to 14.7% of total technological activities inside Belgium, whilst the equivalent proportion is a massive 82% for Dutch-controlled large firms, and a miniscule 0.6% for Japanese-controlled large firms.

By adding up the first two columns, we can see that the relative importance of our large firms varied from around 30% of national technological activities in Canada and Sweden to just over 60% in the Netherlands and Japan, with the remaining seven countries (and Western Europe taken as a whole) in the range from 36% to 54%. By comparing the first and second columns, we see that the relative importance of nationally controlled and foreign-controlled large firms varied more widely amongst countries: national firms from more than 60% in Japan to less than 10% in Belgium, mirrored by foreign firms from nearly 40% in Belgium to just over 1% in Japan. Simple correlation tests show that neither the relative importance of large firms in total technological activities, nor the mix between national and foreign ones, are significantly related to country size as measured by Gross Domestic Product.

The fourth column of Table 5 also shows even greater variation amongst countries in the relative importance of the technological activities of our large firms outside their home countries, from more than 80% of the national total for the Netherlands, to less than 1% for Japan. On the basis of data for 140 large firms, Cantwell and Hodson [1990] have shown that the degree of internationalisation of large firms' technological activities is closely correlated to that of production. The same is true for countries in Table 5.

However, a comparison of the first and fourth columns of Table 5 shows that, in spite of considerable variations amongst the large firms based in different countries, their technological activities remained far from globalised. Only Belgium and Dutch large firms executed more of their technological activities outside their home country than inside. British, Canadian, Swedish and Swiss firms executed between 30% and 42% abroad, whilst firms from the three largest technological countries—FR Germany, Japan and the USA—performed less than 15% outside, as did France and Italy.

Finally, if we compare the second and fourth columns of Table 5, we can conclude that, for most countries, the international technological activities of our large firms are not a dominant feature of national technological systems. Only for the Netherlands and Switzerland did the foreign-executed technological activities controlled by large national firms amount to more than 20% of the national total; and only in Belgium and Canada were foreign large firms relatively more important than national ones. In eight out of the eleven countries in Table 5 (and for Western Europe as a whole), less than 20% of national technological activities were foreign-controlled, and at the same time the technological activities executed abroad by nationally controlled firms performed amounted to less than 20% of national technological activities. This is a long way from any "globalisation" of the world's technological activities.

LARGE FIRM PERFORMANCE AND COUNTRY PERFORMANCE

The third and fourth sections showed that the large firms in our sample produce about half the world's frontier technology, but that their relative importance varies considerably amongst sectors and amongst the eleven frontier countries, as does the degree of internationalisation of their technological activities. In this section, we shall probe more systematically into the links between the technological performance (measured in terms of levels and rates of growth of technological activities) of our large firms, and of these eleven countries. There remain a number of unresolved analytical and policy questions about the effects on a country's technology of the presence of large firms, and about the nature and direction of the interactions between such firms and their home countries. We shall use our data to test a number of simple relationships that have not been tested

12   JOURNAL OF INTERNATIONAL BUSINESS STUDIES, FIRST QUARTER 1991

before. Given the complexities of the real world, we still shall not be able to give complete and conclusive answers.

Structure, Internationalisation and Country Performance

There is a continuing debate about the effects of the structure of industry—and of related technological activities—on a country's technological performance. Some argue that heavy concentration and the prevalence of large firms reduces competition and technological pluralism, and thereby results in a lower level of aggregate technological activities. Others argue the contrary, that large firms can more easily mobilise the required range of skills, reach critical thresholds, and deal with risk, thereby resulting in a higher level of technological activities.

Contradictory arguments are also put forward about the effects on countries' technological activities of multinational firms. For some, a high proportion from foreign-controlled multinationals is likely to augment national activities; for others, it is either the consequence or the cause of deficiencies in nationally controlled activities. Similarly, a high proportion of technological activities undertaken by large firms outside their home countries is for some a sign of strength, and for others a sign of weakness.

Table 6 shows that differences in countries' technological performance, measured in terms of business-funded R&D as a percent of GDP in 1983 (RDGDP), are positively and significantly correlated with differences in performance measured in terms of US patenting per capita (PATPC). Neither performance measure is significantly correlated with shares of national large firms in national technological activities (NLFHSH), nor with shares of foreign large firms (FLFSH). Similarly, neither performance measure is correlated with the extent to which national large firms have internationalised their technological activities (NLFASH). However, improvements in national technological performance, measured as real growth of business-funded R&D between 1967 and 1985, is positively correlated with increasing shares of national large firms, at almost the 5% level of significance.

The considerable amount of remaining variance may be explained by another factor, namely, the influence of sectors of national technological specialisation on the shares of national large firms in national technological activities. This cannot be tested statistically, given insufficient degrees of freedom. However, Table 5 shows that, although they are very different in aggregate technological performance, Canada, Italy and Sweden have in common both a revealed technological advantage[10] in capital goods [Pavitt and Patel 1988], and relatively unconcentrated technological activities as in capital goods (see Table 3). Similarly, Japan and the Netherlands have very different patterns of aggregate performance, but similar relative technological advantages in concentrated sectors: electronics and automobiles (Japan only).

TABLE 6
Structural Correlates of National Technological Performance, 1981-86: 11 OECD Countries

	RDGDP	PATPC	NLFHSH	FLFSH	ONFSH	NLFASH
PATPC	0.765*					
NLFHSH	0.578	0.431				
FLFSH	-0.435	-0.553	-0.724*			
ONFSH	-0.421	-0.097	-0.756*	0.096		
NLFASH	-0.061	-0.036	0.250	0.001	-0.362	
NLFTSH	0.254	0.197	0.701*	-0.373	-0.658	0.865*

Notes:
For each country:
RDGDP = Business-financed Industrial R&D as a percentage of GDP in 1983.
PATPC = Per capita US Patenting, 1981-86.
NLFHSH = Share of National Patenting in USA by National Large Firms: 1981-86.
FLFSH = Share of National Patenting in USA by Foreign Large Firms active in the country: 1981-86.
ONFSH = Share of National Patenting in USA by Other National Firms active in the country: 1981-86.
NLFASH = National Large Firms US patenting from abroad as a percentage of the National Total: 1981-86.
NLFTSH = Total National Large Firms US patenting (home and abroad) as a percentage of the National Total: 1981-86.

*Correlation Coefficient significantly different from zero at the 5% level.

	GRD	GNLFHSH	GFLFSH	GONFSH	GNLFASH
GNLFHSH	0.670				
GFLFSH	-0.393	-0.356			
GONFSH	-0.445	-0.808*	-0.263		
GNLFASH	0.437	0.347	-0.337	-0.146	
NLFTSH	0.636	0.724*	-0.414	-0.487	0.898*

Notes:
For each country:
GRD = Growth of Industry-financed Industrial R&D, defined as the proportionate change between 1967 and 1985.
GNLFHSH = NLFHSH in 1981-86 minus NLFHSH in 1969-74.
GFLFSH = FLFSH in 1981-86 minus FLFSH in 1969-74.
GONFSH = ONFSH in 1981-86 minus ONFSH in 1969-74.
GNLFASH = NLFASH in 1981-86 minus NLFASH in 1969-74.
GNLFTSH = NLFTSH in 1981-86 minus NLFTSH in 1969-74.

*Correlation Coefficient significantly different from zero at the 5% level.

Nationally Controlled Firms and Country Performance

We have shown in an earlier paper the strong correlation between the shares of US patents granted to countries, and the shares granted to their national large firms [Patel and Pavitt 1990]. But this begs the question of causality: Do country characteristics determine the behaviour of their national large firms, or vice-versa?

We have argued elsewhere that firm behaviour may be strongly influenced by country-wide factors: the degree to which the national financial system properly evaluates intangible, firm-specific assets accumulated through technological activities; the national system of basic research, and education and training of management and the work force, that influence the quality of major decisions about technology, and of implementation and learning; and the economic climate—and particularly expectations about growth and profits—that influences firms' propensities to invest in technological activities [Pavitt and Patel 1988]. On the other hand, it can be argued that large firms are not closely coupled to countries: they think and act in terms of world markets, world sources of finance, and world sources of management and worker skills: in typical situations of uncertainty and oligopoly, their discretionary decisions can have major impacts on the rate and direction of countries' technological activities.

Our data can throw some modest empirical light on this debate. We shall typify the competing hypotheses as "country-dominated" and "firm-dominated." In both cases, we would observe a high correlation between country performance and national large firm performance. However, in a country-dominated system, we would also expect a positive and significant correlation between the performance of the two main component parts of national technological activities, namely, the home-based activities of national large firms, and the activities of other national firms. In a firm-dominated system, we would expect instead a high correlation between the performance of the home-based activities of national large firms, and of their foreign activities.

Table 7 shows that aggregate national technological performance is country-dominated rather than firm-dominated. Country performance, measured as business-funded R&D as a percent of GDP in 1983 (RDGDP), or as per capita US patenting in 1981-86 (PATPC), is strongly correlated with the performance of national large firms, measured as per capita US patenting (NLFT), but it is even more strongly correlated with the domestic performance of these large firms (NLFH). It is also significantly correlated with the performance of other national firms (ONF), but not with the foreign performance of national large firms (NLFA). In addition, there is no significant correlation between national large firms' domestic performance, and their foreign performance. Table 7 also shows the same relations hold even more decisively in performance in growth of technological activities.

Sectoral Performance

Whilst differences in countries' aggregate technological performance are closely correlated with differences in the domestically based performance of large firms, the same may not necessarily hold in specific sectors, especially those where technological activities are concentrated in large, multinational firms. We therefore ran correlations, similar to those in Table 7, for each of the thirty-three sectors shown in Table 3, with the performance measures being levels and rates of change of per capita US patenting.

What emerges is a pattern remarkably similar to the one in Table 7. Only in five chemical and chemical-related sectors and in power plant are there

TABLE 7
Correlations between the Technological Performance
of Countries and Nationally Controlled Firms

	RDGDP	NLFH	FLF	ONF	NLFA	NLFT
NLFH	0.811*					
FLF	0.159	0.114				
ONF	0.665	0.825*	0.292			
NLFA	0.313	0.432	0.348	0.459		
NLFT	0.720*	0.909*	0.241	0.797*	0.769*	
PATPC	0.765*	0.941*	0.280	0.965*	0.482	0.890*

Notes:
For each country:
RDGDP = Industry-financed Industrial R&D as a percentage of GDP in 1983.
NLFH = per capita Home-based US patenting of National Large Firms: 1981-86.
FLF = per capita US patenting of Foreign Large Firms active in the country: 1981-86.
ONF = per capita Home-based US patenting of Other National Firms active in the country: 1981-86.
NLFA = per capita US patenting of National Large Firms from abroad: 1981-86.
NLFT = per capita US patenting of National Large Firms (home and abroad): 1981-86.
PATPC = per capita aggregate US patenting for the country: 1981-86.
*Correlation Coefficient significantly different from zero (5% level).

	GRD	GNLFH	GFLF	GONF	GNLFA	GNLFT
GNLFH	0.678					
GFLF	0.121	0.439				
GONF	0.468	0.927*	0.521			
GNLFA	0.595	0.175	0.064	0.103		
GNLFT	0.789*	0.957*	0.416	0.870*	0.453	
GPATPC	0.578	0.980*	0.526	0.982*	0.142	0.929*

Notes:
For each country:
GRD = Growth of Industry-financed Industrial R&D, defined as the proportionate change between 1967 and 1985.
GNLFH = NLFH in 1981-86 minus NLFH in 1969-74.
GFLF = FLF in 1981-86 minus FLF in 1969-74.
GONF = ONF in 1981-86 minus ONF in 1969-74.
GNLFA = NLFA in 1981-86 minus NLFA in 1969-74.
GNLFT = NLFT in 1981-86 minus NLFT in 1969-74.
GPATPC = PATPC in 1981-86 minus PATPC in 1969-74.
*Correlation Coefficient significantly different from zero (5% level).

strong correlations between country performance and the domestic performance of nationally controlled large firms, on the one hand, and their foreign performance, on the other: agricultural chemicals, pharmaceuticals, organic chemicals, dyestuffs and food. Even in the last three of these sectors and in power plant, performance is also correlated with that of other domestic firms.

In twenty-seven out of the thirty-three sectors, country differences are significantly correlated with differences in firms' domestic activities, but not their foreign activities. These sectors comprise all capital goods, materials, transport, and electrical and electronics. In two sectors—hydrocarbons and motor vehicles—national performance is significantly correlated only with the domestic activities of national large firms; and in one sector—textiles—it is significantly correlated only with other domestic firms. In the other twenty-four sectors, national performance is significantly correlated with both. In none of the sectors is national performance significantly correlated with that of foreign large firms.

Finally, the domestic performance of large firms is significantly correlated with that of other domestic firms, in about half the sectors, comprising relatively unconcentrated capital goods sectors, materials, and the concentrated sectors of electronics. Linkages between the performance of these two major elements of national technological activities could be of two types. Horizontally, rivalrous behaviour may lead to imitative increases or decreases in technological activities in certain product fields. Vertically, vigorous technological activities in large users of capital goods may induce a complementary response amongst suppliers.[11]

Firm Specialisation and Country Specialisation

This leads to the last element in our analysis, namely, the interactions between the sectors of technological specialisation of national large—and other categories of—firms, and those of our 11 countries. These are shown in Table 8 which correlates, for each of the eleven countries in Table 5, their revealed technology advantage (RTA) in 1981-86 in each of the thirty-three technological sectors in Table 3, with the RTAs of the various categories of firm in Table 7.[12] The following conclusions emerge.

First, Table 8 shows that the sectoral patterns of technological advantage of large firms and their home countries are significantly similar. The sectoral RTAs of all eleven countries are significantly correlated with those of nationally based large firms (NLFT). This is the dominant relationship between firm and country specialisations in FR Germany, Netherlands, Switzerland, Sweden and the UK; whilst the RTAs of other national firms (ONF) are more highly correlated with countries' specialisations in Canada, France, and Italy.

Second, the sectoral specialisations of foreign large firms (FLF) are strongly correlated with those of their host countries in Belgium and Canada where, as we saw in Table 5, they account for a larger share of national technological activities than national large firms. Otherwise, there are significant correlations between the two in Japan.

Third, the links between the domestic technological specialisations of national large firms (NLFH), and those of other national firms (ONF), are weak: they are significantly correlated only in Canada and Japan.

Finally, the sectoral specialisations of national large firms in foreign countries (NLFA) often reflect those of parent firms (NLFH), with the strong exceptions of France and the USA. In the latter, national large firms are relatively strong in their foreign technological activities in pharmaceuticals,

TABLE 8
Sectoral Specialisations in Technological Activity: Correlations of RTA Indices for 11 Countries across 33 Sectors, 1981-86

Country	Country and NLT	Country and ONF	Country and FLF	NLFH and ONF	NLFA and NLFH	NLFA and Country
United States	0.88*	0.68*	0.32	0.35	-0.11	0.18
Japan	0.89*	0.68*	0.68*	0.54*	0.85*	0.71*
Canada	0.58*	0.94*	0.54*	0.48*	0.33	0.35
Belgium	0.49*	0.48*	0.72*	0.16	0.59*	0.30
FR Germany	0.90*	0.25	0.26	-0.05	0.57*	0.48*
France	0.44*	0.83*	0.06	-0.05	-0.19	-0.30
Italy	0.62*	0.90*	0.29	0.33	0.27	0.15
Netherlands	0.68*	0.29	0.35	-0.16	0.61*	0.51*
Switzerland	0.93*	0.22	-0.08	0.05	0.55*	0.48*
Sweden	0.78*	0.70*	0.02	0.18	0.45*	0.46*
United Kingdom	0.57*	0.27	0.04	0.09	0.44*	0.17

Notes:
See Note 12 for a definition of the RTA Index.

For each country, the RTA Indices are for:
Country = All US Patenting.
NLFH   = Home-based US Patenting of National Large Firms.
FLF   = US Patenting of Foreign Large Firms active in the country.
ONF   = Home-based US Patenting of Other National Firms active in the country.
NLFA   = US Patenting of National Large Firms from abroad.
NLFAT = US Patenting of National Large Firms (home and abroad).

*Correlation Coefficient significantly different from zero (5% level).

machinery, automobiles, and photography and photocopy, all sectors of relative domestic weakness.

CONCLUSIONS

The main conclusion of this paper is that—despite being a critical resource in the global competition and performance of both companies and countries—the production of technology remains far from globalised. Its heavy concentration in the industrialised—as compared to the developing countries—has been recognised for a long time. What we have shown is that, even in the major countries at the world's technological "core," the production of technology remains highly "domesticised" in two senses. First, in most of the countries at the world's technological frontier, the foreign technological activities of large firms are still not the major feature. Second, large firms' technological performance is strongly dependent on the performance of the home country, and not independent of it. These conclusions are very similar to the reported results of Porter's recent research on the sources of competitiveness [1989]. What happens in home countries still matters greatly in the creation of global technological advantage.

Nonetheless, large firms influence countries in other ways. Large firms are particularly important for the production of technology in R&D-intensive

sectors and automobiles. In all our eleven countries, large national firms have a significant influence on sectoral specialisations, whilst other national firms are significant in seven countries, and foreign large firms in three.

Our evidence is in general consistent with the earlier analyses of Dunning [1980], Cantwell [1989] and our own [Pavitt and Patel 1988; Pavitt 1988b]: country-specific factors create both the general conditions that determine the volume of technological activities, and the specific inducement mechanisms that determine their direction. These lead to accumulated firm-specific advantages that are reflected in international patterns of trade, production and related technological activities. It therefore becomes important to understand the nature of the country-specific factors that make up what Andersen and Lundvall [1988] have called "national systems of innovation," including the system of education, training and basic research that forms the infrastructure for firm-specific technological accumulation.

We also need a better understanding of the reasons why large firms keep most of their technological activities at home. Certain key features related to major innovations may help explain the advantages of geographical concentration: the primacy of multidisciplinary and tacit knowledge inputs, and the commercial uncertainties surrounding outputs. Physical proximity facilitates the integration of multidisciplinary knowledge that is tacit and therefore "person-embodied" rather than "information-embodied." It also facilitates the rapid decisionmaking needed to cope with uncertainty. For this reason, it may well be more efficient to have technological activities nationally concentrated, with international "listening posts" and adaptive capabilities maintained through small foreign laboratories, frequent international exchanges often involving what are called "strategic alliances," and proximity to an internationally outward looking system of higher education [Casson 1990; Dosi 1988; Grandstrand and Sjolander 1990; Mowery 1988; Pavitt 1989; Pearce 1990].

In spite of this, we expect to see greater internationalisation of large firms' technological activities in the future, not because it is inherently more efficient, but because it is politically necessary. Uneven technological and competitive developments amongst firms and countries create imbalances, tensions and threats of restrictions on entry into foreign markets. Measures to deal with these threats often involve foreign production and related R&D support, and sometimes independently targeted R&D activities.

In this context, the policies of Swedish large firms are revealing. They perform about 30% of their technological activities outside Sweden (see Table 5), and Hakanson and Nobel [1989] have found that political factors (particularly those related to establishment within European Community countries) have been important in more than 60% of the decisions taken since 1980 to establish R&D activities abroad.

Similar and even more powerful pressures are now on firms in another technologically high performing country, namely Japan, to expand foreign investment and related R&D activities in Europe and North America [Grandstrand et al. 1989]. However, we can see from Table 5 that they have a long way to go before their technological activities become anywhere near "globalised." We can also see that Dutch firms have travelled furthest down this route. It would be intriguing to know whether they intend to continue, or—on simple grounds of managerial efficiency— would prefer in an ideal world to turn back.

NOTES

1. See, in particular, Vernon [1966, 1979], Buckley and Casson [1976], Cantwell [1989].

2. On the reasons why firms internationalise their technological activities, see Hakanson and Nobel [1989], Cantwell and Hodson [1990], and R. Pearce [1990] for recent contributions.

3. For a more detailed discussion of the uses and abuses of patenting statistics as a measure of technological activities, see Pavitt [1988].

4. For a fuller discussion, see Dosi [1988a].

5. For an analysis of the conditions under which this is likely to occur, see Teece [1986].

6. US patenting slightly overestimates technological activities performed in the USA, compared to those performed in other countries, since firms have a higher propensity to patent on home than on foreign markets. It also severely underestimates the considerable volume of R&D undertaken in the USSR and other (former?) centrally planned economies, the efficiency of which is very low in innovation and diffusion, when compared to market economies (see Hanson and Pavitt [1987]).

7. For systematic evidence on intersectoral variations in the relative importance of these barriers, see Levin et al. [1987]; Bertin and Wyatt [1988].

8. On the varying patent practices of firms, see Bertin and Wyatt [1988]. On the skew distribution of the value of patents, see Pakes and Shankerman [1983].

9. Government agencies are granted patents principally in government-funded R&D programmes in defence, aerospace, energy and basic science. Recent studies in Canada and Italy show that, within the category "Individuals" are a significant proportion of commercially active small firms [Amesse et al. 1990; Malerba and Orsenigo 1990].

10. For an analytical discussion, see Dosi [1988b]. Such effects can certainly be observed in the UK: multiple reductions of activity in computers and semiconductors; multiple increases in pharmaceuticals and agricultural chemicals; the positive effects of the National Coal Board in coal-mining machinery [Patel and Pavitt 1987b].

11. Revealed technology advantage (RTA) is defined as the share of a country (or firm, or category of firms) in US patenting in a sector, divided by the share of that country (or firm, or category of firms) in US patenting in all sectors. Some readers will note the similarity to the measure of "revealed comparative advantage" used in analyses of international trade.

12. As an index of specialisation, revealed technology advantage (RTA) corrects for differences amongst countries, categories of firms, and sectors, in the total volume of patenting activity.

REFERENCES

Acs, Zoltan & David Audretsch. 1989. *Small firms and technology*. Directorate of Technology Policy, Ministry of Economic Affairs, the Hague.

Amesse, Fernand, Claude Desranleau, Hamid Etemad, Yves Fortier & Louise Seguin-Delude. 1991. The individual inventor in Canada and the role of entrepreneurship. *Research Policy*, forthcoming.

Andersen, Esben & Bengt-Ake Lundvall. 1988. Small national systems of innovation facing technological revolutions. In Christopher Freeman & Bengt-Ake Lundvall, editors, *Small nations facing technological revolutions*. London: Pinter.

Bertin, Gilles & Sally Wyatt. 1988. *Multinationals and industrial property: The control of the world's technology*. England: Wheatsheaf.

Buckley, Christopher & Mark Casson. 1976. *The future of the multinational enterprise.* London: Macmillan.

Business Week. 1990. The stateless corporation, May 14: 52-60.

Cantwell, John. 1989. *Technological innovation and multinational corporations.* Oxford: Blackwell.

_____ & Christian Hodson. 1990. *The internationalisation of technological activity and British competitiveness: A review of some new evidence.* Mimeo, Economics Department, Reading University.

Casson, Mark. 1990. *Global corporate R&D strategy: A systems view.* Mimeo, Economics Department, Reading University.

Dasgupta, Partha & Joseph Stiglitz. 1980. Industrial structure and the nature of innovative activity. *Economic Journal,* 90: 266-93.

Dosi, Giovanni. 1988a. Sources, procedures and microeconomic effects of innovation. *Journal of Economic Literature,* 26: 1120-71.

_____. 1988b. Institutions and markets in a dynamic world. *The Manchester School,* 56: 119-46.

Dunning, John. 1980. Towards an eclectic theory of international production: Some empirical tests. *Journal of International Business Studies,* Spring/Summer 1: 9-31.

_____. 1989. *Multinational enterprises and the globalisation of technological capacity.* Mimeo, Economics Department, Reading University.

Fagerberg, Jan. 1987. A technology gap approach to why growth rates differ. *Research Policy,* 16: 87-99.

_____. 1988. International competitiveness. *Economic Journal,* 98: 355-74.

Franko, Lawrence. 1989. Global corporate competition: Who's winning, who's losing, and the R&D factor as one reason why. *Strategic Management Journal,* 10: 449-74.

Freeman, Christopher. 1982. *The economics of industrial innovation.* London: Pinter.

Grandstrand, Ove, Soren Sjolander & Sverke Alange. 1989. Strategic technology management issues in Japanese manufacturing industry. *Technology Analysis and Strategic Management,* 1: 259-72.

Grandstrand, Ove & Soren Sjolander. 1990. Managing technology in multi-technology corporations. *Research Policy,* 19: 35-60.

Griliches, Zvi, editor. 1983. *R and D, patents and productivity.* Chicago: Chicago University Press.

Hakanson, Lars & Robert Nobel. 1989. *Overseas research and development in Swedish multinationals* (RP 89/3). Institute of International Business, Stockholm School of Economics.

Hanson, Philip & Keith Pavitt. 1987. *The comparative economics of research, development and innovation in East and West: A survey.* Fundamentals of Pure and Applied Economics, No. 25. New York: Harwood Academic Publishers.

Kleinecht, Alfred. 1987. Measuring R&D in small firms: How much are we missing? *Journal of Industrial Economics,* 36: 253-56.

Levin, Richard, Alvin Klevorick, Richard Nelson & Sidney Winter. 1987. Appropriating the returns from industrial research and development. *Brookings Papers on Economic Activity,* 3: 783-831.

Levin, Richard, Wesley Cohen & David Mowery. 1985. R and D, appropriability, opportunity, and market structure: New evidence on the Schumpeterian hypothesis. *American Economic Review,* 75: 20-24.

Malerba, Franco & Luigi Orsenigo. 1988. *Technological regimes, patterns of innovation and firm variety: A theoretical and empirical investigation of the Italian case.* Mimeo, Institute of Political Economy, Bocconi University, Milan.

_____. 1990. Personal communication.

Mowery, David, editor. 1988. *International collaborative ventures in US manufacturing.* Cambridge, Mass.: Ballinger.

Narin, Francis, Elliot Noma & Ross Perry. 1987. Patents as indicators of corporate technological strength. *Research Policy,* 16: 143-55.

Nelson, Richard & Sidney Winter. 1982. *An evolutionary theory of economic change.* Cambridge Mass.: Belknap.

Pakes, Ariel & Zvi Griliches. 1983. Patents and R and D at the firm level: A first look. In Z. Griliches, editor, *R and D, patents and productivity.* Chicago: Chicago University Press.

Pakes, Ariel & Mark Shankerman. 1983. The rate of obsolescence of knowledge, research gestation lags and the private rate of return to research resources. In Z. Griliches, editor, *R and D, patents and productivity*. Chicago: Chicago University Press.

Patel, Parimal & Keith Pavitt. 1987a. Is Western Europe losing the technological race? *Research Policy*, 16: 59-85.

_____. 1987b. The elements of British technological competitiveness. *National Institute Economic Review*, November, 122: 72-83.

_____. 1988. The international distribution and determinants of technological activities. *Oxford Review of Economic Policy*, Winter, 4: 35-55.

_____. 1989. A comparison of technological activities in FR Germany and the UK. *National Westminster Bank Quarterly Review*, May: 27-42.

_____. 1990. Large firms in Western Europe's technological competitiveness. In L-G. Mattsson & B. Stymne, editors, *Corporate and industry strategies for Europe*. Amsterdam: North Holland.

Pavitt, Keith. 1988. Uses and abuses of patent statistics. In Anthony van Raan, editor, *Handbook of quantitative studies of science and technology*. Amsterdam: Elsevier.

_____. 1988b. International patterns of technological accumulation. In Neil Hood & Jan-Erik Vahlne, editors, *Strategies in global competition*. London: Croom Helm.

_____. 1990. What we know about the usefulness of science: The case for diversity. In Douglas Hague, editor, *The management of science*. London: Macmillan.

_____, Michael Robson & Joe Townsend. 1987. The size distribution of innovating firms in the UK: 1945-83. *Journal of Industrial Economics*, 35: 297-316.

Pearce, Richard. 1990. Communication at seminar at economics department, Reading University, 9th January.

Porter, Michael. 1990. *The competitive advantage of nations*. London: Macmillan.

Rosenberg, Nathan & Claudio Frischtak, editors. 1985. *International technology transfer: Concepts, measures and comparisons*. New York: Praeger.

Soete, Luc. 1978. *Inventive activity, industrial organisation and international trade*. D.Phil thesis, University of Sussex.

_____. 1981. A general test of technological gap trade theory. *Review of World Economics*, 117: 638-66.

_____. 1987. The impact of technological innovation on international trade patterns: The evidence reconsidered. *Research Policy*, 16: 101-30.

_____ & Sally Wyatt. 1983. The use of foreign patenting as an internationally comparable science and technology output indicator. *Scientometrics*, 5: 31-54.

Teece, David. 1986. Profiting from technological innovation: Implications for integration, elaboration, licensing and public policy. *Research Policy*, 15: 285-305.

Vernon, Raymond. 1966. International investment and international trade in the product cycle. *Quarterly Journal of Economics*, 80: 190-207.

_____. 1979. The product-cycle hypothesis in a new international environment. *Oxford Bulletin of Economics and Statistics*, 41: 255-67.

PART THREE

SYSTEMS OF INNOVATION

9

THE JOURNAL OF INDUSTRIAL ECONOMICS   0022-1821 $2.00
Volume XXXV   March 1987   No. 3

THE SIZE DISTRIBUTION OF INNOVATING FIRMS IN THE UK: 1945–1983*

K. PAVITT, M. ROBSON AND J. TOWNSEND

A survey of 4378 significant innovations shows that firms with fewer than 1000 employees commercialised a much larger share than is indicated by their share of R&D expenditures. Innovations per employee have been consistently above average in firms with more than 10 000 employees, and have become so in firms with fewer than 1000. Intersectoral variation in the size distribution of innovating firms can be explained as a function of R&D-based technological opportunities, and of "technological ease of entry" by user firms with principal activities outside the sector.

I. INTRODUCTION

THIS PAPER examines the relationship between firm size and innovative activity, and is based on information on more than 4000 significant innovations commercialised in the UK between 1945 and 1983. Up to the early 1970s, the prevalent assumptions were that the empirical relationship between the volume of innovative activities and firm size was an S-shaped one, with a relatively low share amongst small firms, increasing in the medium and large firms, and then slowing down amongst the very large; or, put another way, the relationship between innovation intensity and firm size was r-shaped (i.e. the first derivative of the S-curve) with a decline amongst the very large. This pattern of innovative activities was often explained by indivisibilities and risk precluding most small firms, and by monopoly power reducing the pressure on the biggest ones.[1]

These assumptions and explanations have since been criticised on two grounds. First, the empirical measures of innovative activity—usually R&D or patenting activities—are inputs or intermediate outputs—and not final outputs of superior products or production processes.[2] Furthermore, if R&D expenditures rather than numbers of patents are taken as the proxy measure of innovative activities, the relationship amongst the largest firms between firm size and innovation intensity is more positive (Soete [1979]).

* This paper has been prepared as part of the research programme of the Designated Research Centre on Science, Technology and Energy Policy, funded by the Economic and Social Research Council (ESRC). It is based on a survey of significant innovations in Britain, funded by the Joint Committee of the ESRC and the Science and Engineering Research Council (SERC). We are indebted to P. Alizadeh, J. Hebden, P. Patel, and S. Wyatt for statistical advice and support, and to two anonymous referees for helpful and detailed comments on an earlier draft.

[1] For summary, see Scherer [1973, Chapter 15].

[2] See Fisher and Temin [1973]. For a summary of the subsequent debate, see Stoneman [1983, pp. 13–14].

Second, the underlying causal model—from firm size and market structure to innovative activity—neglects the reverse causation from innovative activity to firm size and market structure. In particular, it has been postulated that the size distribution of innovating firms is a function of technological opportunity, appropriability, and demand, and that these vary amongst sectors. Starting with this assumption in common, but with very different ones about the behaviour of firms, both Dasgupta and Stiglitz [1980, p. 276] and Nelson and Winter [1982, p. 350] predict a positive association between high technological opportunity and appropriability, on the one hand, and increasing innovation intensity with firm size in markets with elastic demand, on the other.

Based on other models and other evidence, other analysts come to other conclusions. Thus, in his book on the evolution of giant firms in the UK since the Second World War, Prais [1976] compares the size of manufacturing establishments with the size of large firms, and their evolution over time. He concludes that the technical requirements for greater plant sizes do not explain the growth of giant firms in the UK. He also notes considerable increases between 1958 and 1968 in the number of very small establishments in machinery, instruments, electronics, and plastics sectors, all of which were of high technological opportunity.

This paper throws fresh empirical light on the size distribution of innovating firms, and suggests assumptions and hypotheses for further theoretical development and test.[3] Section II describes the nature, scope and limitations of the data. Section III shows a size distribution of innovating firms very different from firms performing R & D, and explores the degree to which this different distribution upsets established assumptions about relationships between firm size and innovation intensity. Section IV shows considerable inter-sectoral variation in the size structure of innovating firms, and explores whether it can be measured and explained by differences in technological opportunity and ease of imitation. Section V concludes with the implications of the analysis for theorising and policy-making about firm size and innovative activities.

II. SCOPE AND LIMITATIONS OF THE SAMPLE

Our data comes from a survey of 4378 innovations, compiled at the Science Policy Research Unit (SPRU) over a period of fifteen years and completed in

---

[3] It does not analyse the relationship between market structure and innovative activity, since our data cannot isolate a market structure variable. The same is true for most previous analyses, which have been based on cross-industry comparisons in one country. As Levin, Cohen and Mowery [1985] have recently shown, the relationship between market structure and R & D intensity disappears, when inter-industry differences in technological opportunity and appropriability are taken into account. Recent improvements in internationally comparable statistics on R & D and on patenting offer better opportunities to isolate a market structure variable.

1984.[4] The procedure was to write to experts in each sector of industry and commerce, asking them to identify significant technical innovations that had been successfully commercialised in the UK since 1945, and to name the firm responsible. Questionnaires were sent to these innovating firms requesting information on a range of parameters, including the employment of the innovating firm and (where appropriate) the innovating unit at the date of the innovation's commercialisation, the firm's and unit's principal product line, and those of the innovation and of its first user. Nearly 400 experts were consulted in the survey, drawn from research and trade associations, government departments, academic institutions, trade and technical journals, individuals and consultants, as well as from firms. It is difficult to prove that such wide consultation has overcome bias towards innovations in large firms. Suffice to say that our results and conclusions suggest that we have, at least in part.

Table I shows that the distribution of the identified innovations over the time period is relatively even. The smaller numbers before 1960 probably reflect the difficulty of finding, in the 1970s, experts who had been active in the earliest period. The consequent variation in coverage over time is therefore likely to be uniform across sectors, and not to result in any change in bias.

The distribution of the sample of innovations according to the principal activity of the innovating firm is shown in Table VI. These sectors are generally at the order level of the 1968 UK Standard Industrial Classification (SIC), together with a number of service sectors. Some orders with few innovations are combined, and some with many are subdivided. Compared to other partial indicators of innovative activities, the sectoral distribution of the sample is closely correlated with that of patenting activity ($r^2 = 0.69$, significant at 99.9% level), and much less closely with industry-financed R&D, followed by total industrial R&D ($r^2 = 0.12$ and 0.10, respectively; non significant). The shares of innovations and patenting in mechanical engineering/machinery and instruments are high, and that of aerospace is low, when compared to R&D activities.[5]

Table VI also shows that about 90 per cent of the significant innovations were commercialised by firms with their principal activity in manufacturing; R&D statistics show a similar degree of concentration (Business Statistics Office, [1979, Table II]). Even greater concentration of innovation is shown in manufacturing product groups: nearly 98 per cent of innovations and nearly 97 per cent of R&D. We shall therefore concentrate our analysis and comparisons on manufacturing firms and products, but recognising that

---

[4] For more complete information, see Townsend *et al.* [1981], Robson and Townsend, [1984]. The tape containing the complete data set, together with the accompanying User Manual, is available from the Survey Archive of the Economic and Social Research Council (ESRC) at the University of Essex in the UK.

[5] For data on UK R&D and patenting, see Townsend *et al.* [1981]. Correlations calculated for 20 matched sectors.

TABLE I
DISTRIBUTION OF IDENTIFIED INNOVATIONS BY YEARS OF FIRST COMMERCIALISATION

Years	1945–49	1950–54	1955–59	1960–64	1965–69	1970–74	1975–79	1980–83	1945–83
Number	226	359	514	684	720	656	823	396*	4378

* Equivalent to 660 over a five year period.
Source: SPRU Innovation Survey [1984].

some innovations were first applied by non-manufacturing firms (e.g. research laboratories, energy and transport utilities, National Coal Board, Ministry of Defence). Whilst such an assumption is valid for hardware innovations, it would not be for software innovations, where service sector firms make a bigger contribution.

The SPRU Innovation Survey has a number of unique attributes for the analysis of firm size and innovative activities that will hopefully become apparent in the course of this analysis.[6] But there are also three sets of limitations. The first relates to the measure of innovative activity. As in previous analyses, we shall assume that data on significant innovations are a visible manifestation of a continuous activity resulting in a stream of incremental innovations and, more infrequently, in significant ones (Pavitt [1984, 1986]; Townsend *et al.* [1981]; Wyatt [1985]). The infrequency of the latter makes them unsuitable for the analysis of individual firms over short periods of time.

Second, unlike previous analyses—mainly about the USA[7]—our sample does not include complete and matched information on both innovative and other economic activity (e.g. sales, net output, employment) of each firm within the large size range. Comparisons of innovation intensity and firm size depend on information from the UK Censuses of Production. On the other hand, our analysis does cover all sizes of firm, and not just the large ones.

Third, analyses of innovative activities in the UK bring out clearly the multinational nature of many innovating firms. For purposes of comparison with UK Census and R & D Survey data, we have collected information on the UK employment of innovating firms; and for purposes of comparing elsewhere their total size, we have also collected information on their world employment.

III. SIZE OF FIRM AND INNOVATIVE INTENSITY

Table II shows that the distribution of innovative activity by firm size is very different according to the SPRU sample of significant innovations, than to the more frequently used proxy of R & D expenditures. In particular, firms with fewer than 1000 UK employees accounted for only 3.3 per cent of the R & D in 1975, but for 34.9 per cent of the identified significant innovations between 1970 and 1979. To what extent do these differences upset previously established evidence on the relationship between firm size and innovation intensity?

III(i). *The first 100 manufacturing firms*

We begin where most similar analyses begin; namely, with the biggest firms in

---

[6] Previous analysis, based on a smaller and partial data set, has been made by Freeman [1971] and Wyatt [1985].
[7] See, for example, Scherer [1965] and Soete [1979].

302   K. PAVITT, M. ROBSON AND J. TOWNSEND

TABLE II
THE DISTRIBUTION OF NUMBERS OF FIRMS* SURVEYED, OF R&D EXPENDITURE AND OF
INNOVATIONS BY FIRM SIZE. ( ) = % AGE

Employment of Firm	R&D, 1975		Innovations 1970–79	
	No. of Firms	Gross R&D Expenditure £M	No. of Firms	No. of Innovations
1–99	—	—	114	173
	—	—	(23.1)	(12.1)
100–499	552	18.6	125	231
	(49.5)	(1.5)	(25.3)	(16.1)
500–999	168	21.8	45	96
	(15.1)	(1.8)	(9.1)	(6.7)
1000–4999	233	90.4	79	172
	(20.9)	(7.5)	(16.0)	(12.0)
5000–9999	68	107.3	34	87
	(6.1)	(8.9)	(6.9)	(6.1)
10 000+	95	971.8	97	677
	(8.5)	(80.3)	(19.6)	(47.1)
All Firms	1116	1209.8	494	1436
	(100.0)	(100.0)	(100.0)	(100.0)

*Both surveys based on firms with R&D or innovations in manufacturing product groups.
Source: SPRU Innovation Survey; Business Statistics Office [1979].

manufacturing. Table III compares cumulative shares and trends in shares of significant innovations of the 100 top manufacturing firms in our sample, ranked by number of innovations, and also by UK employees. Both these indicators of concentration of innovative activity can be compared with the shares of the top 100 manufacturing firms in output and employment, shown in the bottom row.

Beginning with the shares of the top innovating firms, Table III shows a continually higher degree of concentration of production of significant innovations than of output and employment. It also shows no increasing concentration of innovative activities over the period, but a tendency for concentration in each of the ten-year periods to be greater than for the near-forty year period as a whole. This reflects the relative infrequency in the commercialisation of what we have defined as significant innovations, even amongst the top 100 innovating firms in manufacturing. Not all these firms commercialised innovations in each of the 10-year time periods, which is then reflected in a higher degree of concentration than in the near 40-year period as a whole. Even so, the 100-firm concentration ratio for the whole period, of 50.6 per cent, is still well above that for output and employment.

This relatively concentrated innovative activity does not necessarily mean that innovativeness increases with firm size. Unlike Scherer [1965], Soete [1979], and others, we do not have systematic data for each large firm on

THE SIZE DISTRIBUTION OF INNOVATING FIRMS IN THE UK: 1945–1983   303

TABLE III
TRENDS IN CUMULATIVE SHARES OF INNOVATIONS IN MANUFACTURING FIRMS, RANKED
ACCORDING TO NUMBER OF INNOVATIONS AND EMPLOYMENT (%)

Firm with Largest Share of Total		1945–54	1955–64	1965–74	1975–83	1945–83
First 5	Inn	16.3	15.6	14.5	15.1	14.1
	Size	4.5	5.9	4.9	10.9	2.1
First 10	Inn	23.0	22.1	21.4	21.8	20.3
	Size	12.6	11.3	11.6	13.7	7.7
First 20	Inn	33.1	30.3	30.1	31.0	28.4
	Size	21.6	18.7	15.0	19.0	12.9
First 50	Inn	50.4	44.9	44.9	45.9	39.7
	Size	32.1	25.8	24.9	31.9	21.6
First 100	Inn	67.7	60.1	58.8	59.2	50.6
	Size	54.5	41.2	42.8	48.5	30.7
First 100	Output	32(1958)	37(1963)	41(1968)	42(1973)	41(1978)
	Employ.				38(1973)	37(1978)

Inn = Shares of innovations ranked according to number of innovations.
Size = Shares of innovations ranked according to UK employment.
Source: SPRU Innovation Survey, Business Statistical Office [various].

output, employment and innovative activities. We cannot therefore calculate the share of innovations made by the top 100 firms in terms of either employment or output. We can do the next best thing, which is to calculate the cumulative share of innovations made by the innovating firms in our sample, when ranked by their UK employment. This is also shown in Table III. In each of the 10-year periods, the share is higher than for output and employment, but for the 40-year period as a whole it is lower.

This ambiguous result again reflects the relative infrequency of even the largest firms in the commercialisation of significant innovations. It also reflects the even greater infrequency, over the 40-year period, of commercialisation of innovations by firms in the smaller size categories. The employment distribution of innovating firms in the SPRU sample is therefore biassed towards the larger firms. For example, manufacturing firms with more than 50 000 employees account for about 24 per cent of all employment in the SPRU sample, whilst they never account for more than 20 per cent of employment in any of the Censuses of Production since 1945. A more accurate reflection of the innovativeness of firms in the various size categories can be found in a comparison of their shares of innovations in the SPRU sample, with their shares of employment in the Censuses of Production.

III(ii). *All firm size categories*

This is made in Table IV. A ratio of greater than unity shows above-average innovation-intensity, and vice versa. The choice of firm size categories for

TABLE IV

RATIO OF SHARE OF MANUFACTURING FIRMS' INNOVATIONS TO THEIR SHARES OF EMPLOYMENT IN
DIFFERENT SIZE CATEGORIES AND TIME PERIODS

UK Employment	1956–60	1961–65	1966–70	1971–75	1976–80	1981–83
1–99	0.28	0.46	0.64	0.63	0.59	0.63
100–199	0.66	0.77	0.96	1.06	1.19	2.07
200–499	0.80	0.94	0.95	1.32	1.49	1.58
500–999	1.04	1.01	1.10	1.19	1.00	0.92
1000–1999	1.07	1.83	1.50	1.02	0.82	0.34
2000–4999	0.81	0.61	0.69	0.55	0.65	0.91
5000–9999	0.91	0.78	0.73	0.74	0.45	0.37
10 000–19 999	1.68	1.04	1.13	1.15	1.22	1.07
20 000–49 999	1.25	0.86	1.22	1.20	1.00	1.05
50 000+	2.46	2.13	1.27	1.31	1.90	1.90

Note: Employment shares are for 1958, 1963, 1968, 1973, 1978, 1982.
Source: SPRU Innovation Survey, Business Statistical Office [various].

comparison is dictated by those used in the Census, as are the time periods. Table IV shows that, from the late 1950s to the early 1980s, both large firms (with more than 10 000 employees) and small firms (with between 500 and 1000 employees) have usually had above average innovation intensity, whilst medium sized firms (2000–9999 employees) had below average intensity.

Over the period, innovation intensity increased amongst very small firms (1–499 employees), and declined amongst the medium sized ones. These trends are reflected in the shares of firms in different size categories in the commercialisation of innovations: analysis of annual changes between 1945 and 1983 shows a decline in the share of medium sized firms, and an increase in the shares of both small firms and very large firms (more than 50 000 employees).

Thus, although we do not have sufficient data points to make fully satisfactory statistical tests, such information as we have shows above average innovation intensity amongst large firms, and growing innovation intensity amongst small ones. The former result is consistent with earlier analysis of R & D performing firms in the USA (Soete [1979]). The latter result is—as far as we know—quite original,[8] and reflects the already noted ability of the SPRU Innovation Survey to measure innovative activity in small firms. Taken together, they suggest that the relationship between firm size and innovative intensity is not r-shaped, but U-shaped, and becoming increasingly so over time.

III(iii). *Other measures of firm size*

So far, we have measured firm size in terms of UK employment. Table V compares size distributions of innovating firms, according to UK and world

---

[8]It has already been presented by Wyatt [1985].

TABLE V
THE SIZE DISTRIBUTION OF INNOVATING FIRMS IN MANUFACTURING ACCORDING TO DIFFERENT MEASURES

Size Measure	1–199	200–999	1000–9999	10 000–49 999	50 000+	TOTAL
UK Employment	17.0	16.6	24.4	22.6	19.3	100
World Employment	14.5	12.6	19.0	25.2	28.7	100
Unit Employment	22.8	26.1	37.8	11.1 009	2.2	100

Source: SPRU Innovation Survey [1984].

employment, and to employment of innovating unit (i.e. division, subsidiary or independent firm), from 1945–83.

Not unexpectedly, firms with more than 50 000 employees commercialised a much larger share of the innovations, when measured according to world rather than UK employment. The share increased over the period, although the average size of firm did not, probably reflecting the increasing internationalisation of production in firms of this size. Once again, the greatest decline in share over the period was amongst medium sized firms (1000–9999 employees), whilst those with fewer than 200 employees increased their share. These patterns and trends are thus similar to those for firm size measured according to UK employment.

Patterns and trends in the size distribution of innovating units are very different indeed. Table V shows that, for 86.7 per cent of the identified innovations, innovating units had fewer than 10 000 employees, compared to 58.0 per cent of innovating firms according to UK employment, and 46.1 per cent according to world employment. Over the period, the share of innovating units with fewer than 1000 employees increased markedly, at the expense of those with between 1000 and 9999 employees. The contradictory patterns and trends in the sizes of innovating firms and units probably reflects the growing capacity of industrial managers to combine the advantages of increased firm size and internationalisation, with the advantages for the implementation of innovation and other activities of smaller organisational units.

IV. INTERSECTORAL DIFFERENCES

IV(i). *Overview*

Behind these general patterns and trends in innovative activity and innovative intensity, there is considerable variety amongst sectors. This is apparent in Table VI, which shows for each sector the size distribution of innovating firms in terms of world employment, together with average sizes and coefficients of variation (i.e. standard deviation divided by the mean), and the total number of innovations.

TABLE VI
SIZE DISTRIBUTION OF INNOVATING FIRMS ACCORDING TO THEIR PRINCIPAL SECTOR OF ACTIVITY

Principal Sector of Innovating Firm	3/4 Digit SIC's Included	No. of Innovations	% age Distribution of Innovations					Mean Employment	Coefficient of Variation
			1–199	200–999	1000–9999	10000–49000	50000		
Agriculture	001–003	12	8.3	0	66.7	25.0	0	7,652	0.67
Mining	101–104	126	0	2.4	0.8	12.7	84.1	222,974	0.94
Food	211–240	112	3.6	5.4	9.8	57.1	24.1	48,924	1.49
Chemicals	261–279	421	4.8	7.4^I	9.7^D	31.4	46.8^I	69,544	0.98
Metals	311–323	186	0.5	3.8	25.8^D	15.1	54.8^I	111,844	0.95
Machinery	331, 332, 335–339.2, 339.5–339.9	573	26.2^I	27.1	33.7	12.4	0.7	3,801	2.09
Mechanical Engineering	333, 334, 339.3, 339.4 341–349, 391–399	558	14.0	12.0	18.5	38.4^I	17.2	25,881	1.46
Instruments	351–354.2	332	31.6	18.1^I	15.4	16.6	18.4^I	23,665	1.73
Electrical Engineering	361, 362, 368, 369	346	3.2	2.3	4.0^D	15.3^D	75.1^I	111,022	0.68
Electronics	363–367	428	17.5	8.9^I	12.4	27.3	33.9^I	74,203	1.54
Shipbuilding and Offshore Engineering	370	67	13.4^I	14.9^I	46.3^D	23.9	1.5	9,349	1.59
Vehicles	380–382, 384, 385	212	9.4	8.5^I	28.8	27.4	25.9	62,581	1.93

Industry	Code								
Aerospace	383	85	2.4	7.1	17.6[D]	29.4[I]	43.5	38,862	0.77
Textiles, Leather and Clothing	411–450	144	20.1	11.8	32.6[D]	6.9	28.5[I]	30,239	1.46
Bricks, Pottery, Glass, Cement	461–469	157	14.0	7.6	18.5	48.4[I]	11.5	16,824	1.02
Paper	481–484	54	16.7	20.4	13.0	38.9	11.1	17,700	1.25
Printing	485–489	29	6.9	34.5	55.2	3.4	0	4,123	2.17
Rubber & Plastics	471–479, 491–499	91	15.4	27.5	1.1	15.4	40.7	45,485	1.22
Construction	500	39	30.8	7.7	33.3	25.6	2.6	8,362	1.45
Utilities	601–603, 708, 709	44	0	2.3	11.4	11.4	75.0	122,766	0.78
Transport	701–707	38	7.9	5.3	10.5	23.7	52.6	93,283	1.08
Business	810–866	48	12.5	12.5	29.2	35.4	10.4	22,347	1.87
R&D	876	126	42.9	23.8	5.6	27.0	0.8	5,055	1.98
Services	871, 879–899	36	13.9	30.6	2.8	25.0	27.8	26,392	1.17
Defence	901	59	0	1.7	0	0	98.3	288,828	0.17
Other Govt. and Health	872, 874, 901	55	14.5	34.5	18.2	9.1	23.6	76,096	2.23
Total		4378	14.6	12.7	17.9	24.3	30.5	92,579	1.68

I = Increase in annual share, significant at 95% confidence levels.
D = Decrease in annual share, significant at 95% confidence levels.
Source: SPRU Innovation Survey [1984].

First, there is considerable variation amongst sectors in the number of significant innovations produced, from more than 1000 in mechanical engineering/machinery to fewer than 30 in printing. Second, there is the considerable variation amongst sectors in the size distribution of innovating firms. By looking down each of the columns in Table VI, one can identify at least one sector, in each of the size categories, where the share of innovating firms is more than twice the average for the sample as a whole. Firms with fewer than 1000 employees are particularly important in machinery, instruments, and R & D laboratories, where they produced more than 45 per cent of all innovations in the sector; whilst firms with more than 10 000 employees account for more than 75 per cent of all innovations in mining, food, chemicals, electrical products and defence. The average size of innovating firms varies from fewer than 4000 employees in machinery, printing and R & D, to more than 100 000 employees in mining, metals, electrical products, utilities, health and defence. With this considerable variation amongst sectors in the size distribution of innovating firms, Table VI shows that the coefficient of variation for the sample as a whole is 1.68.

Finally, in the thirteen sectors where firms have made more than 100 innovations, we have identified significant trends in the size distribution of innovating firms. In some cases, these reflect trends in the sample as a whole, with an increasing share of innovating firms in both the small and the large categories (chemicals, electronics); and in other cases, they reflect only part of the overall trend, with an increasing share of innovations either for small firms (machinery, instruments, shipbuilding and offshore, vehicles), or for large ones (metals, mechanical engineering, electrical engineering, aerospace, textiles, building materials).

IV(ii). *Measuring technological opportunity and ease of imitation*

How to explain these considerable intersectoral variations? As we saw in section I, recent explanations of the size distribution of innovating firms emphasise technological opportunity and appropriability. High technological opportunity—whether in terms of numbers of significant innovations, patenting or R & D—is associated with the following sectors: chemicals, machinery/mechanical engineering, instruments, electrical machinery and electronics.[9] Tables VI and VII show that firms principally in these sectors have the following similarities and differences.

First, they are not intensively represented in any one part of the size distribution of innovating firms. For example, amongst innovating firms with fewer than 200 employees, 63.7 per cent of the innovations were commercialised by firms principally in machinery/mechanical engineering, instruments and electronics, compared to 43.2 per cent in all size categories. Amongst innovating firms with more than 50 000 employees, 45.2 per cent of

---

[9] See Table VI and Townsend *et al.* [1981].

TABLE VII

CHARACTERISTIC OF INNOVATIVE ACTIVITIES IN MANUFACTURING FIRMS ACCORDING TO PRINCIPAL ACTIVITY AND SIZE OF INNOVATING FIRM

Sector of Principal Activity of Innovating Firms	% Process Innovations	Distribution of Innovations (%)				% Innovations in Sector by Outside Firms
		Specialised	Narrow Diversification	Machinery/Mech. Eng. and Instruments	Other Broad Diversification	
Food	49.1	56.3	1.8	21.5	20.5	20.7
Chemicals	21.0	28.6	37.7	14.2	19.5	22.3
Metals	57.0	57.5	5.9	27.5	9.1	37.6
Machinery/Mech. Eng.	15.6	56.5	19.5	4.7	19.3	32.3
Instruments	19.9	74.1	1.5	5.1	19.3	54.1
Electrical/Electronics	32.6	29.1	41.6	19.2	11.1	26.5
Shipbuilding and Offshore Engineering	50.7	58.2	0	26.8	14.9	71.7
Vehicles	59.9	63.2	8.0	15.1	13.7	33.2
Aerospace	64.7	71.8	0	16.5	11.8	56.7
Textiles	84.0	56.9	4.9	29.9	8.3	39.0
Bricks, Pottery, Glass, Cement	41.4	69.4	0.6	13.4	16.6	25.7
Paper	31.3	46.9	8.3	9.3	35.4	58.3
Printing	79.3	34.5	0	44.8	20.7	33.3
Rubber and Plastics	46.3	46.3	3.7	35.2	14.8	15.3
Total Manufacturing	32.0	49.8	20.2	13.5	16.5	30.0
Size of Innovating Firms						
50 000+	33.9	23.4	33.9	18.5	24.1	n/a
10 000–49 999	31.2	44.3	21.8	13.7	20.1	n/a
1000–10 000	37.2	61.0	12.6	11.1	10.3	n/a
200–9999	27.6	73.7	10.0	7.7	8.7	n/a
1–199	26.9	71.7	7.4	11.4	9.4	n/a

Process Innovations = 2-digit SIC sector of innovating firm is the same as that of the user sector.
Product Innovation = these 2-digit SIC's are different.
Specialised = In principal 3-digit sector of innovating firm.
Narrow Diversification = Elsewhere in principal 2-digit sector of innovating firm.
Broad Diversification = In other 2-digit sectors, other than machinery, mechanical engineering and instruments.
Source: SPRU Innovation Survey [1984].

the innovations were commercialised by firms principally in chemicals and electrical/electronics, compared to 27.3 per cent in firms in all size categories (derived from Table VI).

Second, column one of Table VII shows that they are relatively strongly oriented to product rather than process innovation, particularly those in machinery/mechanical engineering and instruments. It also shows that the relative importance of process innovations increases with firm size, but surprisingly little, especially when compared to the large intersectoral variations. Closer examination shows that, within most sectors of manufacturing, there is no clear positive association between firm size and the relative importance of process innovations.

Third, differing technological opportunities amongst sectors, and amongst firms of different sizes, are reflected in the spread of firms' innovative activities across product groups. Table VII shows that "narrow diversification" in innovative activities (i.e. into other 3-digit product groups within the principal 2-digit activity of the innovating firm) increases steeply with the size of innovating firms, and is heavily concentrated in chemicals and electrical/electronics.[10]

Fourth, Table VII shows that firms principally in machinery/mechanical engineering and instruments differ from those in chemicals and electrical/electronic products, in that they are more specialised in their innovative activities within their principal 3-digit activity; and that, as shown in the last column, a higher proportion of innovations are made by firms with their principal activities outside the sectors. This capacity to make innovations in machinery/mechanical engineering is spread across most sectors. It is linked mainly to innovating firms' activities to improve their own process technologies.

Thus, firms principally in sectors of high technological opportunity can be found heavily represented amongst those that are very large and those that are very small. Those amongst the very large (in chemicals and electrical/electronics) are typified by innovations in technologically related product markets within their principal two-digit SIC sectors, and by the relative difficulty of technological imitation by firms producing principally outside these sectors. Those firms amongst the very small (in machinery/mechanical engineering and instruments) are much more specialised technologically within their principal three-digit SIC sector, and are characterised by relative "ease of technological imitation" by firms producing outside the sectors.

IV(iii). *Some exploratory statistical tests*

These observations, together with an earlier analysis of the SPRU Innovation Survey by one of us (Pavitt [1984]), suggest more systematic relationships

[10] The periodic changes and additions in the two digit categories of "chemicals" and "electrical and electronic" are in fact historical records of technology-based product markets opened up by firms principally in these sectors.

between measurable sectoral characteristics of technical opportunity and ease of imitation, on the one hand, and the size distribution of innovating firms on the other.

(1) We would expect larger sized innovating firms to be positively associated with increasing R & D intensity, with the latter reflecting the level of science based opportunities for diversifying into technologically related product markets.

(2) We would expect little relationship between larger sized innovating firms and large size of production plant, with the latter reflecting technical requirements for increased scale of production. This expectation is based both on the earlier conclusions of Prais [1976], and on the lack of a close relation between firm size and process innovation apparent in Table VII.

(3) We would expect a negative relationship between larger size of innovating firm and increasing "technological ease of imitation", since the latter decreases the likelihood that firms principally in the sector will be allowed to appropriate all technological opportunities.[11]

(4) We would expect a positive relationship between smaller size of innovating firm and a particular form of technological ease of imitation, reflected in bigger contributions to innovation by user firms from outside the sector. A variety of earlier studies have shown that, either through "vertical disintegration" or through untraded flows of information and knowledge, such user firms are an important stimulus for the emergence of small, innovative firms supplying capital goods (Hippel [1978]; Kaplinsky [1983]; Pavitt [1984]; Rosenberg [1976] and [1982]; Rothwell [1983]).

Time, space and data limitations prevent us from translating these expectations into a fully worked-out model encompassing cost and demand characteristics, and the two-way causality between R & D and firm size distribution. However, for the purposes of identifying hypotheses for further elaboration and test, we present briefly in Table VIII the results of some exploratory statistical tests. Reflecting the above four expectations, our independent variables are as follows:

(1) *RD* is the R & D intensity variable, and is the proportion of net output in each sector in 1972 spent on industry funded R & D.

(2) *PS* is the plant size variable, and is the first moment median size of plant employment in the sector in 1963, as calculated by Prais [1976, Table 3.3]. As expected, it never approaches the 90 per cent confidence level, and is excluded from the equations reported in Table VIII.[12]

[11] This concept of "technological ease of imitation" is not exactly the same as the concept of lack of appropriability, as defined by Levin *et al.* [1984]. The latter refers to the ease of all potential entrants to imitate leading innovators, whereas the former refers only to the technological capacity of firms from outside the sector.

[12] The choice of years for *RD* and *PS* is dictated purely by availability. However, the variables do not fluctuate or change greatly over time in sectoral rankings. The plant size variable *PS* has a number of drawbacks: the measure is by employment rather than output; and the data are aggregated for the all-important sectors of machinery/mechanical engineering, instruments, and electrical/electronic products, as well as for aerospace and other vehicles.

312   K. PAVITT, M. ROBSON AND J. TOWNSEND

(3) *EI* is the technological ease of imitation variable, and is the proportion of innovations made in each sector by firms with their principal 2-digit SIC activity outside the sector (see last column of Table VII).

(4) *UI* is the users' innovation variable, and is the proportion of innovations in each sector made by user firms outside the sector (i.e. with the principal 2-digit activity of the innovating firm outside the sector, and the same as the 2-digit sector of use of the innovation).

The dependent variables are measures of the size distribution of innovating firms in each sector and are defined as follows:

VL = percentage of innovations by firms principally in a sector with 50 000 or more employees.

L = percentage of innovations by firms principally in a sector with 10 000 or more employees.

S = percentage of innovations by firms principally in a sector with fewer than 1000 employees.

VS = percentage of innovations by firms principally in a sector with fewer than 200 employees.

The results of the regressions in Table VIII are encouraging, in their support of our expectations. They confirm the importance of R&D intensity (*RD*) in explaining intersectoral differences in the relative importance of very large, large and small innovating firms; and of innovating users (*UI*) in explaining intersectoral differences in the relative importance of small and

TABLE VIII
DETERMINANTS OF THE SIZE STRUCTURE OF INNOVATING FIRMS

Dependent Variables	Independent Variables			F Statistic $(2,11)$	Adjusted R^2
	RD	EI	UI		
VL	4.751[b]	−0.400		4.374	0.342[b]
	(2.712)	(1.309)			
L	4.780[b]	−0.602		3.706	0.294[a]
	(2.261)	(1.623)			
S	−2.357	0.296		2.088	0.143
	(1.699)	(1.216)			
VS	−0.567	0.190		0.947	0.008
	(0.651)	(1.242)			
S	−2.503[b]		1.834[c]	7.449	0.500[c]
	(2.360)		(3.221)		
VS	−0.634		0.960[b]	2.998	0.353[a]
	(0.834)		(2.352)		

[a]Significant at 90% confidence level.
[b]Significant at 95% confidence level.
[c]Significant at 99% confidence level.
t-statistic in brackets.
Note: Based on manufacturing sectors in Table VI, with machinery/mechanical engineering combined and steel excluded (Total = 14 sectors).

very small innovating firms. The technological ease of entry variable (EI) is never significant at the 10 per cent level, but has the predicted sign in all four equations.

V. CONCLUSIONS

There is much more innovative activity amongst firms with fewer than 1000 employees than the R&D statistics had led us to assume. The relationship between innovative activity and firm size may well be increasingly U-shaped, rather than r-shaped. Technological opportunities that result in large firms have less to do with the requirements of scale in production, than with the possibilities for R&D-based diversification into related product markets. Innovative small firms flourish in providing inputs into production, where large user firms also make a contribution.

Assumptions and outcomes reflecting these findings could, at the very least, be built into a variety of traditions of theorising about the relationships between firm size and innovative activities. Thus, useful refinements could be made in models explaining the relationship between firm size and innovative activity in terms of opportunity, appropriability and market. They could recognise, for example, that most of the technological opportunities associated with large firms are R&D-based product innovations rather than process innovations.

More important, they might explore the nature and determinants of varying degrees of appropriation, or ease of imitation, of different types of innovations. On their nature, our results extend those of Levin *et al.* [1984] in showing that, whilst R&D-based innovations tend to be appropriated by large firms, those in producers' goods are so to a lesser extent, given the low technological barriers to entry from user firms. On their determinants, both Mowery [1983] and Teece [1982] have given persuasive explanations for the internalisation and appropriation of R&D-based innovations in terms of transaction cost theories. Nonetheless, it is not possible to explain the incomplete internalisation and appropriation of production/producers' goods-based innovations in the same terms. Other analyses suggest that any satisfactory explanation of the latter will need to move beyond theories of exchange, to the nature of the production technology itself, and in particular the pervasiveness of its application (Lundvall [1984]; Pavitt [1986]).

The findings of this paper could also form the basis of a model to explain observed patterns and trends in the size distribution of innovating firms: very large firms result from the exploitation of continuing and appropriable technological opportunities in related product markets, emerging from R&D, produced in increasingly international markets, and managed through increasingly efficient and smaller units; small firms continue to be innovative in supplying specialised production inputs, in symbiosis with large, innovative users.

314   K. PAVITT, M. ROBSON AND J. TOWNSEND

In the meantime, empirical analysis will benefit from continued improvement in the measurement of technological opportunity and appropriability.[13] It will also benefit from continuous and comparative tracking over time of firms' innovative activities and growth paths. The data sources used in this paper are not suitable for this purpose. Those from other sources— particularly firm-specific statistics on patenting activity[14]—may well be.

Finally, our findings do not point to easy or obvious prescriptions for the policy-maker. Given the high variance in the size distribution of innovating firms both within and between sectors, grand generalisations are often likely to be wrong, and grand policies often likely to be inappropriate. It is tempting to conclude that, under such circumstances, diversity and pluralism should be the only objectives of policy.

However, such a conclusion overlooks some uncomfortable facts emerging from the analysis. In particular, "technological ease of imitation" differs considerably amongst sectors. In some, successful innovating firms will end up being very large. In others, successful innovating firms will be small, but with strong links with often large user firms. In a technological dynamic regime, pluralism and diversity will inevitably include large firms, innovative in their product and process technology. The numbers of innovative small firms will also reflect in part the accumulated skills and strategies of the large user firms on whom they depend for skills, knowledge and markets, rather than any effects of a (difficult to define) "enterprise culture".

KEITH PAVITT, MICHAEL ROBSON AND   ACCEPTED JUNE 1986
JOE TOWNSEND,
Science Policy Research Unit,
University of Sussex,
Mantell Building,
Falmer,
Brighton BN1 9RF,
East Sussex,
UK.

REFERENCES

BUSINESS STATISTICS OFFICE, various dates, *Historical Research of the Census of Production, 1907–1920, Report on the Census of Production, 1973, 1975, 1978 and 1982,* Government Statistical Service (HMSO, London).

BUSINESS STATISTICS OFFICE, 1979, 'Industrial Research and Development Expenditure and Employment 1975', *Business Monitor* No. 14, Government Statistical Service (HMSO, London).

---

[13] In a research project funded by the US National Science Foundation, Levin *et al.* [1984] at Yale University are collecting systematic data on appropriability through a survey of R&D directors in large US firms.

[14] On sources and limitations of patenting statistics, see Pavitt [1985].

DASGUPTA, P. and STIGLITZ, J., 1980, 'Industrial Structure and the Nature of Innovative Activity', *Economic Journal*, Vol. 90, June, pp. 266–293.

FISHER, F. and TEMIN, P., 1973, 'Returns to Scale in Research and Development: What does the Schumpeterian Hypothesis Imply?' *Journal of Political Economy*, Vol. 81, pp. 56–70.

FREEMAN, C., 1971, *The Role of Small Firms in Innovation in the United Kingdom since 1945*, Report to the Bolton Committee of Inquiry on Small Firms, Research Report No. 6 (HMSO, London).

FREEMAN, C., 1982, *The Economics of Industrial Innovation*, Second Edition (Francis Pinter, London).

VON HIPPEL, E., 1978, 'A Customer-Active for Industrial Product Idea Generation', *Research Policy*, Vol. 7, No. 3, July, pp. 240–266.

KAPLINSKY, R., 1983, 'Firm Size and Technical Change in a Dynamic Context', *Journal of Industrial Economics*, Vol. XXXII, No. 1 September, pp. 39–59.

LEVIN, R., KLEVORICK, A., NELSON, R. and WINTER, S., 1984, 'Survey Research on R and D Appropriability and Technological Opportunity Part I: Appropriability', (mimeo), Yale University, Dept. of Economics.

LEVIN, R., COHEN, W. and MOWERY, D., 1985, 'R & D Appropriability, Opportunity, and Market Structure: New Evidence on the Schumpeterian Hypothesis', *American Economic Review*, May.

LUNDVALL, B-å., 1984, 'User-Producer Interaction and Innovation', paper prepared for the Technology Innovation Project (TIP) Workshop, Stanford University, Dept. of Economics, December.

MOWERY, D., 1983, 'The Relationship between Intrafirm and Contractual Forms of Industrial Research in American Manufacturing, 1900–1940', *Explorations in Economic History*, No. 20, pp. 351–374.

NELSON, R. and WINTER, S., 1982, *An Evolutionary Theory of Economic Change* (Belknap Press, Cambridge, Mass.).

PAVITT, K., 1984, 'Sectoral Patterns of Technical Change: Towards a Taxonomy and a Theory', *Research Policy*, Vol. 13, No. 6, pp. 343–373.

PAVITT, K., 1985, 'Patent Statistics as Indicators of Innovative Activities: Possibilities and Problems', *Scientometrics*, Vol. 7, Nos. 1–2, pp. 77–99.

PAVITT, K., 1986, 'Technology, Innovation and Strategic Management', in J. MCGEE, and H. THOMAS, eds., *Strategic Management Research: A European Perspective* (Wiley, New York).

PRAIS, S., 1976, *The Evolution of Giant Firms in Britain* (Cambridge University Press, Cambridge).

ROBSON, M. and TOWNSEND, J., 1984, 'Users Manual for ESRC Archive File on Innovations in Britain Since 1945: 1984 Update' (Science Policy Research Unit, University of Sussex).

ROSENBERG, N., 1976, 'Technological Change in the Machine Tool Industry, 1840–1910', in *Perspectives on Technology* (Cambridge University Press, Cambridge).

ROSENBERG, N., 1982, 'Learning by Using', in *Inside the Black Box: Technology and Economics* (Cambridge University Press, Cambridge).

ROTHWELL, R., 1983, 'Firm Size and Innovation: A Case of Dynamic Complementarity', *The Journal of General Management*, Vol. 8, No. 3, pp. 5–25.

SCHERER, F. M., 1973, *Industrial Market Structure and Economic Performance* (Rand McNally, Chicago).

SCHERER, F. M., 1965, 'Firm Size, Market Structure, Opportunity and the Output of Patented Inventions', *American Economic Review*, Vol. 55, No. 5, pp. 1097–1125.

SOETE, L., 1979, 'Firm Size and Inventive Activity: The Evidence Reconsidered', *European Economic Review*, Vol. 12, pp. 319–390.

STONEMAN, P., 1983, *An Economic Analysis of Technological Change* (Oxford University Press, Oxford).

TEECE, D., 1982, 'Towards an Economic Theory of the Multi-product Firm', *Journal of Economic Behaviour and Organisation*, Vol. 3, No. 1, pp. 39–63.

TOWNSEND, J., HENWOOD, F., THOMAS, G., PAVITT, K. and WYATT, S., 1981, *Innovations in Britain since 1945*, Occasional Paper No. 16, (Science Policy Research Unit, University of Sussex).

WYATT, S., 1985, 'Il Ruolo delle Piccole Imprese nell'attavita innovativa: un'analisi del caso inglese', *Economia e Politica Industriale*, No. 45, pp. 47–82.

10

1 | 'Chips' and 'Trajectories': how does the semiconductor influence the sources and directions of technical change?

Keith Pavitt

Introduction

In this essay, I propose a framework to explain what the 'chip', or the micro-electronics revolution based on the semiconductor, is doing to the sources and directions of technical change in the industrially advanced countries. Christopher Freeman's own research style and achievements suggest why and how such an essay should be written.

First, he has always stressed the importance of describing and understanding present and likely future patterns of technical change.[1] In the 1970s, he jointly led a programme exploring the links between economic growth and global resource depletion, concluding that continuous technical change and social adaptation were essential if the former were to be achieved without the latter.[2] In the 1980s, he and his colleagues have built on the work of Schumpeter, in order to explore the implications of the fundamental and pervasive innovation that is the semiconductor, for trends in the level of economic activity in general, and for employment levels and skills in particular.[3] Like him, I shall assume that the semiconductor does have a significant impact on the rate and direction of technical change in a wide variety of sectors. However, I shall concern myself with the impact on the type of firm that will make innovations, and on the nature of the innovations themselves.

Second, Freeman has always argued that the sources, the rate and the direction of invention and innovation cannot be explained by, or derived from, some simple, general rule. After being asked in the late 1960s to contribute to the debate about the role of small firms in making innovations, he began a painstaking programme of data collection on the sources of more than one thousand significant innovations introduced into the United Kingdom since 1945. He was thereby able to show the considerable variance among sectors in the relative contributions of small and large firms.[4] During the 1970s, J. Townsend and others built up this data, so that in 1981 it comprised more than two thousand, covering more than half of British manufacturing.[5] For each innovation, information was collected on the main knowledge sources, the sectors of production and main use of each innovation, and on the size and principal sector of activity of the innovating firm. These data have enabled this writer to describe and explain intersectoral differences in the sources and directions of technical innovation,[6] as well as to define some of the more general characteristics of technology, innovation and technical change.

These sectoral patterns (or 'trajectories') of technical change are the starting-point of this essay. After briefly describing and explaining them, I suggest tentative hypotheses about the way in which the trajectories are influenced by developments in semiconductor technology. Where possible, I compare these hypotheses with empirical evidence, before reaching some conclusions for theory and for policy.

Technological trajectories

Empirical research shows that technological knowledge in the modern economy has two interconnected properties: it is mainly specific in application, and cumulative in development. Its specificity to particular applications is reflected in three characteristics: first, the heavy concentration of firms' innovation—generating expenditures on development and on production engineering activities, both of which are specific to one product and one production process;[7] second, the heavy reliance of the typical innovating firm on knowledge that is not public, but specific to itself or to other firms in technologically related lines of business; third, the considerable expenditures on technological assimilation and adaptation that are typically incurred when transferring technology from one firm to another, or even from one place to another.[8] Given this specificity in technological knowledge and skills, firms do not make a generalized search when deciding where to move next technologically, but explore zones that use technological knowledge similar to that they already know. Firms at different technological starting-points therefore follow different and cumulative technological trajectories. These can be observed in sectoral patterns of firms' innovative activities, as reflected in either significant innovations made, or R & D resources spent.[9]

Nelson and Winter[10] have proposed three factors that determine technological starting points and trajectories: the available sources of technology, the requirements of users, and the possibilities open to the innovating firm to benefit more than its competitors from any innovations on which it spends resources. This formulation, together with the data compiled by Townsend *et al.*[11] on significant British innovations, has enabled this writer to describe and explain sectoral trajectories. Firms can be divided into three categories: supplier-dominated, production-intensive and science-based. Table 1.1 describes their typical core sectors, as well as the nature, determinants and measured characteristics of their technological trajectories.

Table 1.1 Sectoral technological trajectories: determinants, directions and measured characteristics

Category of firm (1)	Typical core sectors (2)	Determinants of technological trajectories			Technological trajectories (6)	Measured Characteristics			
		Sources of technology (3)	Type of user (4)	Means of appropriation (5)		Source of process technology (7)	Relative balance between product and process innovation (8)	Relative size of innovating firms (9)	Intensity and direction of technological diversification (10)
Supplier-dominated	Agriculture Housing Private services Traditional manufacture	Suppliers Research & extension services Big users	Price-sensitive	Non-technical (e.g. trade marks, marketing, advertising, aesthetic design)	Cost-cutting	Suppliers	Process	Small	Low vertical
Production-intensive: Scale-intensive	Bulk materials (steel, glaze) Assembly (consumer durables & autos)	PE Suppliers R & D	Price-sensitive	Process secrecy & know-how Technical lags Patents Dynamic learning economies	Cost-cutting (Product design)	In-house suppliers	Process	Large	High vertical
Specialized suppliers	Machinery Instruments	Design & Development Users	Performance-sensitive	Design know-how Knowledge of users Patents	Product design	In-house suppliers	Product	Small	Low concentric
Science-based	Electronics electrical Chemicals	R & D Public science PE	Mixed	R & D know-how, patents Process secrecy & know-how Dynamic learning economies	Mixed	In-house suppliers	Mixed	Large	Low vertical / High concentric

Note: PE = Production Engineering Department.
Source: K. Pavitt, 'Sectoral Patterns of Technical Change: Towards a Taxonomy and a Theory', *Research Policy*, 13 (1984), pp. 343–73.

34   Keith Pavitt

Supplier-dominated firms make very little contribution themselves to either their product or their process technology. They can be found mainly in traditional sectors of manufacturing like textiles, in agriculture, in house building, and in many professional, financial and commercial services. Most innovations come from suppliers of equipment and materials, although in some cases contributions are made by government-financed research and extension services, by large customers, or by the relatively few large firms in the sector. Technological trajectories are defined in terms of cost-cutting, based on what is offered by suppliers.

Over time, some firms will evolve from the supplier-dominated to the *production-intensive* category. In 1776 Adam Smith described one of the mechanisms of this evolution: an increasing division of labour and simplification of tasks in production, resulting from an increased size of market, and leading to an increasing substitution of machines for labour. The pressures and incentives to exploit scale economies are particularly strong in what I call scale-intensive firms selling to largely price-sensitive users: those producing standard bulk materials through continuous processes and those producing durable consumer goods and vehicles. In both cases, production processes have become increasingly large, complex and interdependent. Their smooth operation cannot be taken for granted and the costs of failure in any one part of the production system are considerable. Satisfactory operation has therefore come to depend on increasingly professionalized 'production engineering' or 'process engineering' departments, which themselves become an important source of technical change in production processes, and related machinery and instruments.

These scale-intensive firms live in symbiosis with specialized firms supplying production machinery and instrumentation, and who have different technological trajectories from their customers. Given the scale and the interdependence of the production systems to which they contribute, the costs of poor performance of their products are considerable. The technological trajectories of specialized suppliers are therefore strongly focused on the performance and reliability of their products. Specialized suppliers benefit from close and continuous contact with their customers, who often pass onto them skills, information on operating experience, improvements made to equipment in use, and even resources for the design and testing of new equipment. At the same time, each customer benefits from suppliers who have assimilated information and improvements from a large number of users.

In the nineteenth century, *science-based* firms began to emerge, as a result of key scientific discoveries: electromagnetism, radio waves and transistor effects contributed to what is now the electrical and electronics industry, while chemical synthesis and biological synthesis have been the basis of what is now the chemical industry. These key discoveries, and related R & D activities in universities and firms, have been determining influences on the firms' technological trajectories. The pervasive and varied range of applications growing out of these science-based techniques have both enabled rapid growth for successfully innovating firms, and dictated varying emphases on product or process innovation. In bulk synthetic materials and standard consumer products, the same pressures for cost-cutting and increased scale of production have existed as in the production-intensive sectors, whilst in pharmaceuticals, other fine chemicals and in electrical and electronic machinery and equipment, there has been greater pressure for product reliability and performance.

The semiconductor and sectoral directions of technical change

From this analysis, it is clear that the *directions* of technological accumulation in firms are conditioned by the opportunities opened up by fundamental technologies. As Perez has argued,[12] such technologies are defined by the potential that they offer for cost reduction and new products in a wide range of applications. Thus, science-based firms have benefited principally from techniques emerging from research in physics and chemistry; whilst firms in the production-intensive categories have benefited from technologies that have increased the range of potentially useful mechanical products that can be made, and the possibilities of exploiting scale economies in producing them: in particular, from steelmaking, electricity, petroleum and the internal combustion engine.

Similarly, future directions of technical change in firms will be influenced by both their existing technological trajectories and the following specific characteristics of semiconductor technology: first, that it grows out of electronics technology; second, that it contributes to two primary technical functions—information processing, and monitoring and control; third, that both these technical functions have improved at a prodigious rate in the past thirty years, as a result of continuous and rapid improvements in product design, allied to steep dynamic learning economies in the production of semiconductor-based circuitry.[13]

Given these characteristics, we would expect semiconductor technology—like earlier science-based technologies—to have pervasive effects on process technologies in all sectors. Monitoring and control technologies will be adopted particularly rapidly in those sectors where production technologies are large-scale and interdependent, and where firms have the in-house competence to use, to specify and, if necessary, to develop the appropriate production/process control systems: in other words, in firms in our science-based and production-intensive categories. As a consequence, we would also expect improvements in monitoring and control systems to offer opportunities for product innovation more generally in sectors that make production equipment. At the same time, semiconductor technology will help continue the past trajectory of the electronics/electrical sector in creating ample opportunities for product innovations to be adopted in manufacturing, services and households.

Data collected by Northcott, Rogers and Zeilinger on the use of micro-electronics in more than one thousand establishments in British manufacturing in 1981 confirm these expectations.[14] Table 1.2 shows, for each two-digit sector, the proportion of establishments in the sample that had used microelectronics in processes and in products. The third column shows that the sectors that we have defined as *science-based* and *production-intensive* (food and drink, chemicals and metals, mechanical engineering, electrical and instrument engineering, vehicles) have above average use of microelectronics in processes, whilst those that we define as *supplier-dominated* (textiles, clothing and leather) have below average use. Furthermore, the second column shows that the opportunities for product innovations are heavily cononcentrated in electronics and instrument engineering, but are also above average in mechanical engineering and vehicles, again in line with our expectations.

The semiconductor and the sectoral sources of technical change

Thus, with the partial exception of chemical firms, the spread of semi-conductors will not change fundamentally the intersectoral differences in the balance between product and process innovations made by firms, or in the rate of adoption of advanced process technology. But what effect will it have on the sectoral *sources* of semiconductor-related technology? Whilst we can expect an increasing use of electronic skills and electronics products in a wide range of sectors, our analysis of the cumulative nature of technical change suggests that the sectoral sources of electronics-based innovations in production-intensive and supplier-dominated firms will change only marginally, if at all.

Production-intensive firms

We would expect that the main effect of semiconductors in production-intensive firms will be the augmentation of the already existing process-dominated technological trajectories, through the application of monitoring and control technology. As with earlier mechanical and electromechanical production technologies, we would expect significant contributions to their development by both large firms in the user sectors, and small specialized firms supplying equipment incorporating monitoring and control functions.

Evidence from the SPRU data bank on the characteristics of British

Table 1.2 Use of microelectronics by industry

Industry	No. of establishments	Product users	Process users	All users	Non-users	All
		(percentages of the establishments in each industry)				
Food & drink	(125)	0	56	56	44	100
Chemicals & metals	(134)	0	51	51	49	100
Mechanical engineering	(165)	29	43	55	45	100
Electrical & instrument engineering	(133)	58	60	76	24	100
Vehicles	(92)	16	51	54	46	100
Other metal goods	(99)	4	40	40	60	100
Textiles	(97)	0	31	31	69	100
Clothing & leather	(92)	1	21	21	79	100
Paper & printing	(99)	0	52	52	48	100
Other manufacturing	(164)	5	40	40	60	100
TOTAL	(1,200)	13	45	49	51	100

Source: J. Northcott, P. Rogers and A. Zeillinger, *Microelectronics in Industry: Survey Statistics*, London, Policy Studies Institute, 1982.

innovations and innovating firms in instrument engineering between 1945 and 1980 confirms that this is the case. Instrument engineering is the product group that best represents products embodying electronics-based functions for monitoring and control. Significant innovations have mainly been made by both specialized supplier firms and by larger firms, some of which are themselves users of the innovations. As Table 1.3 shows, the sources and directions of innovations in instrument engineering are similar to those in mechanical engineering. Innovating firms with their principal activities in instruments are relatively small (columns 4–6), technologically specialized (column 2) and with a high proportion of their innovations used in other sectors, i.e. product innovations (column 3); at the same time, a relatively high proportion of innovations are made by relatively large firms (including users) principally engaged in other sectors (columns 8–10). Furthermore, Table 1.4 shows in column 4 that users of instrument engineering in process technology are fairly evenly spread across manufacturing, accounting overall for 13 per cent of the process innovations adopted. Column 2 shows that 6 per cent of the innovations produced by manufacturing firms, other than those principally in instruments, use instrument innovations, the percentages being higher in firms typified by continuous processes (iron and steel, cement and glass, paper), as well as in shipbuilding and in electronics and electrically-based firms.

Similar sources of innovation can also be observed in the development of major electronics-based technologies in manufacturing assembly: computer aided design (CAD), numerically controlled machinery (NCM), and robots, where a large number of studies describing and explaining their emergence and diffusion have been completed in the last two years.[15] They all show the major role of users of the technologies in their development. In the 1950s and early 1960s, the major stimuli came from the US Federal Government, for the machining of parts for the high performance products of the aerospace industry; there were similar but smaller programmes in France and the United Kingdom. By the late 1960s, the locus of technological activity began to shift towards the electrical/electronics, automobile and shipbuilding sectors, and towards Western Europe and Japan. According to Gönenc,[16] automobile firms have played a major role in the development and the diffusion of the technologies, given their size, technological resources, and central position in a range of metalworking and metal-using technologies. According to Carlsson, users have been the main driving force in technological change in NCM.[17] The same pattern emerges from a recent Japanese study of the introduction of assembly-based automation.[18]

Users' major involvement results from systemic interdependence in production, and from differentiated technological requirements.

Table 1.3 Characteristics of innovation firms and innovations in instrument engineering and mechanical engineering

Sector	% of innovations produced by firms principally in the sector that are:		Size distribution of employment of innovating firms principally in the sector (Rows add up to 100%)			Innovations in the sector that are made by firms with their principal activities in other sectors			
	Innovations in other product	Used in other sectors	10,000+	1,000–9,999	1–999	As % of all innovations in sector	Size distribution of firms		
							10,000+	1,000–9,999	1–999
(1)	(2)	(3)	(4)	(5)	(6)	(7)	(8)	(9)	(10)
Instrument engineering	19.9	81.4	24.6	21.4	54.0	54.8	54.1	13.8	32.0
Mechanical engineering	16.0	82.5	24.3	36.9	38.8	31.9	67.3	15.2	17.5
All sectors in the sample	31.5	64.0	53.1	21.9	24.9	31.4	53.1	21.9	24.9

Source: SPRU Databank on British innovations since 1945.

40   Keith Pavitt

Table 1.4 Instrument innovations produced and used in manufacturing sectors other than instrument engineering

Sector	Instrument engineering innovations as % of all innovations produced by firms principally in the sector		Instrument engineering innovations as % of all innovations used in the sector	
	% of all innovations produced	All innovations produced	% of all innovations used	All innovations used
(1)	(2)	(3)	(4)	(5)
Food and drink	1.3	78	10.3	68
Chemicals	4.8	78	10.3	71
Metal production	11.9	143	19.2	130
Mechanical engineering	6.5	536	20.1	169
Electronics and electrical	15.2	343	12.0	167
Shipbuilding	12.4	89	10.0	90
Vehicles	3.2	158	8.1	221
Textiles	2.6	77	12.7	377
Leather and footwear	2.0	50	4.4	45
Glass and cement	11.5	87	19.0	63
Paper	11.6	43	10.3	39
Total manufacturing (excluding instruments)	6.0	1,946	13.4	1,452

Source: SPRU Databank on British innovations since 1945.

Symbiotic relationships exist between large-scale users and specialized suppliers of production equipment, with operating experience, specifications, designs and skills flowing from the former to the latter. In NCM, specialized suppliers are on the whole the traditional machinery suppliers that have succeeded in integrating the new electronics technology; whilst in CAD and robots, they are on the whole new entrants spinning off from large-scale users. In a thorough survey, the OECD has recently shown a similar pattern in automation-related software, with significant contributions being made by both specialized suppliers and large-scale users.[19] Many of the latter have established their own automation centres, and some of them are diversifying vertically into automation engineering.

As Rosenberg has shown,[20] this pattern of interdependent users and producers of capital goods has always been central to technological change in the production-intensive sectors. He also identified two other essential features in the pattern, the first of which is the important stimulus to technical change provided by technological imbalances in interdependent production systems. Interdependencies and bottlenecks were particularly important in determining the early rate and direction of development of automation technologies. Numerically controlled machining was essential for aircraft wing configurations proposed in the early 1960s, and CAD was a complement in the design of these configurations. Similarly, CAD/CAM were essential for the design and manufacturing of large-scaled and very large-scaled integrated circuits.

Rosenberg has also stressed the importance of 'vertical disintegration' and 'technological convergence' in the development and diffusion of technologies amongst firms and sectors. However, extension of CAD/CAM to other industries will require the solution of further technical problems. Arnold and Senker[21] have pointed out that the design, machining and assembly of mechanical components that are three-dimensional, irregular and heavy have different and in some ways more difficult technical requirements than those for electronic components that are two-dimensional, regular and light in weight; whilst applications in clothing require, as Hoffman and Rush[22] pointed out, the solution of the long-standing problem of the automated handling of non-rigid materials. Many of these technologial bottlenecks are non-electronic in nature, and often require improvements in mechanical technology: for example, Rendeiro[23] points out that improvements in performance resulting from NCM have depended in part on the finishing and accuracy of lead screws and other mechanical components.

As and when technological bottlenecks have been overcome, the benefit of adopting electronics-based assembly technology have been multiform. Labour saving has remained a major trajectory. Gönenc[24] has suggested—as Rosenberg[25] has done for the nineteenth century—that the stimulus comes not just from increasing labour costs, but from the desire to decrease dependence on an unreliable labour-force. He also suggests that—at least in Scandinavia—it has come from the pressure to eliminate labour from tedious and dangerous work. However, most analysts identify another set of benefits which they give at least equal importance. These are improved product performance and quality, increased speed of product development, reduced holdings of inventory, and increased flexibility in production that can result from the combined uses of CAD, NCM and robots.[26]

Supplier-dominated firms

The channels of diffusion of electronics-based production into the supplier dominated sectors in traditional manufacturing are similar to those of vertical disintegration and technological convergence described by Rosenberg.[27] The development and use of the technology in the leading user sectors (i.e. aerospace, automobiles, electrical/electronics) improves performance and reduces costs to a level where specialist firms staffed with personnel previously trained in these user firms can begin to adapt and simplify the technologies for use in supplier-dominated firms. Hoffman and Rush[28] have described and analysed in some detail how this process has affected the clothing industry. CAD has begun to revolutionize the design process, and both NC sewing machines and robots are to be used in the industry. Most of these innovations have come from both traditional and new equipment suppliers, although a few large user firms in clothing have also made a contribution, and large customers of the clothing industry continue to put pressure on the many small and technically understaffed firms in the industry to adopt the latest technology.

In other words, the sources of technical innovation in clothing remain largely outside the industry, even though the nature and the pace of such innovation are being profoundly modified. Whilst this is likely to remain true for most of traditional manufacturing and for many services, a significant shift in the sources of innovation is under way in certain firms with heavy and complex operations in information processing. Decreasing costs and increased performance of equipment have opened new opportunities for increasingly automated and interdependent systems of information processing, progressively replacing the free-standing information processing 'tools' (typewriters, computers, filing systems) that had existed previously. These organizations have developed professionalized 'systems engineering' groups, whose functions are similar to 'production engineering' and 'process engineering' groups in scale-intensive manufacturing, namely, to specify, assimilate and operate increasingly complex and interdependent 'production' systems. Over time, these user organizations have developed the capacity to write their own software; to modify, design and specify their own interdependent systems; and eventually to build their own hardware. As a result, the sources of technical change move progressively from complete reliance on general purpose equipment suppliers to an increasing contribution by large-scale user firms themselves, and by specialized suppliers of hardware and software, many of which are 'spin-offs' from users.

The recent OECD report confirms this trend in data processing software.[23] Over time, the share of production has shifted markedly from computer suppliers to computer users, and to newly formed software houses. The study by Barras and Swann of the British insurance industry shows the same trend, and explains that it results from incompatibilities amongst bits of equipment, from lack of appropriate software from general purpose suppliers, and from product differentiation.[29] Both studies confirm the importance of specialized data-processing groups in user firms.

However, some of the findings suggest that this trajectory of technological development may not continue. Trends in hardware and software are inevitably dominated by the wide range of products offered by IBM.

Deviation by users from their range as a result of in-house development generally leads to big adjustment costs back to the dominant (i.e. IBM) trajectory. Specialized data-processing groups in user firms can be harmful to the extent that they reduce the degree to which the design of products and production systems are compatible with package software provided by suppliers; and the role of such groups may diminish with the advent of distributed processing. Equipment supplier firms, on the other hand, may reverse the trend and increase their contribution to software production, as a result of technical change in hardware and software, and of changes in their strategies.

The semiconductor and the size structure of innovating firms

We would expect that the size structure of innovating firms will reflect both earlier technological trajectories, and the specific effects of semi-conductor technology. In the electronic/electrical sector, successfully innovative firms will continue to grow through product market diversi-fication. Those that have successfully exploited semiconductor tech-nology since the 1950s have already become very big. Freeman[30] pointed out that employment in IBM increased from 22,000 in 1946 to 205,000 in 1971, and by now it is more than 350,000. However, experience over the past thirty years shows that these large organizations do not succeed in exploiting commercially all of the considerable number of product market opportunities emerging from their technological activities; some of the 'spin-off' firms started by their former personnel can become very big.

We would also predict that developments in microchip technology will enable the emergence of an increasing number of small, specialized and technologically progressive firms. Some of these—like suppliers of elec-tronic chip-making machinery—will emerge in symbiosis with the mass production of electronic products.[31] In addition, just as the provision of cheap and high-quality steel in the nineteenth century enabled the forma-tion of myriads of small machinery firms, capable of matching steel to a great variety of differentiated mechanical functions, so the provision of cheap information processing capacity in the microchip will enable an increasing number of small firms to match this capacity to an equal variety of differentiated information and control functions. The formation of specialized firms in instrumentation engineering, described earlier, is an early manifestation of this trend. Just as the typical technological activity of the small machinery firm is machine design, so the typical technological activities of the small information-based firms will be soft-ware, and the design of special chips and of interfaces with bigger systems. Thus, the semiconductor may result in a bimodal size distribu-tion of innovating firms in the electronics/electrical industry, with both increasingly large, innovative and diversified firms, as well as an increasing number of small and specialized ones.

Some writers have argued that quite the reverse will happen in the production-intensive categories of firms: on the one hand, assembly-based firms can survive being somewhat smaller, whilst machinery suppliers can survive only by getting bigger. In the major assembly-based industry—automobiles—Altshuler et al.[32] have documented the adoption of modern and flexible automation. In contrast to the earlier dedicated automation, it can produce and assemble components for a variety of models and variants. It will reduce the volume and increase the skill of labour required, reduce economies of scale, and increase flexibility. The authors argue that medium-sized automobile producers will not in future suffer from cost disadvantages, or from over-commitment to the one model or variant.

At the other extreme, it is often argued that specialized firms supplying capital goods will have to increase in size in the future, given that increasing flexibility will enable standard machines to be manufactured in series or large batches, and then to be programmed for specific uses.[33] It seems to this writer that such a trend is unlikely for two reasons. First, it ignores the enormous *mechanical* diversity of machining requirements. It is unlikely that the same machine could be used efficiently for, say, manufacturing gearboxes and cylinder blocks. What flexible automation allows is the manufacture of more than one type of gearbox on the same machine, or of cylinder block on another machine. Second the argument ignores the opportunities open to the machine builders themselves, as a result of flexible automation: CAD, NCM and robots enable small and specialized supplier firms both to adapt their products to the specific requirement of users, and to take advantage of any economies of scale in component production or purchase. As a result, we would expect small and specialized machinery firms to continue to be able to design and sell equipment specifically adapted to the requirements of the large and sophisticated users.

However, larger firms amongst equipment suppliers might gain another advantage as a result of the application of semiconductor technology. Flexible automation is opening up new markets in small-firm traditional manufacture and services, because it has become relatively cheap, and because it can increasingly be used in small batch or one-off production.[34] These user firms are unlikely themselves to become important sources of new technology, even if automation increases their relative efficiency. Larger equipment-producing firms will have an advantage in these markets, since a strong capability will be necessary for the design and engineering of complete production systems, and the education of users.[35]

Creative destruction or creative accumulation?

To what extent has the pervasive diffusion of semiconductor technology been a process of 'creative destruction;, where the new displaces the old, or of 'creative accumulation', where the new builds on the old? The diffusion of what Freeman and his colleagues have called a 'new technology system'[36] involves—almost by definition—the infusion of new skills, and a speeding-up of the rate of advance of frontier, state-of-the-art technology. These factors in themselves might appear to be generating a process of creative destruction by speeding up the growth and decline of firms. As Nelson and Winter[37] have shown, industrial structures are more volatile when the rate of technical change is faster, and when competitive advances in skills and technology are more easily appropriated by innovating firms. However, these more rapid rates of change of structures and market shares—or what Klein[38] has called 'fast history'—are not synonymous with our definition of 'creative destruction', which requires in addition that growth be associated with new skills and organizations independent from old ones. Our theory of technical change, which stresses the cumulative and differentiated nature of technical change, suggests that accumulation and complementarity of skills in existing organizations typify the diffusions of a new technology system rather than destruction and displacement. Most of the evidence suggests that this is the case.

To begin with, there is the stability within the fast-moving electrical/ electronic sector itself. Table 1.5 identifies seventeen American and West European firms as leaders in American patenting activity in key areas of electronics in the 1970s. Ten of these had aready been identified by Freeman (in *The Economics of Industrial Innovation*) as leaders in British patenting in electronic capital goods and office machinery in the 1930s, late 1940s and 1950s: IBM, Philips, RCA, Westinghouse, Bell, General Electric, Siemens, Burroughs, Sperry, Rand and ITT. The remaining seven can be considered as new entrants that have grown very rapidly as a consequence of relatively high rates of innovation: Texas Instruments, Motorola, Honeywell, Hewlett-Packard, Xerox, Data General and Rockwell. Whilst this rate of entry of new firms is probably higher than in most other sectors, it cannot be described as 'creative destruction', since old-established firms continue to create technically at a higher rate than the new entrants. However, the semiconductor has changed the shape of the firms and the competition within the industry, since it has led to growing tehnological convergence and interdependence among office machinery, consumer electronics, telecommunications, and industrial automation.[39]

Stability also typifies the applications of the semiconductor in

Table 1.5 Shares of top ten organizations in American patenting in selected areas of microelectronics since 1969

Patenting organization	Integrated circuit structure patents		Patenting organization	CPUs and other systems patents		Patenting organization	Digital logic circuits patents		Patenting organization	Semi-conductor memories patents		Patenting organization	Speech analysis and synthesis patents	
	%	No.		%	No.		%	No.		%	No.		%	No.
IBM	12.2	289	IBM	12.2	73	IBM	12.6	199	IBM	16.9	194	Bell	12.2	85
Texas Instrs.	8.2	194	Texas Instrs.	12.0	72	RCA	5.9	93	Texas Instrs.	6.7	77	IBM	4.2	29
US Philips	6.9	163	Burroughs	5.7	34	Motorola	5.4	85	RCA	5.7	65	NEC	3.1	23
RCA	6.5	153	Motorola	4.7	28	Bell	4.1	65	Bell	5.7	65	US Navy	2.7	19
Hitachi	5.5	131	Honeywell	4.3	26	Hitachi	3.7	58	Siemens	4.4	50	Texas Instrs.	2.3	16
Motorola	5.1	121	Hewlett-Pack.	2.8	17	Texas instrs.	3.2	51	Gen. Elec.	3.3	38	US Army	1.9	13
Westinghouse	3.4	80	Bell	2.5	15	Rockwell	3.0	47	Motorola	3.2	37	US Philips	1.9	13
Bell	3.3	79	Sperry Rand	2.5	15	Toshiba	2.6	41	Hitachi	3.1	35	Hitachi	1.4	10
Gen. Elec.	3.3	78	Xerox	2.3	14	Siemens	2.5	39	Westinghouse	2.8	32	ITT	1.4	10
Siemens	2.6	61	Data General	2.2	13	Westinghouse	2.1	33	Sperry Rand	2.6	30	Sharp	1.3	9
Top 10 above	57.0	1,349		51.3	307		45.0	711		53.4	623		32.7	227
Total	100	2,367		100	599		100	1,579		100	1,147		100	695

Note: For IC structure and CPUs, organization patents granted 1/1969–12/1982. Total from 1/1969–12/1979. Total from 1/1969 to 12/1982. For digital logic circuits, semiconductor memories and speech analysis, organization patents granted 1/1969–6/1982. Total from 1/1969 to 12/1982.
Source: Office of Technology Assessment and Forecast, *Patent Profiles: Microelectronics I, and II*, Washington: US Department of Commerce, 1981, 1983.

production-intensive sectors. Large-scale and established users in sectors typified by continuous processes and mass assembly have played a major role in developing control instrumentation technology and flexible automation. As with earlier generations of capital goods, development and diffusion have happened in symbiosis with small and specialized supplier firms. In the case of NCM, these suppliers are in general the long-established ones who have successfully assimilated electronics technology and skills into their machine design and operation; Rendeiro[40] shows that, in NC machine tools, manufacturers of numerical controls have on the whole remained separate and distinct from the machine builders. Even in the case of CAD and robots, where equipment suppliers have often been new entrants, they have generally emerged from large-scale users.

There is, none the less, case evidence of abrupt and discontinuous changes resulting from microelectronics: for example, watches, meters, calculators, printing and publishing. One possible explanation is that they are products that have a high informational content, and where there has been a radical switch from electromechanical to electronics-based technology. However, if this were a universal law, IBM would have been destroyed in the office machinery market in the 1950s by the established electronics companies. Similarly, established printing machinery companies would have been displaced by electronics-based ones; but Haywood[41] shows that some of them have successfully diversified into the electronics-based technology.

An alternative explanation is that it is not the *type*, but the *level* of technology in established firms that matters. If firms are in sectors that do have a strong technological component, and if they are at the frontier of established technology, they are more likely to be successful and efficient in assimilating the new technologies than firms that are not. This is a stronger hypothesis than one allowing accumulation from one technological regime to the next. It says that being good at the old is a necessary condition for being good at the new. It is consistent with the stabilities observed in electronics and scale-intensive sectors, and with the estimate of Cranfield Institue of Technology that 60 per cent of robotics specialists in the United Kingdom have their previous experience and training in mechanical and production engineering.[42] It is also consistent with Mowery's observations after examining large American firms' R & D expenditures in the period from 1921 to 1946:

> research investment seems to have acted as a form of insurance for the already large firms, reinforcing the position of dominant firms ... rather than precipitating significant turnover. Such an interpretation is consistent with the evidence ... concerning reductions in the rate of turnover among the largest American firms since 1920.[43]

48   Keith Pavitt

Mowery stresses the development of an oligopolistic market structure as the primary explanation of these trends. However, there are at least three other explanations related to the characteristics of technological accumulation.

First, firms accustomed to managing professional scientists and engineers in the old technologies may be more likely to learn how to attract and to use scientists and engineers in the new technologies. In his study of textile machinery, Rothwell[44] found that firms with better developed R & D laboratories were better able to assimilate and exploit advances from non-mechanical technologies, especially aerodynamics, fibres and electronics.

Second, as we have aready pointed out, the full exploitation of rapid progress in electronics technology often depends on upstream or downstream improvements in more conventional technologies. Particularly revealing evidence of this comes from two countries which, if the level of adoption of robots is any guide, are particularly successful in the exploitation of electronics-based technology, namely, Sweden and Japan. Thus, based on a detailed study of a Swedish engineering workshop, Eliasson concludes that the introduction and spread of electronics will be a relatively slow and piecemeal process, given 'unsatisfactory precision and reliability of measurement and sensory equipment and crude mechanical installations that lay behind in development'.[45] Similarly, a recent Japanese survey of the spread of flexible automation identifies '[the] development of high precision in mechanical technology'[46] as one of four major preconditions for its development and adoption. Compared to conventional mass production systems, flexible automation requires greater process accuracy, given more restricted opportunities for learning by doing.

Third, there is the evidence that effective assimilation of electronics-based technology depends on more general competence in management. Eliasson concludes that one of the most important features determining the assimilation of electronics-based technology in Sweden is 'the lack of centralized knowledge of the production process itself'.[47] Arnold and Senker have called this the 'computerisation effect': according to Arnold 'computers impose a need for orderly, clearly defined systems: they are fast but stupid, unlike people who are slow but intelligent enough to muddle their way through ill-defined procedures'.[48]

From such a perspective, Japanese management innovations introduced in mass assembly can be seen as an essential *prerequisite* for the introduction of flexible automation, since they help make explicit the essential features of conventional production technology, as well as increase the incentives for quality control. According to Altshuler *et al.*, the Japanese approach

assumes that if production workers are given the skills and responsibilities to diagnose problems, repair equipment and spot defects, then
the ranks of supervisors and machine repairmen can be greatly thinned
even as quality is improved ... [and] asserts that high buffers hide problems rather than providing time for their repair; that defect prevention
by workers is far superior to defect detection by supervisors; and that
ideas for improving the production process can come largely from the
line workers who know the system best.[49]

Conclusions

The empirical data on which the above interpretations are based are
highly imperfect. They are also incomplete, given that the proposed
taxonomy in Table 1.1 does not cover what might be called a 'state
capitalist' category of firm, whose products tend to be large-scale and
sophisticated capital goods used by organizations whose purchasing
decisions are heavily influenced by governments: in particular, those
related to defence, energy, transport and communication, the first and last
of which are of central importance in the diffusion of semiconductor-
based technology. Although the data are tentative and incomplete, their
conclusions are clear and straightforward. Sectoral differences in the
sources and directions of technical change are similar both before and
after the assimilation of semiconductor technology; and the adoption of
electronics-based technology is largely a cumulative process, with the
new building firmly on the old.

Thus, in electronics we continue to see the growth of large firms out of
technology-based diversification into new and growing product markets;
in production-intensive firms, symbiotic links remain between large-
scale users, and small and specialized suppliers of capital goods for the
development of electronics-based production technology; and in
supplier-dominated firms, dependence on others continues for applications of semiconductor technology. Compared to existing 'trajectories',
the main changes are a more rapid growth of electronics than chemicals
firms; the growing importance of small and specialized electronics-based
firms providing equipment and software; greater flexibility in scale-
intensive firms; and an emerging technological competence in service
firms that are large-scale processors of information. A number of other
conclusions, both practical and conceptual, also emerge from the
analysis.

First, policies to encourage the introduction and diffusion of microelectronics technology should go beyond support for large firms to
include small electronics-based firms, and both large and small-scale
users of equipment embodying electronics-based functions. It should also

go beyond electronics technologies to include other technologies that are complementary, and potential bottlenecks. According to a British industrialist, R. Curry, writing in 1982:

> I can purchase 'chips' for pennies but I cannot purchase reliable components to interface them to mechanical mechanisms. This means that while it is relatively easy to design calculators and TV games, it is not easy to use microelectronics in the automotive or engineering capital equipment industries. ... Research into microelectronics without corresponding research into, for example, the more mundane problems of the electronic–mechanical interface means that micro-electronics cannot successfully or easily be used in product design.[50]

Second, given this interdependence between established and radical technologies, it is important to distinguish a 'fundamental technology' (i.e. one enabling the emergence of a new, pervasive technological family) from a 'strategic technology' (i.e one whose control enables a firm or a country to achieve competitive success). Thus, whilst the semiconductor is certainly a fundamental technology, it is not necessarily strategic, but will influence strongly the strategic opportunities that do emerge. For example, firms at the frontier of flexible automation do not necessarily control the relevant semiconductor technology itself, but the interface between it and conventional mechanical and production engineering.

Third, it is unrealistic to expect that—as has been suggested by certain analysts[51]—developing and newly industrializing countries will be able to 'leapfrog' mature industrial countries in electronics technologies, because of the latters' strong commitment and competence in older and slower-moving technologies. Where—as is often the case—there are strong complementarities between old and new technologies, developing and newly industrializing countries will be constrained to follow similar technological trajectories to mature industrialized countries. Countries like Japan (and earlier Germany) that have caught up and overtaken leading countries have done so through rapid technological accumulation in both old and new technologies.

Fourth, the exploitation of even radical technological change—whether at the level of the firm, sector or country—is unlikely to be random but to build upon, or grow out of, previously existing competence. Thus, the rapid rate of development and adoption of robots in Swedish industry[52] reflects in part a high pre-existing level of technological competence in mechanical and production engineering. For similar reasons, we would predict that firms with strong technological competence in pharmaceutical products are likely to become strong in future in the exploitation of biotechnology.

Finally, no simple model is likely to describe and explain satisfactorily the sources and directions of technical change in the emergence and diffusion of a major new technology. Thus, the 'product cycle' model,[53] where fast-moving technologies emerge through new small firms and product innovation and then gradually stabilize technologically into large firms concentrating on process innovations, is inadequate. Although new small firms have played an important role in the development of electronics technology, major contributions have been made from the beginning by large firms—in electronics, office machinery and automobiles—and by their spin-off suppliers. Furthermore, a series of *process* innovations in the manufacture of semiconductor devices have been fundamental from the beginning, and have been the basis of subsequent product innovations.

Similarly, explanations of the rate and direction of technical change based entirely on economic signals and institutional constraints have limited explanatory power. For example, they neglect accumulated mechanical and production engineering skills in explaining international variations in the adoption of robots. Or they are tempted to explain the advent of flexible automation as a rapid response in the 1970s to the requirements of slow and volatile economic growth, and to neglect the contribution of the semiconductor, and of twenty years of accumulated experiment and learning in CAD, NCM and robots.

The concepts of technological 'trajectories' or 'paradigms' as developed by Nelson and Winter, Rosenberg, Dosi[54] and this writer may be a more promising analytical path to follow, since they recognize the cumulative and differentiated nature of technology. They also allow that the logic of technology itself does strongly influence—if not determine—the rate and direction of technical change.

Notes

I wish to thank the following colleagues for helpful suggestions and criticisms on an earlier draft of this paper: J. Gershuny, K. Hoffman, D. Jones, M. McLean, J.-J. Salomon, P. Senker.

1. S. Cole, C. Freeman, M. Jahoda and K. Pavitt, *Thinking about the Future: A Critique of 'The Limits to Growth'*, London, Chatto and Windus, 1973.
2. C. Freeman and M. Jahoda, *World Futures: The Great Debate*, London, Martin Robertson, 1978.
3. C. Freeman, J. Clark and L. Soete, *Unemployment and Technical Innovation: A Study of Long Waves and Economic Development*, London, Frances Pinter, 1982.
4. C. Freeman, *The Economics of Industrial Innovation*, London, Frances Pinter, 2nd edn., 1982.
5. J. Townsend *et al.*, *Innovations in Britain since 1945*, SPRU Occasional Paper No. 16, University of Sussex, 1981.

6. K. Pavitt, 'Sectoral Patterns of Technical Change: Towards a Taxonomy and a Theory', *Research Policy*, **13** (1984), pp. 343–73.

7. J. Kamin *et al.*, 'Some Determinants of Cost Distributions in the Process of Technological Innovation', *Research Policy*, **11** (1982), pp. 83–94.

8. K. Pavitt and L. Soete, 'International Differences in Economic Growth and the International Location of Innovation', in H. Giersch, (ed.), *Emerging Technologies: Consequences for Economic Growth, Structural Change, and Employment*, Tübingen, J. C. B. Mohr, 1982; D. Teece, *The Multinational Corporation and the Resource Cost of International Technology Transfer*, Cambridge, Mass., Ballinger, 1977.

9. K. Pavitt, 'Some Characteristics of Innovative Activities in British Industry', *Omega*, **11** (1983), pp. 113–30.

10. R. Nelson and S. Winter, *An Evolutionary Theory of Economic Change*, Cambridge, Mass., Harvard University Press, 1982.

11. Townsend, *et al.*, op. cit. note 5.

12. C. Perez, 'Structural Change and the Assimilation of New Technologies in the Economic and Social Systems', *Futures*, **15** (1983), pp. 357–75.

13. J. McLean and H. Rush, *The Impact of Microelectronics in the UK: A Suggested Classification and Illustrative Case Studies*, SPRU Occasional Paper No. 6, University of Sussex, 1978.

14. J. Northcott, P. Rogers and A. Zeilinger, *Microelectronics in Industry: Survey Statistics*, London, Policy Studies Institute, 1982. A similar and more recent survey in 1983 confirms these results: J. Northcott and P. Rogers, *Microelectronics in British Industry: The Pattern of Change*, London, PSI, 1984.

15. E. Arnold, *Computer-Aided Design in Europe*, Sussex European Papers No. 14, University of Sussex European Research Centre, 1984.

16. R. Gönenc, *Électronisation et Ré-organisations Verticales dans l'Industrie*, Unpublished Thèse de Troisième Cycle, University of Paris X, Nanterre, 1984.

17. B. Carlsson, 'The Machine Tool Industry—Problems and Prospects in an International Perspective', *The Industrial Institute for Economic and Social Research*, Stockholm, December 1983, Working Paper No. 96.

18. 'Preconditions for Flexible Manufacturing System', *Mechatronics News*, **1**, No. 1 (1983), pp. 3–4.

19. OECD, *Software: A New Industry*, Committee for Information, Computer and Communications Policy, document ICP (84)4, February, 1984. I am grateful to Mr H. P. Gassman for bringing this excellent synthesis to my attention. The main contributor to its preparation was R. Gönenc.

20. N. Rosenberg, *Perspectives on Technology*, Cambridge, Cambridge University Press, 1976; N. Rosenberg, *Inside the Black Box: Technology and Economics*, Cambridge, Cambridge University Press, 1983.

21. E. Arnold and P. Senker, *Designing the Future: The Implications of CAD Interactive Graphics for Employment and Skills in the British Engineering Industry*, Engineering Industry Training Board, Occasional Paper No. 9, 1982.

22. K. Hoffman and H. Rush, 'Microelectronics and Clothing: The Impact of Technical Change on a Global Industry', Science Policy Research Unit, University of Sussex, mimeo. 1983.

23. J. Rendeiro, *Technical Change and Strategic Evolution in the Machine Tool Industry*, Sussex European Research Centre, mimeo. 1984.

24. Gönenc, op. cit., note 19.

25. Rosenberg, op. cit., note 20.

26. P. Senker, 'Some Problems in Implementing Computer-Aided Engineering—A General Review', *Computer Aided Engineering Journal*, **1** (1983), pp. 25–31.
27. Rosenberg op. cit., note 20.
28. Hoffman and Rush, op. cit., note 22.
29. R. Barras and J. Swann, *The Adoption and Impact of Information Technology in the UK Insurance Industry*, London, Technical Change Centre, 1982.
30. Freeman, op. cit., note 4.
31. M. McLean, 'Chip Makers Report Economic Lift', *Electronics Times*, 2 June 1983, p. 6.
32. A. Altshuler, M. Anderson, D. Jones, D. Roos and J. Womack, *The Future of the Automobile: Report of MIT's International Automobile Programme*, London, Allen and Unwin, 1984.
33. Gönenc, op. cit., note 19.
34. T. Sasaki, 'Microcomputers in the Japanese Consumer Durables Industry—Status and Prospects', in M. McLean (ed.), *The Japanese Electronics Challenge*, London, Frances Pinter, 1982; Carlsson, op. cit., note 17.
35. Carlsson, ibid.
36. Freeman *et al.*, op. cit., note 3.
38. B. Klein, *Dynamic Economics*, Cambridge, Mass., Harvard University Press 1977.
39. Sasaki, op. cit., note 33.
40. Rendeiro, op. cit., note 23.
41. B. Haywood, *Technical Change and Employment in the British Printing Industry*, World Employment Programme Research, International Labour Organisation, Geneva, September, 1982.
42. Gönenc, op. cit., note 19, Figure 26.
43. D. Mowery, *The Emergence and Growth of Industrial Research in American Manufacturing*, Unpublished Ph.D. dissertation, Stanford University, 1980, p. 7.
44. R. Rothwell, *Innovation in Textile Machinery: Some Significant Factors in Success and Failure*, SPRU Occasional Paper No. 2, University of Sussex, 1976.
45. G. Eliasson, 'Electronics, Economic Growth and Employment—Revolution or Evolution', in Giersch, op. cit., note 8, pp. 77–95.
46. Op. cit., note 18, p. 1.7.
47. Arnold and Senker, op. cit., note 21.
48. Ibid., p. 6.
49. Altshuler *et al.*, op. cit., note 31.
50. R. Curry, Written submission to the House of Lords Select Committee on Science and Technology, *Sub Committee on Engineering Research and Development*, Vol. III—Written Evidence, London, HMSO, 1983, pp. 75–7.
51. L. Soete, 'International Diffusion of Technology, Industrial Development and Technological Leapfrogging', *World Development*, **13** (1985), pp. 409–22.
52. 'Robots: The Users and the Makers', *OECD Observer*, No. 123 (July 1983), pp. 11–17.
53. R. Vernon, 'International Investment and International Trade in the Product Cycle', *Quarterly Journal of Economics*, **80** (1966), pp. 190–207; J. Utterback

54   Keith Pavitt

and W. Abernathy, 'A Dynamic Model of Product and Process Innovation', *Omega*, **3** (1975), pp. 639–56.

54. G. Dosi, 'Technological Paradigms and Technological Trajectories', *Research Policy*, **11** (1982), pp. 147–62.

11

Uneven (and Divergent) Technological Accumulation among Advanced Countries: Evidence and a Framework of Explanation*

PARIMAL PATEL and KEITH PAVITT

(Science Policy Research Unit, University of Sussex, Falmer, Brighton BN1 9RF, UK)

We present evidence for the advanced OECD countries of uneven and divergent patterns of technological accumulation. We show that 'global' firms will not smooth out the differences, since their technological activities are strongly influenced by conditions in their own countries. We suggest that—in addition to diversity in cumulative technological trajectories—the divergent patterns reflect international differences in the capacities of management, financial and training institutions properly to evaluate— and exploit—the learning benefits of technological investments. For these reasons, we conclude that technological gaps among the advanced OECD countries are here to stay.

1. The Persistence of Technology Gaps

The renewed interest over the past 10 years in the nature and determinants of international patterns of economic growth has confirmed that international 'catch up' in technology and productivity is neither automatic nor easy, since it depends on investment in tangible capital, and intangible capital in the form of education and training and—at least in the industrially advanced countries —of business expenditures on R&D and related activities (Fagerberg, 1987, 1993). These factors explain why some developing countries have been success- ful in reducing the technology and productivity gap, while others have not.

* This paper draws heavily on the results of research undertaken in the ESRC (Economic and Social Research Council)-funded Centre for Science, Technology, Energy and the Environment Policy (STEEP) at the Science Policy Research Unit (SPRU), University of Sussex. We are grateful to two anonymous referees for comments on an earlier draft.

——————— *Uneven (and Divergent) Technological Accumulation* ———————

This is because the international diffusion of technology is neither automatic nor easy (see Bell and Pavitt, 1993). Both material artifacts and the knowledge to develop and operate them are complex involving multiple dimensions and constraints that cannot be reduced entirely to codified knowledge, whether in the form of operating instructions, or a predictive model and theory. Tacit knowledge—underlying the ability to cope with complexity—is acquired essentially through experience, and trial and error. It is misleading to assume that such trial and error is either random, or a purely costless by-product of other activities like 'learning by doing' or 'learning by using'. Tacit (and other forms of) knowledge are increasingly acquired within firms through deliberately planned and funded activities in the form of product design, production engineering, quality control, education and staff training, research, or the development and testing of prototypes and pilot plant. Differences among countries in the resources devoted to such deliberate learning—or 'technological accumulation'—have led to international technological gaps which, in turn, have led to international differences in economic performance.

But while uneven and divergent development is readily acknowledged among the developing countries, the same is not true for the advanced (OECD) countries. Until recently, it was commonly assumed that the open trading system would allow the rapid international diffusion of technology, so that the catching up of Western Europe and Japan to the levels of technology and efficiency of the world's leading country (the US) would be relatively smooth. In fact, there has also been uneven development among industrial countries. Some (e.g. the UK, see Pavitt, 1980) have caught up only very partially, while others (e.g. FR Germany and Japan) have actually overtaken the world's technological leading country—the US—in certain important sectors (Nelson, 1990). At a more aggregate level, Soete and Verspagen (1993) have shown recently that productivity convergence in the OECD countries stopped at the end of the 1970s.

Thus, technology gaps among the industrial countries have not been eliminated. Hence the continuing relevance of the 'neo-technology' theories of trade and growth, that were pioneered by Posner (1961) and Vernon (1966), and confirmed by Soete (1981) and Fagerberg (1987, 1988), as well as by the company-based analyses of Cantwell (1989), Franko (1989) and Geroski *et al.* (1993) Hence, also, the growing interest in the implications of international technology gaps for policy (Ergas, 1984) and for theory (Dosi *et al.*, 1990).

We shall now present statistical evidence of uneven and divergent technological accumulation in the 1980s, and shall argue that technological gaps among the OECD countries will not be eliminated in the 1990s. Given that the activities contributing to technological accumulation are complex and varied, all statistical measures are bound to be imperfect. However, as a result of the growing demands from public and private policy makers for better data, progress has been made in both measurement and conceptualization. The advantages and drawbacks of the various measures have been reviewed extensively elsewhere (Freeman, 1987; van Raan, 1988; Grilliches, 1990; Patel and Pavitt, 1994a). In particular, we have shown in our earlier work that the combined use of data on R&D activities, and on patenting in the USA by country of origin, gives a plausible and consistent picture of technological activities at the world's technological frontier.[1]

2. *The Evidence of Uneven (and Divergent) Technological Accumulation*

Among Countries in the Volume of Technological Activities

The data on R&D and US patenting activities show no evidence of convergence in national capacities for technological accumulation since the early 1970s, and some evidence of divergence in the 1980s.

Table 1 presents trends in the percentage of gross domestic product (GDP) spent by business on R&D activities in 17 OECD countries since 1967.[2] These show a certain stability in the rankings throughout the period at the two ends of the spectrum: Switzerland has remained with the highest share, and Ireland, Spain and Portugal with the three lowest shares. Otherwise there are countries who started near the top but have moved down the rankings: Canada, The Netherlands and—above all—the UK, there are also countries that have improved their positions: FR Germany, Sweden, Japan and—above all—Finland. In general, stability in the rankings of the countries is confirmed by a statistically significant (positive) correlation between their ranks in 1967 and in 1991.[3]

Overall, there are no statistical signs of convergence in the industry-

---

[1] R&D is a better measure of rates of change in real resources over time, but it measures technological activities in small firms only very imperfectly. US patenting is a better measure of technological activities in small firms and can be broken down quite finely by specific firms and specific technical fields. Neither measure is satisfactory for software technology, but no alternative yet exists. And neither measure captures all the activities that lead to product and process innovations, such as design, management, production engineering, marketing and learning by doing.

[2] Government funded R&D performed in industry is excluded. This is concentrated in defense-related activities, and in few countries: principally the US, UK, France and FR Germany, where it has clearly stimulated accumulation in defense-related technologies (see Tables 5, 9 and 10 below). Its wider effects on technological accumulation are a matter of debate. Our own conclusion is that defense R&D has considerable opportunity costs, particularly in electronics, where leading-edge technologies and markets have shifted to civilian applications.

[3] The correlation coefficient for the 17 countries is 0.82, which is significant at the 5% level.

TABLE 1. Trends in Industry Financed R&D as a Percentage of GDP in 17 OECD Countries: 1967 to 1991

	1967	1969	1971	1975	1977	1979	1981	1983	1985	1987	1989	1991
Belgium	0.66	0.64	0.71	0.84	0.91	0.95	0.96	1.02	1.09	1.16	1.14	1.16
Canada	0.40	0.39	0.38	0.33	0.32	0.39	0.49	0.45	0.56	0.57	0.54	0.59
Denmark	0.34	0.39	0.41	0.41	0.41	0.42	0.46	0.53	0.60	0.66	0.71	0.85
Finland	0.30	0.32	0.44	0.44	0.49	0.53	0.62	0.72	0.89	0.98	1.07	1.07
France	0.60	0.64	0.67	0.68	0.69	0.75	0.79	0.88	0.92	0.92	0.98	0.99
FR Germany	0.94	1.03	1.13	1.11	1.12	1.32	1.40	1.48	1.65	1.80	1.78	1.57
Ireland	0.19	0.23	0.30	0.23	0.22	0.23	0.26	0.27	0.35	0.40	0.45	0.58
Italy	0.33	0.38	0.44	0.43	0.37	0.40	0.43	0.42	0.49	0.49	0.56	0.61
Japan	0.83	1.00	1.09	1.12	1.11	1.19	1.38	1.59	1.81	1.82	2.05	2.13
The Netherlands	1.12	1.04	1.02	0.97	0.87	0.86	0.83	0.89	0.96	1.11	1.07	0.91
Norway	0.35	0.39	0.41	0.49	0.49	0.50	0.50	0.61	0.80	0.88	0.81	0.77
Portugal	0.04	0.06	0.09	0.05	0.04	0.09	0.10	0.11	0.11	0.11	0.14	0.14
Spain	0.08	0.08	0.11	0.18	0.18	0.18	0.18	0.22	0.25	0.29	0.34	0.38
Sweden	0.71	0.69	0.80	0.96	1.07	1.11	1.24	1.45	1.71	1.74	1.68	1.71
Switzerland	1.78	1.78	1.67	1.67	1.71	1.74	1.68	1.67	2.16	2.13	2.07	2.07
UK	1.00	0.92	0.81	0.80	0.80	0.82	0.91	0.86	0.95	1.02	1.04	0.94
US	0.99	1.03	0.97	0.98	0.98	1.05	1.17	1.31	1.42	1.37	1.36	1.36
Standard deviation:												
All countries	0.46	0.46	0.43	0.43	0.45	0.47	0.48	0.52	0.61	0.61	0.60	0.58
Excluding US	0.47	0.46	0.43	0.44	0.45	0.47	0.48	0.52	0.62	0.62	0.62	0.59

Source: OECD.

——————— *Uneven (and Divergent) Technological Accumulation* ———————

funded shares over time, since the standard deviation of the distribution has not decreased over time. On the contrary, it has increased markedly in the 1980s, suggesting technological divergence among countries. In this context, it is worth noting that the US share began slipping progressively below that of FR Germany, Japan, Sweden and Switzerland in the 1970s, and that the gap grew much larger in the 1980s.

Table 2 shows trends in per capita national patenting in the US for the same 17 OECD countries. At first sight the evidence about divergence is more ambiguous. When the US is included, the standard deviation of the population increases between the late 1960s and the early 1970s—thereby suggesting divergence—but then decreases until the mid-1980s, after which it increases again to its original level. However, there are well-known reasons for excluding the US from such a comparison, since for firms in the US, we are measuring domestic patenting, whereas for firms in other countries we are measuring foreign patenting. Given the propensity of firms to seek patent protection more intensely in their home country (Bertin and Wyatt, 1988), the rate of technological accumulation in the US is overestimated. At the same time, given the tendency of firms to give increasing attention to patenting in foreign markets, trends over time will tend to overestimate any

TABLE 2. Trends in Per Capita Patenting in the US from 17 OECD Countries

	1963–68	1969–74	1975–80	1981–85	1986–90
Switzerland	138.0	197.1	207.1	179.2	193.6
US	236.6	244.7	181.3	156.7	177.9
Japan	9.9	38.9	56.0	82.5	139.4
Germany	54.7	86.6	91.8	97.7	122.2
Sweden	64.2	94.7	100.9	87.1	99.6
Canada	41.3	55.8	49.0	45.8	63.1
The Netherlands	35.9	48.8	46.8	47.0	60.4
Finland	5.1	14.8	22.3	30.8	50.4
France	26.1	41.0	39.7	38.8	49.6
UK	43.3	55.8	46.6	40.1	47.8
Denmark	18.4	31.9	29.7	27.9	35.7
Belgium	16.3	29.3	26.4	23.8	30.4
Norway	12.7	20.3	23.0	19.4	27.2
Italy	7.8	13.1	13.0	14.0	20.2
Ireland	2.0	6.5	5.1	6.7	12.9
Spain	1.2	2.1	2.3	1.6	3.1
Portugal	0.3	0.6	0.3	0.4	0.6
Standard deviation					
All countries	60.56	67.56	59.44	52.05	59.26
Excluding US	35.05	48.91	51.44	46.13	53.58

Source: Based on data supplied to SPRU by the US Patent and Trademark Office.

decline in US performance (Kitti and Schiffel, 1978) and thereby show a spurious degree of convergence.

When the US is excluded, the evidence in Table 2 on the whole confirms that in Table 1. Throughout the period, Switzerland stays at the top and Ireland, Spain and Portugal at the bottom. Britain's relative position declines, while Finland, FR Germany and Japan improve. In general, stability in the rankings of the countries is confirmed by a significant and positive correlation between their ranks in 1963–68 and 1986–90.[4] At the same time there is an indication of international divergence in that the standard deviation increases over the period. The one anomaly is the reduction in the standard deviation in 1981–85, but this may reflect the reduction in the overall number of patents granted, following a reduction in the number of patent examiners (see Grilliches, 1990).

Finally, Table 3 presents recent trends in patenting in the US by a number of developing countries. We are very much aware of the inadequacies of US patenting as a measure of the largely imitative activities in technological accumulation that are performed in developing countries. Studies using other approaches have shown the superior performance of East Asian countries, compared to those of Latin America and to India (see for example Dahlman *et al.*, 1987). Table 3 simply shows that, while most of the developing countries have continued with a very low level of US patenting, Taiwan and South Korea have both seen massive increases. This indicates that technology in Taiwan and South Korea is now attaining world best practice levels in an increasing number of fields—a striking example of technological catch-up compared to the advanced countries,[5] and of technological divergence compared to other developing countries.

Among Countries in Workforce Education and Training

International comparisons of education and training over the past 10 years have moved well beyond the average numbers of years of schooling that used to be common usage. Greater attention is now paid to the distribution of education levels among different groups in the working population, and to quality as measured through educational attainment. One of the main pioneers has been Prais (together with his colleagues) at the National Institute of Economic and Social Research in London.[6] Some of the major results

———

[4] The correlation coefficient for the 17 countries is 0.80, which is significant at the 5% level.

[5] If the level and rate of increase of US patenting is a reliable guide to technology levels, South Korea and Taiwan are now at the level of technology in Japan about 35 to 40 years ago.

[6] See, most recently, Prais (1993).

—————— *Uneven (and Divergent) Technological Accumulation* ——————

TABLE 3. US Patenting Activities of Selected Developing Countries: 1969–1992

Country	1969	1970	1971	1972	1973	1974	1975	1976	1977	1978	1979	1980	1981	1982	1983	1984	1985	1986	1987	1988	1989	1990	1991	1992
Taiwan	0	0	0	0	1	0	23	28	52	29	38	65	79	88	65	97	174	208	343	457	592	732	904	1000
South Korea	0	3	2	7	5	7	11	7	5	12	4	8	15	14	26	29	38	45	84	97	159	225	402	538
People's Republic of China	5	6	15	8	10	22	1	6	1	0	2	1	3	0	1	2	1	9	23	47	52	47	52	41
Hong Kong	7	8	19	7	15	9	10	20	9	21	13	27	33	18	14	24	25	30	34	41	48	52	50	60
Mexico	67	43	63	43	42	51	66	78	42	24	36	41	43	35	32	42	32	37	49	44	39	32	28	39
Brazil	18	17	14	16	18	21	17	18	21	24	19	24	23	27	19	20	30	27	34	29	36	41	61	40
Venezuela	6	3	13	7	5	7	0	0	0	2	11	11	12	10	5	11	15	21	24	20	23	17	16	20
Argentina	17	23	22	29	27	24	24	24	20	21	24	18	25	18	21	20	11	17	18	16	20	17	—	—
Singapore	2	0	4	4	7	6	1	3	3	2	0	3	4	3	5	4	9	3	11	6	18	—	—	—
India	18	16	10	19	21	17	13	17	13	14	14	4	6	4	14	12	10	18	12	14	14	23	22	24

Source: Based on data supplied to SPRU by the US Patent and Trademark Office.

TABLE 4. Vocational Qualifications of the Workforce in Britain, The Netherlands, Germany, France and Switzerland

Level of qualification	Britain 1988	The Netherlands 1989	Germany 1987	France 1988	Switzerland 1991
University degrees	10	8	11	7	11
Higher technician diplomas	7	19	7	7	9
Craft/lower technical diplomas	20	38	56	33	57
No vocational qualifications	63	35	26	53	23
Total	100	100	100	100	100

Source: Prais, 1983.

of their work are summarized in Table 4, which uses census data to compare the vocational qualifications of the workforce in five European countries.

It shows striking similarities between countries in the proportion with university degrees (7–11%) but even more striking differences in the proportions with intermediate qualifications (66% in Switzerland–27% in the UK) and with no vocational qualifications (63% in the UK–23% in Switzerland). Although there had been some improvement in the UK position in the 1980s, the qualifications gap between Germany and France, on the one hand, and the UK, on the other, actually widened (Patel and Pavitt, 1991b). These skill levels in the workforce are reflected in productivity differences resulting from differences in machine maintenance, consistency in product quality, workforce flexibility and learning times on new jobs.

These studies have been supplemented by comparisons of educational attainment across countries, and which tend to confirm their findings: thus, Dutch adolescents are 2–3 years ahead of their English counterparts in mathematical attainment (Mason *et al.*, 1992). Similar differences are found over a broader geographical area, with adolescents from Japan, the four East Asian 'tiger' countries, and continental Europe (including Hungary) clearly outperforming their counterparts from the US (and often the UK) in mathematics (Newton *et al.*, 1992).

Among Countries in the Sectoral Composition of National Technological Activities

So far, we have compared countries' aggregate technological performance. Table 5 shows the sectoral patterns of technological advantage of 19 OECD countries. On the basis of the US patent classification, technologies have been divided into 11 fields. The content of most of them will be clear from their titles: technologies for extracting and processing raw materials are

—————— *Uneven (and Divergent) Technological Accumulation* ——————

related mainly to food, oil and gas; defense-related technologies are defined as aerospace and munitions. For each country–region and technological field, we have calculated an index of 'Revealed Technology Advantage' (RTA) in 1963–68 and 1985–90.[7]

Table 5 shows markedly different patterns and trends among the three main, technology-producing regions of the world—US, Europe and Japan— in their fields of technological advantage and disadvantage. The US has seen rapid decline in motor vehicles and consumer electronics; growing relative strength in technologies related to weapons, raw materials and telecommunications; and an improving position in chemicals. In Japan, almost the opposite has happened: growing relative strength in electronic consumer and capital goods and motor vehicles, together with rapid relative decline in chemicals, and continued weakness in raw materials and weapons. In Western Europe, the pattern is different again, and very close to that of its dominant country—FR Germany: continuing strength in chemicals, growing strength in weapons, continued though declining strength in motor vehicles, and weakness in electronics.

Table 6 examines the similarities and differences among countries' technological specializations in greater and more systematic detail.[8] It uses correlation analysis to measure both the stability over time of each country's sectoral strengths and weaknesses in technology (first row), and the degree to which they are similar to those of other countries (correlation matrix). The first row shows that, with five exceptions (Australia, Ireland, Italy, Portugal and the UK) most OECD countries have a statistically significant degree of stability in their technological strengths and weaknesses between the 1960s and the 1980s: 10 at the 1% level, and a further four at the 5% level, thereby confirming the path-dependent nature of national patterns of accumulation of technological knowledge.

The correlation matrix also confirms the differentiated nature of technological knowledge, with the very different strengths and weaknesses in Japan, the US and Western Europe: each is negatively correlated with the other two; and significantly so in two cases out of three (the USA with the other two regions). More generally, it confirms that countries tend to differ markedly in their patterns of technological specialization.[9] Of the 171 correlations among pairs of countries in Table 6, only 31 (18%) are positively

---

[7] RTA is defined as a country's or region's (or firm's) share of all US patenting in a technological field, divided by its share of all US patenting in all fields. An RTA of more than one therefore shows a country's or region's relative strength in a technology, and less than one its relative weakness. These measures correspond broadly to the measures of comparative advantage used in trade analyses.

[8] For this analysis we use a more detailed breakdown than that used in Table 5. Again on the basis of the US Patent Classification we have divided technologies into 34 fields.

[9] Archibugi and Pianta (1992) also show that OECD countries' degree of technological specialization is increasing over time.

————————— Uneven (and Divergent) Technological Accumulation —————————

TABLE 5. Sectoral Patterns of Revealed Technological Advantage:* 1963–68 to 1985–90

		Fine chemicals	Industrial chemicals	Materials	Mechanical engineering	Vehicles	Electrical machinery	Electronic capital goods	Telecommunications	Electronic consumer goods	Raw material related	Defense related
USA	1963–68	0.89	0.93	1.04	1.01	0.89	1.00	1.02	1.03	0.94	1.08	0.99
	1985–90	0.97	0.98	0.95	0.99	0.55	1.01	0.97	1.04	0.65	1.28	1.15
Europe**	1963–68	1.34	1.29	0.86	0.99	1.48	1.00	0.92	0.91	1.26	0.61	1.14
	1985–90	1.33	1.19	0.83	1.13	1.02	0.92	0.61	0.94	0.59	0.83	1.40
Japan	1963–68	2.95	1.62	1.02	0.77	0.83	1.17	1.47	1.06	1.99	0.44	0.36
	1985–90	0.72	0.92	1.42	0.85	2.21	1.08	1.65	0.97	2.50	0.37	0.09
Australia	1963–68	1.05	0.69	0.80	1.16	1.44	0.74	0.19	0.72	1.48	1.02	0.31
	1985–90	0.80	0.65	0.42	1.21	1.35	0.59	0.30	0.54	0.34	1.82	1.63
Austria	1963–68	1.41	0.80	1.13	1.25	1.21	0.62	0.28	0.38	1.69	0.39	0.26
	1985–90	0.84	0.73	0.68	1.39	1.94	0.82	0.32	0.43	0.41	0.90	1.96
Belgium	1963–68	1.23	1.38	3.99	0.71	0.44	0.92	0.69	1.02	3.97	0.42	0.78
	1985–90	1.85	1.79	2.21	0.77	0.19	0.98	0.21	0.50	1.24	1.10	0.80
Canada	1963–68	0.71	0.81	0.75	1.11	1.35	0.78	0.56	0.85	0.36	1.43	0.71
	1985–90	0.68	0.74	0.67	1.15	0.65	0.83	0.40	1.38	0.47	1.69	1.13
Denmark	1963–68	3.05	0.77	0.97	1.11	0.39	0.87	0.47	0.59	1.17	0.92	0.12
	1985–90	2.38	0.91	0.49	1.19	0.20	0.90	0.21	0.46	0.54	0.82	0.39
Finland	1963–68	0.00	0.62	0.00	1.30	2.17	0.72	0.00	0.31	0.20	1.45	0.47
	1985–90	0.88	0.76	0.76	1.54	0.71	0.68	0.10	0.51	0.15	1.13	0.82

———————— *Uneven (and Divergent) Technological Accumulation* ————————

France	1963–68	1.86	1.02	1.05	1.02	2.11	1.23	0.84	1.15	0.81	0.49	1.11
	1985–90	1.34	1.03	0.88	1.04	0.57	1.17	0.85	1.56	0.49	1.03	1.55
Germany	1963–68	1.12	1.49	0.68	0.96	1.43	0.78	0.92	0.74	1.88	0.54	1.04
	1985–90	1.14	1.37	0.85	1.18	1.42	0.87	0.52	0.79	0.54	0.61	1.58
Ireland	1963–68	0.00	0.39	0.00	1.29	3.00	0.00	0.00	0.65	0.00	0.55	0.00
	1985–90	1.57	1.18	0.55	0.87	0.63	1.38	0.94	0.72	0.99	1.66	0.00
Italy	1963–68	1.29	1.93	0.51	0.93	1.26	0.68	0.87	0.69	0.53	0.71	0.78
	1985–90	1.77	1.10	0.62	1.15	1.21	0.73	0.76	0.85	0.41	0.78	0.92
The Netherlands	1963–68	1.71	1.46	1.24	0.75	0.15	1.34	1.90	1.11	1.95	1.15	0.15
	1985–90	0.54	1.13	0.89	0.87	0.26	1.27	1.24	1.25	1.82	1.07	0.33
Norway	1963–68	0.94	0.63	0.00	1.25	0.36	1.16	0.65	0.47	0.29	0.91	0.46
	1985–90	0.83	0.62	0.27	1.14	0.33	0.70	0.24	0.92	0.43	2.30	1.79
Portugal	1963–68	10.58	1.41	0.00	0.99	0.00	0.67	0.00	0.00	0.00	0.98	3.46
	1985–90	1.90	1.66	0.00	1.11	0.00	0.43	0.00	0.00	0.00	2.97	0.00
Spain	1963–68	0.84	0.56	0.48	1.20	3.00	0.86	0.13	1.02	0.47	0.47	1.93
	1985–90	1.93	0.76	0.37	1.22	1.84	0.75	0.12	0.55	0.07	0.85	2.72
Sweden	1963–68	0.94	0.44	0.44	1.22	1.16	1.14	0.72	1.37	0.38	0.68	2.35
	1985–90	0.72	0.57	0.58	1.43	0.93	0.89	0.33	0.79	0.23	0.99	1.86
Switzerland	1963–68	2.60	2.01	0.27	0.90	0.56	0.85	0.56	0.73	0.63	0.47	1.45
	1985–90	1.81	1.48	0.63	1.12	0.51	0.78	0.41	0.64	0.43	0.72	1.16
UK	1963–68	0.88	1.03	1.09	1.02	1.99	1.22	1.04	1.03	0.93	0.70	1.28
	1985–90	1.83	1.05	0.92	1.04	0.86	0.88	0.71	1.07	0.73	0.99	1.37

Notes: * For the definition of the Revealed Technology Advantage Index see footnote 7 in the text.
      ** Europe is defined as the 15 European countries included in this table.
Source: Based on data supplied to SPRU by the US Patent and Trademark Office.

———————— Uneven (and Divergent) Technological Accumulation ————————

TABLE 6. Stability and Similarities among Countries in their Sectoral Specializations:

	Australia	Austria	Belgium	Canada	Denmark	Finland	France	Germany	Ireland
Stability: correlations over time: 1963–68 to 1985–90									
	0.28	0.76*	0.54*	0.67*	0.47*	0.59*	0.82*	0.35*	0.05
Similarities: correlations among countries: 1985–90									
Austria	0.36*								
Belgium	−0.09	−0.14							
Canada	0.52*	0.47*	0.05						
Denmark	0.18	−0.03	0.33*	0.32					
Finland	0.47*	0.45*	0.20	0.54*	0.45*				
France	−0.27	−0.16	0.10	−0.14	0.10	−0.15			
Germany	0.27	0.05	0.22	−0.18	0.21	0.32	0.29		
Ireland	0.07	−0.10	0.09	0.21	0.14	0.03	−0.09	−0.31	
Italy	0.28	0.28	0.06	0.34*	0.30	0.53*	−0.23	0.22	0.28
Japan	−0.43*	−0.07	0.06	−0.44*	−0.22	−0.26	−0.44*	−0.20	−0.13
The Netherlands	−0.24	−0.18	−0.03	0.06	−0.04	0.07	−0.33*	−0.38*	0.27
Norway	0.36*	0.36*	−0.20	0.62*	0.22	0.28	0.02	−0.12	0.02
Portugal	0.32	0.48*	0.17	0.31	0.11	0.43*	−0.09	0.15	0.11
Spain	0.32	0.13	−0.11	0.34*	0.68*	0.38*	0.00	0.28	−0.07
Sweden	0.25	0.46*	−0.05	0.38*	0.40*	0.53*	0.36*	0.26	−0.07
Switzerland	0.35*	−0.12	0.11	−0.21	0.01	0.07	0.06	0.72*	−0.07
UK	0.08	−0.15	−0.04	−0.10	−0.10	0.40*	0.03	0.23	−0.02
US	0.22	−0.03	−0.20	0.42*	−0.02	−0.01	0.23	−0.37*	0.25
Western Europe	0.26	0.04	0.21	−0.08	0.33*	0.34*	0.49*	0.93*	−0.19

Notes: For the definition of the Revealed Technology Advantage Index see foonote 7 in the text.
Europe is defined as the 15 European countries included in this table.
* Denotes correlation coefficient significantly different from zero at the 5% level.
Source: Based on data supplied to SPRU by the US Patent and Trademark Office.

and significantly correlated at the 5% level. Among these we find FR Germany similar to Switzerland (chemicals and machinery), and Canada similar to Australia, Finland and Norway (raw material-based technologies). Japan has a unique pattern of specialization, with no significant positive correlations with other countries but plenty of negative ones.

Implications

The above comparisons show some striking differences—even divergences— in the rate and direction of technological accumulation in the industrial countries. Those related to fields of technological specialization reflect inevitable diversity in stages of economic and technological development, or a desirable diversity in fields of national scientific and technological specialization. Those related to differences in the overall volume of investment in

———————— *Uneven (and Divergent) Technological Accumulation* ————————

Correlations of Revealed Technology Advantage Indices across 34 Sectors

Italy	Japan	The Netherlands	Norway	Portugal	Spain	Sweden	Switzerland	UK	US
0.32	0.45*	0.66*	0.35*	0.25	0.53*	0.73*	0.83*	0.23	0.56*
−0.13									
0.08	0.24								
−0.03	−0.50*	−0.23							
0.35*	−0.20	−0.04	0.06						
0.38*	−0.23	−0.28	0.41*	0.20					
0.19	−0.38*	−0.35*	0.30	0.30	0.45*				
0.17	−0.19	−0.26	−0.14	0.04	0.13	−0.04			
−0.03	−0.34*	−0.20	0.17	0.00	0.32	0.10	0.20		
−0.09	−0.81*	−0.03	0.50*	0.06	−0.02	0.13	−0.26	0.11	
0.27	−0.41*	−0.37*	−0.02	0.17	0.35*	0.39*	0.73*	0.45*	−0.19

technological accumulation should be causes of disquiet because—if allowed to persist—they are likely to reinforce any uneven and divergent rates of national technological and economic development in future.

The most striking international differences and technological divergencies in the overall rate of accumulation are in the core, namely, in the volume of change-generating (including R&D) activities supported by business firms, and in the skills of the workforce that they employ and that they (unequally) train.[10] Japan and FR Germany are the major countries with high company R&D and workforce skills, with the UK (and probably the US) among the major industrial countries at the other end of the spectrum.

[10] National differences in basic research also exist but appear—in the long term—to adjust to the level of demand for skills and knowledge from the business sector. For further discussion see Patel and Pavitt (1944b).

3. *The Effects of 'Global Corporations'*

The Importance of Home Countries for Technological Activities

The unfettered behavior of large corporations is unlikely—in and of itself—to smooth out these international differences in technological accumulation, for three sets of reasons.

First, large firms do not play a major role in the development and control of some significant fields of technology: in particular, capital goods, components, and measuring and control instruments (Patel and Pavitt, 1991a). As the recent experience of the Japanese automobile industry shows, large firms can stimulate technological accumulation in these fields through their supplier networks, including those outside Japan. But this depends on the autonomous development of skills and technological capabilities among the suppliers themselves.

Second, technology-generating activities remain among the most domesticated of all corporate activities.[11] Table 7 shows that, in the second half of

TABLE 7. Geographic Location of Large Firms' US Patenting Activities, According to Nationality: 1985–90

Firms' nationality	Percentage shares					
	Home	Abroad	Of which:			
			US	Europe	Japan	Other
Japan (13)	98.9	1.1	0.8	0.3	–	0.0
US (249)	92.2	7.8	–	6.0	0.5	1.3
Italy (7)	88.1	11.9	5.4	6.2	0.0	0.3
France (26)	86.6	13.4	5.1	7.5	0.3	0.5
Germany (43)	84.7	15.3	10.3	3.8	0.4	0.7
Finland (7)	81.7	18.3	1.9	11.4	0.0	4.9
Norway (3)	68.1	31.9	12.6	19.3	0.0	0.0
Canada (17)	66.8	33.2	25.2	7.3	0.3	0.5
Sweden (13)	60.7	39.3	12.5	25.8	0.2	0.8
UK (56)	54.9	45.1	35.4	6.7	0.2	2.7
Switzerland (10)	53.0	47.0	19.7	26.1	0.6	0.5
The Netherlands (9)	42.1	57.9	26.2	30.5	0.5	0.6
Belgium (4)	36.4	63.6	23.8	39.3	0.0	0.6
All firms (587)	89.0	11.0	4.1	5.6	0.3	0.9

Note: The parentheses contain the number of firms based in each country.
Source: Based on data supplied to SPRU by the US Patent and Trademark Office.

[11] We use data on the address of inventors patenting in the US as a proxy measure of the international location of large firms' technological activities. These data are consistent with the available (but less comprehensive) studies based on corporate R&D expenditures. See Patel (1994).

the 1980s, 89% of the technological activities of the world's largest firms continued to be performed in their home country—a 1% increase over the previous five-year period. Not unsurprisingly, the share performed in foreign countries by large firms based in smaller countries tends to be higher, although the proportion for Finnish firms (18.3%) compared to that for British firms (45.1%) shows that factors other than size are at work.

In particular, Cantwell (1992) has shown that the share of foreign in total production is the most important factor explaining differences among firms in the location of their technological activities, but that foreign patenting shares are smaller than foreign production shares. In other words, the technology intensity of foreign production is consistently and significantly less than that of home production and concerned mainly with product and process adaptation to local conditions.

Third, Table 7 also shows that the foreign technological activities of large firms are not globalized but are concentrated almost exclusively in the 'triad' countries—especially the US and Europe (and, more specifically, Germany). More detailed data shows that the largest proportionate increases in foreign technological activities in the 1980s were in British and Swedish firms— especially in the US, and in a number of smaller European countries outside the EC—Switzerland, Finland, Norway—all increasing their share within the EC. The firms based in countries with the largest technological activities— the US, Japan and FR Germany—had among the lowest proportionate increases in foreign technological activities. Most of any increase in foreign technological activities came as a by-product of take-overs and divestitures, rather than an explicit relocation of technological activities.

Finally, this pattern suggests that, globalization of markets and (increasingly) production notwithstanding, there remains at least one compelling reason for companies to concentrate a high proportion of their technological activities in one location. The development and commercialization of major innovations requires the mobilization of a variety of often tacit (person-embodied) skills and involves high uncertainties. Both are best handled through intense and frequent personal communications and rapid decision making—in other words, through geographical concentration (see Porter, 1990). In this context, it is worth noting that the rapid product development times in Japanese firms (Clark *et al.*, 1987) have been achieved from an almost exclusively Japanese base, while the strongly globalized R&D activities of the Dutch Philips company are said to have slowed down product development.

This is consistent with the inter-industry differences in Table 8, which shows that firms making products with the highest technology intensities are among those with the lowest degrees of internationalization of their

──────────── *Uneven (and Divergent) Technological Accumulation* ────────────

TABLE 8. Geographic Location of Large Firms' US Patenting Activities, According to Product Group: 1985–90

Firms' nationality	Abroad	Percentage shares			
		Of which:			
		US	Europe	Japan	Other
Drink and tobacco (18)	30.8	17.5	11.1	0.4	1.8
Food (48)	25.0	14.8	8.5	0.1	1.7
Building materials (28)	20.6	9.1	9.8	0.1	1.6
Other transport (5)	19.7	2.0	6.8	0.0	10.9
Pharmaceuticals (25)	16.7	5.5	8.3	1.1	1.7
Mining and petroleum (47)	15.0	9.7	3.5	0.1	1.6
Chemicals (72)	14.4	8.0	5.1	0.3	1.0
Machinery (68)	13.7	3.5	9.1	0.1	1.1
Metals (57)	12.8	5.4	5.7	0.1	1.6
Electrical (58)	10.2	2.6	6.8	0.3	0.4
Computers (17)	8.9	0.1	6.6	1.1	1.1
Paper and wood (34)	8.1	2.4	4.9	0.1	0.7
Rubber and plastics (10)	6.1	0.9	2.4	0.4	2.4
Textiles etc. (18)	4.7	1.4	1.8	0.8	0.6
Motor vehicles (43)	4.4	0.9	3.2	0.1	0.2
Instruments (20)	4.4	0.4	2.8	0.5	0.8
Aircraft (19)	2.9	0.3	1.8	0.1	0.7
All firms (587)	11.0	4.1	5.6	0.3	0.9

Note: The parentheses contain the number of firms in each product group.
Source: Based on data supplied to SPRU by the US Patent and Trademark Office.

underlying technological activities: those producing aircraft, instruments, motor vehicles, computers and other electrical products are all below the average for the population of firms as a whole. In all these products, links between R&D and design, on the one hand, and production, on the other, are particularly important in the launching of major new products and benefit from geographical proximity.[12]

By contrast, we see in Table 8 a high proportion of foreign R&D in industries, where some localized technological activities are required, either to adapt products to differentiated local tastes, or to exploit local natural resources: food, drink and tobacco, building materials, mining and petroleum. All in all, these data suggest a general theorem: localized 'low-tech' products require globalized R&D; global 'high-tech' products do not.

[12] The one technology intensive exception is pharmaceutical products, where the share of foreign R&D is high, but where R&D and production are unimportant compared to the links with high quality basic research, and with nationally based agencies for testing and validation.

———————— *Uneven (and Divergent) Technological Accumulation* ————————

Uneven Development among Firms Reflects Uneven Development among Countries

They also suggest that conditions in the home country remain preponderant in the technological performance of large firms. This is confirmed in Table 9, which presents evidence of uneven technological development since the late 1960s among the world's largest firms.[13] It shows, in the same 11 technological fields as in Table 5, trends in the total US patenting of the world's top 20 firms in the period 1985–90. Two major conclusions emerge from this table.

First, the technological strengths and weaknesses of each region, shown in Table 5, are in general reflected in the number of nationally based large firms appearing in the top 20 in Table 9.[14] Thus, as summarized in Table 10, Japanese firms make up 11 of the top 20 firms in motor vehicles and 14 in consumer electronics and photography, US firms make up 16 of the top 20 in raw materials and 15 in defense, while European firms have their largest numbers in chemicals.

Second, in addition to uneven development of large firms in each technological field according to their nationality, there has also been an uneven degree of stability (or instability) in the firms' shares and rankings within each technological field. A casual reading of Table 9 shows that in some fields, the leaders of the early 1970s continued to be so into the late 1980s, while in others new leaders emerged during the period. This is shown statistically in the final column of Table 10, which presents the correlation of the shares of the top 20 firms in 1985–90 with their shares in 1969–74.

Thus the low (and statistically insignificant) correlations in motor vehicles and in electronic consumer goods, mainly reflect the emergence of Japanese firms as technological leaders in these fields, while the high (and statistically significant) correlations in electrical machinery and telecommunications reflect mainly a re-enforcement of the dominance of established US and some European firms. The more stable shares in industrial chemicals reflect the continuing strength of mainly European firms.

4. *Possible Causes of Uneven Development*

The above evidence suggests that the behavior of large firms will not, in and of itself, lead to a wider international spread of R&D and related activities.

---

[13] It is based on data that we have compiled on more than 600 of the world's largest, technologically active firms, as measured by their patent activity in the US (see Patel and Pavitt, 1991a).

[14] For a more systematic statistical proof, see Patel and Pavitt (1991a).

——————— *Uneven (and Divergent) Technological Accumulation* ———————

TABLE 9. Shares of US Patenting for Top 20 Firms in 11 Technical Fields: Sorted According to Shares in 1985–90

	Nationality	1969–74	1985–90
Fine chemicals			
1 Bayer	FRG	2.84	3.70
2 Hoechst	FRG	1.61	2.53
3 Merck	US	2.57	2.44
4 Ciba-Geigy	Switzerland	4.33	2.27
5 Imperial Chemical Industries	UK	1.98	1.98
6 E. I. Du Pont De Nemours	US	1.48	1.79
7 Warner-Lambert	US	0.71	1.58
8 Eli Lilly Industries	US	1.67	1.46
9 Dow Chemical	US	1.21	1.24
10 BASF	FRG	0.58	1.19
11 Pfizer	US	1.17	1.09
12 American Cyanamid	US	2.43	1.04
13 Johnson + Johnson	US	0.36	1.03
14 Boehringer Mannheim	FRG	1.00	1.01
15 Hoffmann-La Roche	Switzerland	1.59	0.96
16 SmithKline Beckman	US	1.40	0.96
17 Monsanto	US	2.87	0.88
18 Squibb	US	1.03	0.88
19 Takeda	Japan	1.21	0.88
20 Beecham	UK	0.23	0.82
Other chemicals			
1 Bayer	FRG	2.57	2.77
2 Dow Chemical	US	2.40	2.67
3 Hoechst	FRG	2.37	2.45
4 BASF	FRG	1.40	2.31
5 Ciba-Geigy	Switzerland	2.54	1.92
6 General Electric	US	1.39	1.86
7 E. I. Du Pont De Nemours	US	3.29	1.68
8 Imperial Chemical Industries	UK	2.14	1.16
9 Shell Oil	The Netherlands	0.85	1.09
10 Eastman Kodak	US	1.33	1.03
11 Union Carbide	US	1.11	0.86
12 Exxon	US	0.79	0.84
13 Allied-Signal	US	1.62	0.81
14 Henkel	FRG	0.35	0.80
15 Rhone-Poulenc	France	0.63	0.69
16 Phillips Petroleum	US	1.46	0.66
17 Sumitomo Chemical	Japan	0.55	0.65
18 Texaco	US	0.41	0.64
19 3M	US	0.52	0.62
20 Monsanto	US	2.22	0.61
Materials			
1 3M	US	1.36	2.26
2 Fuji Photo Film	Japan	0.31	2.24
3 Ppg	US	2.72	1.81

TABLE 9. (*continued*)

	Nationality	1969–74	1985–90
Materials (continued)			
4 General Electric	US	2.32	1.76
5 E. I. Du Pont De Nemours	US	3.48	1.57
6 Hitachi	Japan	0.20	1.43
7 Corning Glass Works	US	2.77	1.13
8 Dow Chemical	US	1.26	1.11
9 Hoechst	FRG	1.01	1.08
10 Saint-Gobain Industries	France	0.96	0.95
11 Emhart	US	0.46	0.93
12 TDK	US	0.10	0.91
13 Owens-Corning Fiberglas	US	1.83	0.89
14 Allied-Signal	US	0.65	0.86
15 Toshiba	Japan	0.38	0.80
16 Sumitomo Electric Industries	Japan	0.06	0.75
17 Kimberly-Clark	US	0.61	0.75
18 W. R. Grace	US	0.69	0.71
19 Bayer	FRG	0.59	0.69
20 GTE	US	0.28	0.65
Non-electrical machinery			
1 General Motors	US	1.23	0.91
2 Hitachi	Japan	0.18	0.88
3 General Electric	US	1.22	0.80
4 Canon	Japan	0.05	0.71
5 Toshiba	Japan	0.08	0.69
6 Siemens	FRG	0.29	0.65
7 Philips	The Netherlands	0.39	0.54
8 United Technologies	US	0.47	0.54
9 Nissan Motor	Japan	0.12	0.52
10 Westinghouse Electric	US	0.56	0.51
11 Honda	Japan	0.03	0.51
12 Allied-Signal	US	0.78	0.50
13 Toyot Jidosha Kogyo	Japan	0.10	0.50
14 Fuji Photo Film	Japan	0.09	0.41
15 Mitsubishi Denki	Japan	0.03	0.38
16 IBM	US	0.46	0.38
17 ITT Industries	US	0.35	0.35
18 Robert Bosch	FRG	0.24	0.34
19 ATT	US	0.56	0.34
20 Aisin Seiki	Japan	0.10	0.31
Vehicles			
1 Honda	Japan	0.91	9.12
2 Nissan Motor	Japan	1.74	5.73
3 Toyota Jidosha Kogyo	Japan	0.76	4.84
4 Robert Bosch	FRG	3.79	4.27
5 Mazda Motor	Japan	0.78	2.93
6 General Motors	US	5.21	2.79
7 Mitsubishi Denki	Japan	0.21	2.78

—————— *Uneven (and Divergent) Technological Accumulation* ——————

TABLE 9. (*continued*)

	Nationality	1969–74	1985–90
Vehicles (continued)			
8 Nippondenso	Japan	1.27	2.57
9 Fuji Heavy Industries	Japan	0.06	2.20
10 Hitachi	Japan	0.38	1.91
11 Yamaha Motor	Japan	0.32	1.85
12 Daimler-Benz	FRG	2.71	1.50
13 Ford Motor	US	2.41	1.50
14 Brunswick	US	0.42	1.10
15 Aisin Seiki	Japan	0.34	1.10
16 Lucas	UK	0.95	0.87
17 Porsche	FRG	0.42	0.86
18 Outboard Marine	US	0.70	0.84
19 Caterpillar	US	1.99	0.76
20 Kawasaki Jukogyo	Japan	0.08	0.76
Electrical machinery			
1 General Electric	US	5.77	2.99
2 Westinghouse Electric	US	3.22	2.68
3 Philips	The Netherlands	1.44	2.15
4 Amp	US	1.10	2.02
5 Mitsubishi Denki	Japan	0.20	1.97
6 Hitachi	Japan	0.53	1.91
7 Siemens	FRG	1.47	1.85
8 Toshiba	Japan	0.41	1.53
9 General Motors	US	1.81	1.30
10 GTE	US	1.20	1.29
11 Motorola	US	0.51	0.97
12 Matsushita Electric Industrial	Japan	0.85	0.88
13 Asea Brown Boveri Ab	Switzerland	0.83	0.80
14 United Technologies	US	0.93	0.67
15 NEC	Japan	0.22	0.63
16 ATT	US	1.25	0.53
17 Robert Bosch	FRG	0.49	0.53
18 Allied-Signal	US	0.72	0.52
19 Honeywell	US	0.76	0.51
20 Canon	Japan	0.07	0.51
Electronic capital goods and components			
1 Toshiba	Japan	0.53	5.29
2 IBM	US	8.83	5.25
3 Hitachi	Japan	1.71	4.79
4 Motorola	US	2.15	2.88
5 Texas Instruments	US	1.97	2.88
6 NEC	Japan	0.97	2.75
7 Mitsubishi Denki	Japan	0.13	2.73
8 Fujitsu	Japan	0.38	2.59
9 General Electric	US	6.77	2.50
10 Philips	The Netherlands	2.81	2.43
11 ATT	US	6.05	2.09

———————— *Uneven (and Divergent) Technological Accumulation* ————————

TABLE 9. (*continued*)

	Nationality	1969–74	1985–90
Electronic capital goods and components (continued)			
12 Siemens	FRG	1.75	1.79
13 Honeywell	US	2.23	1.08
14 Unisys	US	3.47	1.03
15 Sharp	Japan	0.05	1.03
16 Canon	Japan	0.04	0.97
17 General Motors	US	1.36	0.92
18 Tektronix	US	0.26	0.89
19 Thomson-Csf	France	0.32	0.81
20 Sony	Japan	0.43	0.79
Telecommunications			
1 ATT	US	5.97	4.24
2 Siemens	FRG	2.22	3.25
3 General Electric	US	4.29	2.89
4 Philips	The Netherlands	1.54	2.55
5 Motorola	US	1.02	2.55
6 NEC	Japan	0.61	2.43
7 Westinghouse Electric	US	3.30	1.79
8 Toshiba	Japan	0.26	1.64
9 Mitsubishi Denki	Japan	0.21	1.49
10 ITT Industries	US	3.55	1.41
11 Hitachi	Japan	0.43	1.41
12 General Motors	US	1.57	1.30
13 Thomson-Csf	France	0.82	1.26
14 GTE	US	1.49	1.21
15 IBM	US	1.19	1.12
16 Northern Telecom	Canada	0.54	1.02
17 Fujitsu	Japan	0.20	0.86
18 Rockwell International	US	0.63	0.77
19 CGE	France	0.58	0.75
20 Alps Electric	Japan	0.13	0.74
Electronic consumer goods			
1 Canon	Japan	0.95	6.51
2 Fuji Photo Film	Japan	2.12	6.21
3 Eastman Kodak	US	6.24	3.32
4 Toshiba	Japan	0.42	3.27
5 General Electric	US	3.97	3.06
6 Philips	The Netherlands	2.38	3.04
7 Sony	Japan	1.02	2.94
8 Hitachi	Japan	0.62	2.89
9 Minolta Camera	Japan	0.88	2.47
10 Xerox	US	3.79	2.29
11 Konica	Japan	0.52	1.95
12 Ricoh	Japan	1.00	1.87
13 Matsushita Electric Industrial	Japan	1.20	1.61
14 Sharp	Japan	0.00	1.60
15 IBM	US	2.92	1.44

——————— *Uneven (and Divergent) Technological Accumulation* ———————

TABLE 9. (*continued*)

	Nationality	1969–74	1985–90
Electronic consumer goods (continued)			
16 Pioneer Electronic	Japan	0.18	1.44
17 Olympus Optical	Japan	0.14	1.22
18 Mitsubishi Denki	Japan	0.07	1.11
19 NEC	Japan	0.36	1.05
20 Siemens	FRG	0.59	1.01
Technologies for extracting and processing raw materials			
1 Mobil Oil	US	2.17	4.91
2 Exxon	US	3.00	2.25
3 Halliburton	US	0.60	1.62
4 Chevron	US	2.66	1.47
5 Philip Morris	US	1.32	1.42
6 Baker Hughes	US	0.41	1.38
7 Texaco	US	2.49	1.36
8 Phillips Petroleum	US	2.35	1.36
9 Nabisco Brands	US	0.32	1.29
10 Amoco	US	0.18	1.28
11 Shell Oil	The Netherlands	2.13	1.26
12 Allied-Signal	US	3.00	1.17
13 Atlantic Richfield	US	0.89	1.10
14 Deere	US	1.04	0.89
15 Union Oil Of California	US	0.57	0.83
16 E. I. Duk Pont De Nemours	US	0.87	0.67
17 Nissan Motor	Japan	0.03	0.61
18 Schlumberger	US	0.85	0.55
19 British-American Tobacco	UK	0.19	0.55
20 Nestle	Switzerland	0.15	0.55
Defense-related technologies			
1 Boeing	US	1.06	4.29
2 MBB	FRG	1.18	2.50
3 General Electric	US	1.58	1.44
4 Oerlikon-Buhrle Ag	Switzerland	0.89	1.35
5 British Aerospace	UK	0.66	1.28
6 Morton Thiokol	US	0.93	1.11
7 Feldmuhle	FRG	1.56	1.08
8 General Dynamics	US	0.29	1.08
9 Imperial Chemical Industries	UK	1.33	0.95
10 Honeywell	US	0.31	0.93
11 United Technologies	US	1.04	0.82
12 Aerospatiale	France	0.35	0.82
13 General Motors	US	0.54	0.80
14 Westinghouse Electric	US	0.15	0.80
15 Olin	US	1.16	0.71
16 Lockheed	US	0.79	0.69
17 Grumman	US	0.02	0.66
18 Ford Motor	US	0.08	0.51
19 Sundstrand	US	0.02	0.51
20 Rockwell International	US	0.83	0.46

Source: Based on data supplied to SPRU by the US Patent and Trademark Office.

—————— *Uneven (and Divergent) Technological Accumulation* ——————

TABLE 10. Nationalities of the Top 20 Firms in US Patenting: 1985–90

	Japan	US	Western Europe	Correlation of shares of the top 20: 1969–74 to 1985–90
Defense related technologies	0	14	6	0.37
Fine chemicals	1	12	7	0.54
Industrial chemicals	1	11	8	0.66*
Raw materials based technologies	1	16	3	0.45
Materials	4	13	3	0.41
Electrical machinery	6	10	4	0.68*
Telecommunications	6	10	4	0.70*
Electronic capital goods	8	9	3	0.51
Non-electrical machinery	9	8	3	0.41
Motor vehicles	11	5	4	0.15
Electronic consumer goods	14	4	2	0.27

Note: *Denotes a correlation coefficient significantly different from zero at 5% level.
Source: Based on data supplied to SPRU by the US Patent and Trademark Office.

On the contrary, our statistical analysis confirms Porter's conclusion (1990) that the conditions in large firms' home countries have a major impact on the rate and direction of their technological activities. We shall now propose a framework of analysis that might eventually explain international differences in the rate and direction of technological accumulation. At this stage, it does not lend itself to rigorous modelling and statistical analysis, although it is consistent with the conclusions of a wide range of more qualitative analyses, as well as with our own data.

International Differences in the Volume of Technological Activities: 'Institutional Failure' in the Competence to Evaluate and Benefit from Technological Learning

Some of the observed international differences in the volume of technological activities may reflect differences in the degree of market failure.[15] However, we would also stress the importance of institutional failures in the competence to evaluate and benefit from investments in technology that are increasingly specialized and professionalized in nature (e.g. industrial R&D laboratories employing highly qualified specialists in a variety of fields of science and engineering), and are long term and complex in their economic impact (e.g. from research on photons, through the laser, to the compact

[15] In particular, the effects of labor mobility on the incentive on business firms to train their workforce; and the effects of intellectual property rights on the incentive to invest in innovative activities.

——————— *Uneven (and Divergent) Technological Accumulation* ———————

disc, over a period of 25 years). For purposes of exposition, we have found it useful to distinguish between national systems of innovation that we define as 'myopic', and those that we define as 'dynamic' (Pavitt and Patel, 1988).

Briefly stated, 'myopic' systems treat investments in technological activities just like any conventional investment: they are undertaken in response to a well-defined market demand, and include a strong discount for risk and time. As a consequence, technological activities often do not compare favorably with conventional investments. 'Dynamic' national systems of innovation, on the other hand, recognize that technological activities are not the same as any other investment. In addition to tangible outcomes in the form of products, processes and profits, they also entail the accumulation of important but intangible assets, in the form of irreversible processes of technological, organizational and market learning, that enable them to undertake subsequent investments, that they otherwise could not have made.[16] The archetypal dynamic national systems of innovation are those of FR Germany[17] and Japan, while the myopic systems are the UK and the US. The essential differences can be found in three sets of institutions:

(i) First, in the financial system underlying business activity: in Germany and Japan, these give greater weight to longer-term performance, when the benefits of investments in learning begin to accrue. And they generate both the information and the competence to enable firm-specific intangible assets to be evaluated by the providers of finance (Hu, 1975; Corbett and Mayer, 1991).

(ii) Second, there are the methods of management, especially those employed in large firms in R&D-intensive sectors: in the UK and US, the relatively greater power and prestige given to financial (as opposed to technical) competence is more likely to lead to incentive and control mechanisms based on short-term financial performance, and to decentralized divisional structures insensitive to new and longer term technological opportunities that top management is not competent to evaluate (Abernathy and Hayes, 1980; Lawrence, 1980).

(iii) Third, there is the system of education and training: the German and Japanese systems of widespread yet rigorous general and vocational education provide a strong basis for cumulative learning. The British and US systems of higher education have performed relatively well, but the other two-thirds of the labor force are less well trained and educated than their counterparts in continental Europe and East Asia (Newton *et al.*, 1992; Prais, 1993).

[16] In other words, investments in technology nearly always have an 'option value'. See Myers (1984) and Mitchell and Hamilton (1988).

[17] Sweden and Switzerland have many 'dynamic' institutional characteristics similar to Germany.

International Differences in Fields of Technological Specialization: Local Inducement Mechanisms and Cumulative Trajectories

Further, we propose that the observed international differences in the sectoral patterns of technological accumulation emerge from the localized nature of technological accumulation, and the consequent importance of the local inducement mechanisms that guide and constrain firms along cumulative technological trajectories. We know from earlier debates about the relative importance of 'technology push' and 'demand pull' that these inducement mechanisms are numerous, and that their relative importance varies among sectors. It is nonetheless possible to distinguish three mechanisms.

Factor endowments. Examples include the stimulus of scarce labor for labor-saving innovations in the US; and the different technological trajectories followed by the automobile industries of the US, and of Europe and East Asia, as consequence of very different fuel prices.

Directions of persistent investment. Especially those with strong inter-sectoral linkages: examples include the extraction and processing of natural resources (North America, Australia and Scandinavia), defense (US, France, UK), public infrastructure (France) and automobiles (Japan, Germany, Italy).

The cumulative mastery of core technologies and their underlying knowledge bases. Examples include Germany in chemicals and machinery, Sweden in machinery, Switzerland in fine chemicals, The Netherlands in electronics; Japan in electronics and automobiles; the US in chemicals and electronics; the UK in chemicals.

The relative significance of these mechanisms change over time. In the early stages, the directions of technical change in a country or region are strongly influenced by local market inducement mechanisms related to scarce (or abundant) factors of production and local investment opportunities. At higher levels of development, the local accumulation of specific technological skills itself becomes a focusing device for technical change. At this stage, firms become less dependent on the home country for creating the appropriate market signals, and more so for its provision of high quality skills and knowledge bases that local firms can exploit on world markets.

5. Conclusions

In their essay on the rise and fall of US technological leadership, Nelson and Wright (1992) conclude that two sets of factors—both related to the

───────── *Uneven (and Divergent) Technological Accumulation* ─────────

increasing interdependence of the world economy—led to the erosion of the massive US technological (and productivity) lead held before and immediately after World War II.

(i)   Together with massive social changes, increasing international openness of markets after World War II eroded the US advantages in market size, natural resource availability and more egalitarian income distribution, that were of central importance to the US lead in mass production.

(ii)   The highly educated and increasingly international nature of technological communities has eroded the US lead in 'high-technology' product groups.

If these mechanisms are dominant and can be generalized internationally, we could expect technology gaps to disappear among the industrially advanced countries. But both our data and our proposed framework of explanation suggest otherwise.

First (as Nelson and Wright themselves recognize), there exist international disparities in education that will influence (and for a long time to come) the 'human capital endowments' from which firms can benefit. Countries with a strong endowment of science and engineering graduates, but a badly educated workforce are likely to be constrained to specialize in fields like drugs and software, where the skills of the general workforce are not critical. Countries with a skilled general workforce will have a wider range of opportunities, including assembly and process industries, where production-related skills are of central importance.

Second, our evidence suggests that the distinction between myopic and dynamic systems of finance and management does matter, and influences not only the size of the overall commitment of resources to technological accumulation, but the capacity to maintain and develop competences in core technologies that open a range of potential future product opportunities. Thus, relative overall decline in the US has been accompanied by a major loss of competence in automobiles and in the UK in automobiles and electronics.

The UK decline is of long standing, and its nature and causes have been widely documented and debated elsewhere. For the US, Chandler (1992) identifies a number of factors changing US corporate behavior toward technology and innovation since the 1960s. In particular, the growing influence of business school graduates with universal recipes for management problems, the difficulties facing corporate management in making informed entrepreneurial judgments over a large number of often disparate product divisions, and the changing role of the investment banks from underwriting

─────────── *Uneven (and Divergent) Technological Accumulation* ───────────

long-term corporate investments to trading in corporate control, all conspired to change the bases of strategic decisions:

> ROI data were no longer the basis for discussion between corporate and operational management as to performance, profit and long-term plans. Instead, ROI became a reality in itself—a target sent down from the corporate office for division managers to meet ... ROI too often failed to incorporate complex, non-quantifiable data as to the nature of specific product markets, changing production technology, competitor's activity and internal organizational problems—data that corporate and operating managers had in the past discussed in their long person-to-person evaluation of past and current performance and the allocation of resources. Top management decisions were becoming based on numbers, not knowledge (Chandler, 1992, pp 277–278).

The same features exist in other OECD countries, and there is no reason to believe that they will diminish in future. In our view, uneven and divergent technological development among the industrially advanced countries is here to stay.

References

Abernathy, W. and R. Hayes (1980), 'Managing Our Way to Economic Decline,' *Harvard Business Review*, (July/August) pp. 67–77.

Archibugi, D. and M. Pianta (1992), *The Technological Specialisation of Advanced Countries*. Kluwer Academic Publishers: Dordrecht.

Bell, M. and K. Pavitt (1993), 'Technological Accumulation and Industrial Growth: Contrasts between Developed and Developing Countries,' *Industrial and Corporate Change*, 2, 157–210.

Bertin, G. and S. Wyatt (1988), *Multinationals and Industrial Property*. Harvester-Wheatsheaf: Hemel Hempstead.

Cantwell, J. (1989), *Technological Innovation and Multinational Corporations*. Blackwell: Oxford.

Cantwell, J. (1992), 'The Internationalisation of Technological Activity and its Implications for Competitiveness,' in O. Grandstrand, L. Hakanson and S. Sjolander (eds), *Technology Management and International Business*. Wiley: Chichester.

Chandler, A. (1992), 'Corporate Strategy, Structure and Control Methods in the United States in the 20th Century,' *Industrial and Corporate Change*, 1; 263–284.

Clark, K., T. Fujimoto and W. Chew (1987), 'Product Development in the World Auto Industry,' *Brookings Papers on Economic Activity*, 3.

Corbett, J. and C. Mayer (1991), *Financial Reform in Eastern Europe*. Discussion Paper No. 603, Centre for Economic Policy Research (CEPR): London.

Dahlman, C., B. Ross-Larsen and L. Westphal (1987), 'Managing Technological Development: Lessons from Newly Industrialising Countries,' *World Development*, 15, 759–775.

Dosi, G., K. Pavitt and L. Soete (1990), *The Economics of Technical Change and International Trade*. Wheatsheaf: Hemel Hempstead.

Ergas, H. (1984), *Why Do Some Countries Innovate More Than Others?* Paper No. 5, Centre for European Policy Studies: Brussels.

Fagerberg, J. (1987), 'A Technology Gap Approach to Why Growth Rates Differ,' in C. Freeman (ed.), *Output Measurement in Science and Technology: Essays in Honour of Y. Fabian*. North-Holland: Amsterdam.

─────────── *Uneven (and Divergent) Technological Accumulation* ───────────

Fagerberg, J. (1988), 'International Competitiveness,' *Economic Journal*, 98, 355–374.

Fagerberg, J. (1993), 'Technology and International Differences in Growth Rates' (mimeo), Norwegian Institute of International Affairs (NUPI): Oslo.

Franko, L. (1989), 'Global Corporate Competition: Who's Winning, Who's Losing, and the R&D Factor as One Reason Why,' *Strategic Management Journal*, 10, 449–474.

Freeman, C. (ed.) (1987), *Output Measurement in Science and Technology: Essays in Honour of Y. Fabian*. North-Holland: Amsterdam.

Geroski, P., S. Machin and J. van Reenen (1993), 'The Profitability of Innovating Firms,' *RAND Journal of Economics*, 24, 198–211.

Grilliches, Z. (1990), 'Patent Statistics as Economic Indicators: A Survey,' *Journal of Economic Literature*, 28, 1661–1707.

Hu, Y.-S. (1975), *National Attitudes and the Financing of Industry*. Political and Economic Planning, Vol. XLI, Broadsheet No. 559, London.

Kitti C. and Schiffel, D. (1978), 'Rates of Invention: International Patent Comparisons,' *Research Policy*, 7, 323–340.

Lawrence, P. (1980), *Managers and Management in W. Germany*. Croom Helm: London

Mason, G. S. Prais and B. van Ark (1992), 'Vocational Education and Productivity in the Netherlands and Britain,' *National Institute Economic Review*, (May), 45–63.

Mitchell, G. and W. Hamilton (1988), 'Managing R&D as a Strategic Option,' *Research-Technology Management*, 31, 15–22.

Myers, S. (1984), 'Finance Theory and Finance Strategy,' *Interfaces*, 14, 126–137.

Nelson, R. (1990), 'US Technological Leadership: Where Did It Come From, and Where Did It Go?' *Research Policy*, 19, 117–132.

Nelson, R. and G. Wright (1992), 'The Rise and Fall of American Technological Leadership: The Postwar Era in Historical Perspective,' *Journal of Economic Literature*, 30, 1931–1964.

Newton, K., P. de Broucker, G. McDougal, K. McMullen, T. Schweitzer and T. Siedule (1992), *Education and Training in Canada*. Canada Communication Group: Ottawa.

Patel, P. (1995), 'Localised Production of Technology for Global Markets,' *Cambridge Journal of Economics* (in press).

Patel, P. and K. Pavitt (1991a), 'Large Firms in the Production of the World's Technology: An Important Case of 'Non-Globalisation', *Journal of International Business Studies*, 22, 1–21.

Patel, P. and K. Pavitt (1991b), 'Europe's Technological Performance,' in C. Freeman, M. Sharp and W. Walker, *Technology in Europe's Future*. Pinter: London.

Patel, P. and K. Pavitt (1994a), 'Patterns of Technological Activity: Their Measurement and Interpretation,' in P. Stoneman (ed.), *The Economics of Innovation and Technical Change*. Basil Blackwell: Oxford, (forthcoming).

Patel, P. and K. Pavitt (1994b), 'National Innovation Systems: Why They Are Important and How They Might be Measured and Compared,' *Economics of Innovation and New Technology* (forthcoming).

Pavitt, K. (1980), *Technical Innovation and British Economic Performance*. Macmillan: London.

Pavitt, K. and P. Patel (1988), 'The International Distribution and Determinants of Technological Activities,' *Oxford Review of Economic Policy*, 4, 35–55.

Porter, M. (1990), *The Competitive Advantage of Nations*. Macmillan: London.

Posner, M. (1961), 'International Trade and Technical Change,' *Oxford Economic Papers*, 13, 323–341.

Prais, S. (1993), *Economic Performance and Education: The Nature of Britain's Deficiencies*. Discussion Paper No. 52, National Institute for Economic and Social Research: London.

van Raan, A. (ed.), *Handbook of Quantitative Studies of Science and Technology*. North-Holland: Amsterdam.

Soete, L. (1981), 'A General Test of Technological Gap Trade Theory,' *Weltwirtschaftliches Archiv.*, 117, 638–666.

Soete, L. and B. Verspagen (1993), 'Technology and Growth: The Complex Dynamics of Catching Up, Falling Behind, and Taking Over,' in A. Szirmai, B. van Ark and D. Pilat, *Explaining Economic Growth.* Elsevier: Amsterdam.

Vernon, R. (1966), 'International Investment and International Trade in the Product Cycle,' *Quarterly Journal of Economics*, **80**, 190–207.

What makes basic research economically useful? *

Keith Pavitt

Science Policy Research Unit, University of Sussex, Brighton BN1 9RF, U.K.

Final version received May 1990

Economic analysis has helped us understand the strong economic dimension in the explosive growth of science, and (more recently) the reasons for continuing public subsidies. However, the growing domination of the "market failure" approach has led to the analytical neglect of two major questions for policy-makers. How does science contribute to technology? Are the technological benefits from science increasingly becoming international?

On the former, too much attention has been devoted to the relatively narrow range of scientific fields producing knowledge with direct technological applications, and too little to the much broader range of fields, the skills of which contribute to most technologies. On the latter, national systems of science and of technology remain closely coupled in most major countries, in spite of the technological activities of large multinational firms.

Empirical research is needed on concentration, scale and efficiency in the performance of basic research, where techniques and insights from the applied economics of industrial R&D are of considerable relevance. There is no convincing evidence so far of unexploited economies of scale in basic research.

This evidence shows that many policies for greater "selectivity and concentration" in basic research have been misconceived. Economists and other social scientists could help by formulating more persuasive justifications for public subsidy for basic research, and by making more realistic assumptions about the nature of science and technology.

* This paper draws on a presentation made at Section F (Economics) of the Annual Meeting of the British Association for the Advancement of Science (Sheffield, 1989), and subsequently published in Hague [15]. It is based on the research programme of the ESRC-funded Centre for Science, Technology and Energy Policy, within the Science Policy Research Unit. I am grateful for criticisms and suggestions to Diana Hicks, John Irvine, Ben Martin, Frances Narin, Geoffrey Oldham, Pari Patel, Jackie Senker, Margaret Sharp, and to two anonymous referees. The usual disclaimers hold.

Research Policy 20 (1991) 109–119
North-Holland

1. Introduction

In this paper, I discuss what we know about the economic usefulness of basic research, and its implications for public policy. It is has become commonplace to argue that we need better knowledge about the economic impact of basic science to assist the agonising choices that must be made, given the "steady state" in the inputs that society is now able and willing to devote to it. Such a steady state of zero growth in science was first raised in the public consciousness by Derek de Solla Price [42] in his classic essay *Little Science, Big Science*, published in 1963. Some analysts say that it has now happened, or is about to happen (see, for example, Ziman [53]).

However, the available data suggest otherwise. At least since the mid-1960s, civilian R&D has grown in the OECD area, both in real terms and as a percentage of Gross Domestic Product. Growth has been particularly rapid in Japan, and slow in the UK. After a deceleration in the 1970s, the rate of growth has in fact increased in the 1980s [37,38]. Furthermore, data published by the National Science Board [32, p. 52] show that the total employment of scientists and engineers in the USA increased annually by 6 percent between 1976 and 1986, and is expected to increase by a further 36 percent by the year 2000. In Japan, the numbers of science and engineering graduates continue to increase, and, as in the USA, a growing proportion are finding employment outside manufacturing in professional services, finance and insurance [23]. These are not symptoms of a stationary state, but rather of vigorous growth in demand for professional and research skills in science and technology.

For basic research, the picture is more complicated. According to OECD estimates R&D in higher education grew in real terms in all OECD countries until the mid-1970s, when it stabilised and even declined slightly in the Federal Republic of Germany, Italy and the UK [37, p. 44]. According to a recent study by Irvine et al. [20], there has been real growth in academic science since then in most major OECD countries, but at a slower rate than Gross Domestic Product. Within the national totals, separately budgeted academic research has grown more rapidly than the share embedded within general university funds.

Thus, for most countries policies for basic science evolve within a regime for growth in demand for research skills and knowledge, no doubt reflecting the importance of technological change in economic efficiency and welfare (see, for example, Fagerberg [10,11]). The case for improving the knowledge-base for policies for basic science must therefore be the old-fashioned one of "timeliness and promise". It is timely to improve understanding of an activity that consumes considerable resources (approximately 0.3–0.5 percent of Gross Domestic Product), and that has a major influence on society's capacity to respond to economic and social demands. And there is considerable promise of such improvement, given our qualitative understanding of the nature and determinants of science, technology and the links between them [29], and recent advances in quantitative, bibliometric techniques.

In section 2, I identify two major contributions by (political) economists to better understanding: the growth of sciences as a factor of production, and the economic case for the public subsidy of basic research. I argue that unwarranted emphasis by contemporary economists on the "public good" and information-like properties of science (and sometimes even of technology) has led to the neglect of two centrally important problems of contemporary science policy: the contribution of science to technology (discussed in section 3) and the supposed "internationalisation" of science and technology (section 4). I then argue that economists can make a useful contribution to better understanding of the properties of basic research-performing institutions, and that there is no convincing empirical evidence of unexploited economies of scale in basic research (section 5). I conclude in section 6 that policies for greater selectivity and concentration have been misconceived, and that we need better theorising on the nature of science and technology, and on why basic science deserves public subsidy.

2. Some contributions by economists

Economic analysis has made two major contributions to the policy debate about the management of science. It has shown, first, that the growth of basic science must be understood and justified mainly by its contribution to economic and social progress; and, second, that basic science should be supported mainly through public subsidy.

2.1. Science as an economic activity

The importance of science as an economic activity was in fact recognised very early on. In Chapter One of *The Wealth of Nations*, Adam Smith pointed out that technical advances were made not only at the point of production, but also by suppliers of capital goods, and by "philosophers and men of speculation", which is what scientists were then called. Perhaps less well known in the UK are the predictions of Alexis de Tocqueville in *Democracy in America*. He observed the down-to-earth and problem-solving nature of US society early in the 19th century, and predicted the rapid expansion of science, for three reasons: first, as a form of conspicuous intellectual consumption, funded from the great accumulations of private wealth that de Tocqueville rightly predicted the US system would produce; second, as a foundation for the education of the large number of applied scientists that de Tocqueville predicted (again rightly) that modernising society would require; and third, as a source of fundamental knowledge needed to facilitate and guide the solving of practical problems.

Thus, the rapid growth of modern science must be seen as part of a more general process of the specialisation and professionalisation of productive activities in modernising societies. To this, we must add Marx's important insights into the major influence of economic and social demands on the rate and direction of scientific advance, through the problems that they pose, the empirical data and techniques of measurement that they generate, and the financial resources that they make available (see Rosenberg [43]). For all these reasons, economists are right to argue that large expenditures on science can be neither understood nor justified solely on cultural and aesthetic grounds; they inevitably have important economic and social dimensions. However, we shall see in section 3 below that the links today between science and technological practice are far from straightforward.

2.2. The public subsidy of science

In the meantime, another major contribution of economics to the management of science has been the analytical justification for regular and large-scale government funding of basic research. As is often the case, principle followed practice rather than led to it, since governments in some countries (and most notably Germany) had already been funding basic research for a very long time. After World War II, the USA followed suit, and in the early 1950s established the National Science Foundation. In 1959, Nelson published his pioneering paper entitled "The Simple Economics of Basic Research" [33], in which he argued that – left to itself – a competitive market will invest less than the optimum in basic research. This is because a profit-seeking firm cannot not be sure of capturing all the benefits of the basic science that it sponsors, given major uncertainties about the benefits for the sponsoring firm, and the difficulties it faces in extracting compensation from subsequent imitators. At the same time, a policy of secrecy aimed at stopping such imitation would be sub-optimal, since it would restrict applications with small marginal cost. If, in addition, profit-seeking firms are risk-averse, or have short-term horizons in their decisions to allocate resources, private expenditures on basic research will be even more sub-optimal.

Nelson's insights have been developed and modified over the past thirty years, notably by Arrow [1] and Averch [2] in the USA, and by Kay and Llewellyn Smith [22], Dasgupta [6], Stoneman [47] and – most recently – Stoneman and Vickers [48] in the UK. Risk aversion, low or zero marginal cost of application, and difficulties in appropriating benefits, have become standard explanations for the public subsidy of science. At the same time there has been a subtle shift in emphasis. Nelson's original paper was grounded in research on the development of the transistor

[34], and his paper is spliced with examples of the development and application of science. Over time, progressively fewer references have been made to the empirical evidence, and more to the standard theorems of welfare economics. Whilst it might be advantageous in the economics classroom to stress the "public good" characteristics of science, and to minimise or ignore the distinctions and interactions between science and technology, this has effectively excluded economists from two of the major debates of contemporary science policy: the nature and extent of the contributions of science to technology, and the impact of national science on national technology [27].

2.3. Technology as science?

It is comfortable as well as convenient to treat science and technology as the same thing, given the similarities in their inputs (scientists, engineers, laboratories) and their outputs (knowledge), and given the well-known examples of outstanding science performed in corporate laboratories. However, this neglects the very different nature and purpose of the core activities of university and business laboratories. In universities, basic research seeks generalisations based on a restricted number of variables, and results in publications and reproducible experiments. In business, a combination of research, and (more important) development, testing, production engineering and operating experience accumulates knowledge on the many critical operating variables of an artefact, and result in knowledge that is not only specific, but partly tacit (uncodifiable) and therefore difficult and costly to reproduce.

Given these differences, basic research is more likely to meet the conditions for private under-investment, as defined by Nelson and others, which explains the higher proportion of public funding in basic research than in development in all OECD countries. Economists conscious of the distinction between science and technology have made a major contribution to the policy debate by stressing the complementary nature of private and public investments in science and technology, with the former concentrating on the short-term and specific, and the latter on the long-term and the general. They have also warned of the dangers and inefficiencies of heavy public funding of commercial development activities [9,21]. However, insufficient attention has in general been directed by economists to the interface between science and technology.

2.4. Science as a "free good"?

One other reason for this lack of attention has been a common confusion between the reasonable assumption that the results of science are a "public good" (i.e. codified, published, easily reproduced and therefore deserving of public subsidy), and the unreasonable assumption that they are a "free good" (i.e. costless to apply as a technology, once read). In a paper entitled "Why do Firms do Basic Research (with Their Own Money)", Rosenberg [45] argues that basic research financed and performed in (mainly large) firms often grows out of practical problem-solving, and that the two are highly interactive. He also argues that in-house basic research is essential in order to monitor and evaluate research being conducted elsewhere:

> "This point is important...in identifying a serious limitation in the way economists reason about scientific knowledge and research in general....such knowledge is regarded by economists as being "on the shelf" and costlessly available to all comers once it has been produced. But this model is seriously flawed because it frequently requires a substantial research capability to understand, interpret and the appraise knowledge that has been placed on the shelf – whether basic or applied. The cost of maintaining this capability is high, because it is likely to require a cadre of in-house scientists who can do these things. And, in order to maintain such a cadre, the firm must be willing to let them perform basic research. The most effective way to remain plugged in to the scientific network is to be a participant in the research process."

This has implications for the way we view the impact of science on technology, and for the reasons for public subsidy. We shall take them up in sections 3 to 6 below.

3. The impact of science on technology

The impact of science on technology is bound to be of central concern to science policy-makers.

I summarise below what we already know from earlier studies, and identify subjects for future research.

3.1. Calculating the economic return from basic science?

Resource-starved basic scientists no doubt welcome studies demonstrating a high economic return to basic research. On such study has just been completed in the USA, by a distinguished economic expert on R&D – E. Mansfield [25]. It is one of the most ingenious and persuasive of its kind but, as pointed out by David and his colleagues [7], calculations of this kind do not satisfactorily reflect the nature of the impact of science and technology:

> "The outputs of basic research rarely possess intrinsic economic value. Instead, they are critically important inputs to other investment processes that yield further research findings, and sometimes yield innovations,...Policies that focus exclusively on the support of basic research with an eye to its economic payoffs will be ineffective unless they are also concerned with these complementary factors.
>
> The alternative conceptualization...that we have developed focuses on basic research as a process of learning about the physical world that can better inform the processes of applied research and development. Rather than yielding outputs that are marketed commercially, basic research interacts with applied research in a complex and iterative manner to increase the productivity of both basic and applied research. The development of links between the basic and applied research enterprises are critical to the productivity and economic payoffs of both activities" (pp. 68–69).

3.2. The complexity of science's impact on technology

We know from the results of past research that these links between basic science and technology are in fact complex along at least four dimensions [49].

(i)   The intensity of direct transfers of knowledge from basic science to application varies widely amongst sectors of economic activity, and amongst scientific field. The most systematic analyses have been made in the USA, on the basis of patent citations to journals [4,30], and of a survey of industrial R&D directors [36]. They both confirm strong links in chemicals and drugs firms to basic research in biology, whilst the links of electronics firms are also intense but to more applied research activities in physics. In mechanical and transport technologies, on the other hand, the links to science are weak.

(ii) The nature of the impact of basic research on technology also varies widely from the generation of epoch-making new technologies (e.g. electricity, synthetic materials, semi-conductors; see Freeman et al. [13]), through accumulated improvements in continuous flow industries resulting from routine chemical analysis [44], to insights and methods for dealing with applied problems. In all cases, operationally viable technology requires combinations with knowledge from other sources, including design and production engineering.

(iii) Basic science has an impact on technology not just through direct knowledge transfers, but also through access to skills, methods and instruments [40].

(iv) Knowledge transfers are mainly person-embodied, involving personal contacts, movements, and participation in national and international networks [14].

3.3. Is basic research a growing source of technology?

Some analysts (for example, Martin and Irvine [26]) claim that we are now witnessing a significant increase in the direct use in technology of the results of basic research. Others claim that such "strategic" areas of science should receive priority support from government. In my view, the evidence is ambiguous and incomplete (see also, Williams [50,51]).

Narin and Frame [31] have produced the most persuasive quantitative evidence so far. They have shown sharply upward trends in the frequency with which US patents, originating in a number of countries, contain citations to publications other than patents: from about 0.2 cites to "other publications" per US patent in 1975, to between 0.9 cites for US patents of US origin – and 0.4 cites for US patents of Japanese origin – in 1986. On this basis they claim that the technology reflected in US patents is much more "science-dependent" than ten years ago. They further show that the time-lags in the citations from patents to other publications are diminishing rapidly, and that science-intensive patents are relatively highly cited.

Whilst suggestive, this evidence has its limitations. It is not yet clear to what extent the "other publications", cited in patents, reproduce basic or applied research, from universities or from corporate laboratories. In addition, a high proportion of technology is not patented, because it is kept secret (e.g. process technology), because it is tacit and non-codifiable know-how, or because – as in the increasingly important case of software technology – it is very difficult to protect through patenting. This non-patented technology is likely to be less dependant on science, and more on cumulative design and engineering skills. Together with a number of colleagues, I have argued elsewhere that it is increasing as a proportion of total technological activity [46].

In addition, it is worth noting that, in the USA, the recent report from MIT *Made in America* [8] has claimed that it is precisely because of deficiencies in these engineering skills that US firms are not capturing the full economic benefits from exploiting scientific advances. They further claim that engineering education in the USA has become too science-based.

More generally the evidence from US R&D statistics are ambiguous. Whilst there are signs of increasing corporate commitment to basic research in the 1980s, this follows an extended period of decline, and it has only just regained its share of the early 1960s (National Science Board [32], Appendix Tables 1–40 and 5–1). According to Mowery [28], the increasingly generous provision of funds for academic research by the Federal Government after World War II had led to a reduction in the direct funding by business firms, and:

" ...(b)oth the recent upsurge in state funding of applied research and the proliferation of collaborative research relationship between universities and industry thus represent a partial revival of earlier relationship that were sundered by the dramatic changes in the structure of the

U.S. national research system during and after World War II" (pp. 23–24).

Even if certain fields of basic research make increasingly important direct knowledge inputs into technology, it is misleading to assume that only they contribute to technology, and other fields do not. There are at least two other influences of science and technology that are equally, if not more, important: research training and skills; and unplanned applications.

3.4. The broad demands for research skills

One important function of academic research is the provision of trained research personnel, who go on to work in applied activities and take with them not just the knowledge resulting from their research, but also skills, methods, and a web of professional contacts that will help them tackle the technological problems that they later face.

In one of their less well-known studies, Irvine and Martin [19] have shown that Masters and Doctoral graduates from British radio-astronomy benefited in subsequent non-academic careers from the research skills – rather than the research knowledge – that they obtained during their post-graduate training. A more comprehensive survey

Table 1
The relevance of scientific fields to technology (USA)

Field of science	Number of industries (out of 130) ranking scientific field at 5 or above (out of 7) in relevance to its technology of:	
	Science (i.e. skills)	Academic research (i.e. know-ledge)
Biology	14	12
Chemistry	74	19
Geology	4	0
Mathematics	30	5
Physics	44	4
Agricultural Science	16	17
Applied Maths & OR	32	16
Computer Science	79	34
Materials Science	99	29
Medical Science	8	7
Metallurgy	60	21

Source: Nelson and Levin [36]; Nelson [35].

undertaken at Yale University suggests that this is the rule rather than the exception.

The relevant results are summarised in table 1. They show the responses of 650 US industrial research executives, spread across 130 industries, who were asked to rank the relevance to their technology of a number of fields of pure and applied science. Table 1 lists the number of industries in which each scientific field was given high ranking according to two criteria: first, the relevance of the skill base in the science to the technology; second, the relevance of the academic research knowledge to the technology. As the authors point out:

"Industrial scientists and engineers almost always need training in the basic scientific principles and research techniques of their field, and providing this training is a central function of universities. Current academic research in a field, however, may or may not be relevant to technical advance in industry, even if academic training is important" (Nelson and Levin [36]).

Table 1 shows that, in most scientific fields, whether pure or applied, academic training and skills are relevant over a far larger number of industrial technologies than is academic research. The expectations are the pure and applied biological sciences where we know from other studies that academic research is at present very close to technology [30]. These results show clearly that most scientific fields are much more strategically important to technology than data on direct transfers of knowledge would lead us to believe.

3.5. Unplanned applications

Another major influence of science on technology is through unplanned applications, where useful knowledge emerges from research undertaken purely out of curiosity, without any strategic mission or expectation of application. Two US studies – one undertaken in the late 1960s and the other some 20 years later – both show the importance of such research for achieving relatively short-term technological objectives [18,25]. In both cases, important innovations would have been substantially delayed without contributions from unprogrammed research performed in the ten years preceding the commercial launch of the innovations. Furthermore, in both studies, unprogrammed research contributed about 10 percent of the important knowledge inputs.

One implication of these findings is that programmed R&D should be built on a wider spread of non-programmed research. Analysts like Nelson [33], and Kay and Llewellyn Smith [22], have gone further and used various examples to suggest that more useful knowledge is produced in the long term by allowing basic scientists to pursue their own interests, than by fixing practical objectives for their work. It is a view that needs to be considered seriously by analysts in future (see, for example, Council for Science and Society [5]).

4. Is the application of science (at last) being internationalised?

The analytical apparatus developed by economists to justify public subsidy to basic research has in general assumed a closed economy. This is paradoxical given that the main stimulus for public policies for science and technology have not come from any notions of (national) market failure, but from what is perceived as best practice in a world system of international competition where technological leads and lags are of central importance. It is also perhaps fortunate that the subject has not been pursued too often within the mainstream analytical framework: if we assume that basic research is a "free good", an open international world would in principle permit any one country to live off the rest of the world's basic research [22].

But the real world is more complicated. As Rosenberg has pointed out, the ability to assimilate the results of other people's basic research depends in part on the performance of basic research, oneself. An active national competence in basic research is therefore a necessary condition for benefiting from research undertaken elsewhere in the world; indeed it can be viewed as a national scientific intelligence system. And since most transfers of knowledge and skills between science and technology are person-embodied, the constraints of distance and language have meant that nation-based transfers between science and technology have been the rule rather than the exception.

Now, it is argued, conditions are changing. The barriers of distance and language are lower than

Table 2
Foreign controlled domestic technology compared to nationally controlled foreign technology (based on US patenting, 1981–86)

Home country	US patenting from inside country by foreign firms (as % of country's total US patenting)	US patenting by national firms from outside home country (as % of country's total US patenting)
Belgium	45.7	16.5
France	11.8	3.8
FR Germany	11.5	8.5
Italy	11.2	3.0
Netherlands	9.5	73.4
Sweden	5.4	16.7
Switzerland	12.5	27.8
UK	22.3	24.5
W. Europe	7.4	9.3
Canada	28.1	12.5
Japan	1.2	0.5
USA	4.2	4.4

Source: Patel and Pavitt [39].

they used to be. And firms are increasingly internationalising their R&D activities, which enable them more easily to benefit from academic science in foreign countries, through personal contacts and the hiring of scientists and engineers with research experience from local systems of higher education. Does this mean that linkages between science and technology will become internationalised? Does it mean that an increasing proportion of the benefits of national governments' investments in basic research will "leak away" through foreign-controlled firms to other countries?

This is a subject that deserves further research. Suffice to suggest at this stage that the degree of "leakage" depends, as a first approximation, on the proportion of a country's corporate technological activity that is controlled by foreign firms, which reflects their capacity to monitor and absorb local basic research skills and knowledge. Similarly, the importance of the foreign technological activities of nationally owned firms will reflect a country's capacity to benefit from basic research undertaken in other countries.

Table 2 is a first attempt to measure and compare these variables across countries. The first column compares the proportion of each country's US patenting originating from foreign-controlled firms. It shows that, in most countries, large foreign firms still play a relatively small role in national technological activities; only in Belgium, Canada and the UK do they account for more than 20 percent of the total. The second column compares the US patenting of nationally-controlled firms from outside their home country, as a proportion of total national patenting in the USA. For the Netherlands, this amounts to more than 70 percent of the national total, and more than 20 percent for Switzerland and the UK.

Taken together, the two measures show that most national technological systems are relatively self-contained. Both measures of internationalisation are less than a quarter of total technological activities in eight out of the 11 countries. In Belgium and Canada, foreign-controlled domestic technological activities are much greater than domestically controlled foreign technological activities, whereas for the Netherlands, Sweden and Switzerland the opposite is the case. When Western Europe is considered as a whole, the degree of internationalisation is much less than for most of the European countries taken individually, but still greater than that of either Japan or the USA.

These results show that complete internationalisation of links between science and technology is not at all likely in the immediate future. In most countries, national science will still be feeding into largely nationally controlled technology, and close links with foreign science through personal contacts and recruitment will in most cases be small compared to national links. Contrary to conventional wisdom, Japan is not well positioned to benefit from foreign countries' basic science, since their firms undertake such a small proportion of their technological activities outside Japan. Our data suggest that the Dutch are much better at it.

5. The properties of basic research-producing institutions

In addition to the links between science and application, we need a better understanding of the properties of basic research-producing institutions, particularly universities and university departments. Public policy in the UK (and perhaps in other countries) increasingly assumes that there are advantages to greater scale and concentration

in basic research activities, although there is no systematic evidence that this is the case.

In this context, policy would be better informed as the result of a research programme that combines recent advances in bibliometric methods, with accumulated experience in industrial economics in understanding the links between technological activities, firm size and industrial concentration. There would no doubt be similar room for debate over the adequacy of the various measures used. But similarly useful results would probably emerge, showing considerable variations amongst scientific fields in concentration and economies of scale.

5.1. What are the unexploited economies of scale in basic research?

Partly as what they would consider as legitimate acts of academic self-defence, British scholars have been among the first to identify the problems to be clarified. As Hare and Wyatt [16] have recently pointed out, little systematic evidence is available on economies of scale in basic research. In the USA, Frame and Narin [12] found no economies of scale in biomedical research, when output was measured by numbers of publications. In the UK, Hicks and Skea [17] have come to similar conclusions to. Frame and Narin in a preliminary analysis of 45 physics departments in Britain: no unexploited economies of scale, when output is measured in numbers of publications. Williams [51] came to the same conclusion in an unpublished study of chemistry departments.

This type of analysis should be extended to other scientific fields, and down into sub-fields. The sensitivity of results to various measures of inputs and outputs should also be tested: for example, McAllister and Narin [24] found no economies of scale in US biomedical research in terms of the number of publications, but they did find higher citation rates amongst the larger institutions. This might reflect the greater quality of large institutions' basic research. but it might also reflect their greater visibility.

5.2. How do basic research institutions evolve?

Just as in the analysis of firms' technological activities, cross-sectional comparisons of size, concentration and efficiency, while useful, will also raise further important questions for theory and policy: in particular, how and why do the existing patterns come about? This leads on to four further questions, each of which is also central to the analysis of the dynamics of technical change, concentration and efficiency in industry:

- are large and productive research institutions good because they are big, or big because they are good?
- what are the characteristics of productive institutions? To what extent do they grow out of accumulated scientific and managerial skills?
- what is the appropriate organisational unit in which such skills are accumulated? Preliminary analysis by Platt [41] suggests that it is not at the level of a university as a whole, but (if at all) in closely related subjects;
- what are the mechanisms through which good research practice and productivity are diffused (or not) throughout the research community?

6. Conclusions

6.1. Gaps in empirical knowledge

Conclusions for policy are bound to be tentative, given the still shaky theoretical and empirical base, which is why I have signalled throughout the paper where further research is required. The three most important subjects (in my view) are:

- the economic and social benefits of "unstrategic" science, particularly the development of useful research skills and networks, and unplanned applications;
- the nature and effects of the internationalisation of scientific and technological activities;
- the structure, efficiency and dynamics of national systems of basic research.

6.2. Misguided policies seeking "relevance"

In the meantime, our analysis suggests that the objectives of many policies seeking to make basic research more useful may turn out to have been badly misconceived. Policies of high priority for basic research that are directly and obviously applicable ignore the considerable indirect benefits across a broad range of scientific fields resulting from training and from unplanned discoveries.

Policies for concentration in larger units in basic research are based on the unproven premise that big is necessarily beautiful (i.e. efficient). Both policies neglect the all-important fact that the application of basic research depends overwhelmingly on the size and persistence in investment in downstream activities by business firms. Dealing with deficiencies in business R&D by making basic research more "relevant" is like pushing a piece of string.

6.3. A revised case for public subsidy for basic research

Our analysis suggests that the justification for public subsidy, in terms of complete inappropriability of immediately applicable knowledge, is a weak one. In fact, the results of basic research are rarely immediately applicable, and making them so also increases their appropriability, since – in seeking potential applications – firms learn how to combine the results of basic research with other firm-specific assets, and this cannot be imitated overnight. In three other dimensions, the case for public subsidy is stronger.

The first justification was originally stressed strongly by Nelson [33], but has been neglected since then: namely, the considerable uncertainties before the event in knowing if, when and where the results of basic research might be applied. We now know from transaction cost theory that high uncertainty is one reason why markets are not necessarily efficient [52]. The probabilities of application will be greater with an open and flexible interface between basic research and application, which implies public subsidy for the former. The case for such a subsidy is strongest for "unstrategic" fields of curiosity driven research, the application of which cannot be foreseen.

A second, and potentially new, justification grows out of internationalisation of the technological activities of large firms, discussed in section 4. Facilities for basic research and training can be considered as an increasingly important part of the infrastructure for downstream technological and production activities. Countries may therefore decide to subsidise them, in order to attract foreign firms or even to retain national ones. Recent interest in so-called "science parks" might sometimes be one manifestation of this trend. Clearly there are dangers of competitive subsidy, the implications of which should keep game and trade theorists busy for some time.

The final and most important justification for public subsidy is training in research skills, since private firms cannot fully benefit from providing it when researchers, once trained, can and do move elsewhere. There is, in addition, the important insight of Dasgupta [6] that, since the results of basic research are public and those of applied research and development often are not, training through basic research enables more informed choices and recruitment into the technological research community.

6.4. Better conceptualisations of science and technology

This last justification illustrates a broader conclusion emerging from this paper (see also Mowery and Rosenberg [29]): economists and other social scientists will benefit enormously in both the accuracy and impact of their analyses, if they drop their conceptualisations of science and technology as activities producing easily transmissible and applicable "information", and recognise them instead as search processes and skills embodied in individuals and institutions. In this context, they would more easily appreciate the importance of basic research as both training and a cumulative body of knowledge. As we have seen, this was clear to de Tocqueville a long time ago. It was also clear to one of the major figures in the development of modern policies for basic science, Vannevar Bush, who pointed out in 1945 that "(t)he responsibility for the creation of new scientific knowledge – and for most of its application – rests on that small body of men and women who understand the fundamental laws of nature and are skilled in the techniques of scientific research" [3, p. 7].

References

[1] K. Arrow, Economic Welfare and the Allocation of Resources for Invention, in: R. Nelson (ed.), *The Rate and Direction of Inventive Activity* (Princeton University Press, New Jersey, 1962).

[2] H. Averch, *A Strategic Analysis of Science and Technology Policy* (John Hopkins Press, Baltimore, 1985).

[3] V. Bush, *Science: the Endless Frontier* (National Science Foundation, Washington, 1945) (reprinted, 1960).

[4] M. Carpenter, *Patent Citations as Indicators of Scientific and Technological Linkages* (Computer Horizons Inc., New Jersey, 1983).

[5] Council for Science and Society, *The Value of "Useless" Research: Supporting Science and Scholarship for the Long Run*, London, 1989.

[6] P. Dasgupta, The Economic Theory of Technology Policy: an Introduction in: P. Dasgupta and P. Stoneman, *Economic Policy and Technological Performance* (Cambridge University Press, Cambridge, 1987).

[7] P. David, D. Mowery and W. Steinmuller, The Economic Analysis of Payoffs from Basic Research-An Examination of the Case of Particle Physics Research, *CEPR Publication No. 122*, Center for Economic Policy Research, Stanford University, California, 1988.

[8] M. Dertouzos, R. Lester and R. Solow (eds.), *Made in America: Regaining the Productive Edge* (MIT Press, Cambridge, MA, 1989).

[9] G. Eads and R. Nelson, Governmental Support of Advanced Civilian Technology: Power Reactors and the Supersonic Transport, *Public Policy* 19 (1971) 405–428.

[10] J. Fagerberg, A Technology Gap Approach to Why Growth Rates Differ, *Research Policy* 16 (1987) 87–99.

[11] J. Fagerberg, International Competitiveness, *Economic Journal* 98 (1988) 355–374.

[12] J. Frame and F. Narin, NIH Funding and Biomedical Publication Output, *Federation Proceedings* 35 (1976) 2529–2532.

[13] C. Freeman, J. Clark and L. Soete, *Unemployment and Technical Innovation* (Pinter, London, 1982).

[14] M. Gibbons and R. Johnston, The Roles of Science in Technological Innovation, *Research Policy* 3 (1974) 220–242.

[15] D. Hague (ed.), *The Management of Science* (Macmillan, 1990).

[16] P. Hare and G. Wyatt, Modelling the Determination of Research Output in British Universities, *Research Policy* 17 (1988) 315–328.

[17] D. Hicks and J. Skea, Is Big Really Better?, *Physics World* 2 (12) 31–34.

[18] Illinois Institute of Technology Research Institute, *Report on Project TRACES* (National Science Foundation, Washington, 1969).

[19] J. Irvine and B. Martin, The Economic Effects of Big Science: the Case of Radio-Astronomy, in: *Proceedings of the International Colloquium on Economic Effects of Space and Other advanced Technologies*, Ref. ESA SP 151 (European Space Agency, Paris, 1980).

[20] J. Irvine, B. Martin and P. Isard, *Investing in the Future* (Edward Elgar Publishing, Aldershot, Hampshire, UK, 1990).

[21] J. Jewkes, *Government and High Technology* (Institute of Economic Affairs, Occasional Paper 37, London, 1972).

[22] J. Kay and C. Llewellyn Smith, Science Policy and Public Spending, *Fiscal Studies* 6 (1985) 14–23.

[23] F. Kodama, Some Analysis on Recent Changes in Japanese Supply and Employment Pattern of Engineers, (mimeo), National Institute of Science and Technology Policy, Tokyo, 1989.

[24] P. McAllister and F. Narin, Characterization of the Research Papers of US Medical Schools, *Journal of the*

American Society for Information Science 34 (1983) 123–131.

[25] E. Mansfield, The Social Rate of Return from Academic Research, (mimeo), University of Pennsylvania, Philadelphia, 1989.

[26] B. Martin and J. Irvine, *Research Foresight: Priority-Setting in Science* (Pinter, London, 1989).

[27] D. Mowery, Economic Theory and Government Technology Policy, *Policy Sciences* 16 (1983) 27–43.

[28] D. Mowery, The Growth of U.S. Industrial Research, (mimeo), Haas School of Business, University of California, Berkeley, 1990.

[29] D. Mowery and N. Rosenberg, *Technology and the Pursuit of Economic Growth* (Cambridge University Press, Cambridge, 1989).

[30] F. Narin and F. Noma, Is Technology becoming Science?, *Scientometrics* 7 (1985) 369–381.

[31] F. Narin and J. Frame, The Growth of Japanese Science and Technology, *Science* 245 (1989) 600–604.

[32] National Science Board, *Science and Engineering Indicators–1987* (Washington, 1987).

[33] R. Nelson, The Simple Economics of Basic Scientific Research, *Journal of Political Economy* 67 (1959) 297–306.

[34] R. Nelson, The Link between Science and Invention: the Case of the Transistor, in: R. Nelson (ed.), *The Rate and Direction of Inventive Activity* (Princeton University Press, New Jersey, 1962).

[35] R. Nelson, *Understanding Technical Change as an Evolutionary Process* (North-Holland, Amsterdam, 1987).

[36] R. Nelson and R. Levin, The Influence of Science University Research and Technical Societies on Industrial R&D and Technical Advance, *Policy Discussion Paper Series Number 3*, Research Program on Technology Change, Yale University, Newhaven, Connecticut, 1986.

[37] OECD, *Technical Change and Economic Policy* (Paris, 1980).

[38] OECD, *R&D, Production and Diffusion of Technology*, OECD Science and Technology Indicators Report, No. 3, Paris, 1989.

[39] P. Patel and K. Pavitt, Do Large Firms Control the World's Technology?, (mimeo), Science Policy Research Unit, University of Sussex, Brighton, 1989.

[40] K. Pavitt, The Objectives of Technology Policy, *Science and Public Policy* 14 (1987) 182–188.

[41] J. Platt, Research Policy in British Higher Education and Its Sociological Assumptions, *Society* 22 (1988) 513–529.

[42] D. de Solla Price, *Little Science, Big Science* (Columbia University Press, New York, 1963).

[43] N. Rosenberg, Karl Marx on the Economic Role of Science, in: N Rosenberg, *Perspectives on Technology* (Cambridge University Press, Cambridge, 1976) pp. 126–138.

[44] N. Rosenberg, The Commercial Exploitation of Science by American Industry, in: K. Clark, A. Hayes and C. Lorenz (eds.), *The Uneasy Alliance: Managing the Productivity–Technology Dilemma* (Harvard Business School Press, Boston, 1985).

[45] N Rosenberg, Why do Firms do Basic Research (with Their Own Money)?, *Research Policy* (1990) (forthcoming).

[46] L. Soete, B. Verspagen, P. Patel and K. Pavitt, *Recent Comparative Trends in Technology Indicators in the OECD Area*, International Seminar on Science, Technology and Economic Growth (OECD, Paris, 1989).

[47] P. Stoneman, *The Economic Analysis of Technology Policy* (Oxford, University Press, Oxford, 1987).

[48] P. Stoneman and J. Vickers, The Assessment: the Economics of Technology Policy, *Oxford Review of Economic Policy* 4 (1988) i–xvi.

[49] B. Williams, The Direct and Indirect Role of Higher Education in Industrial Innovation: What Should We Expect?, *Minerva* 2–3 (1986) 145–171.

[50] B. Williams, Applied Research in Higher Education and Technology Transfer, in: *The Economics of Higher Education* (Pergamon Press, Oxford, 1990).

[51] B. Williams, Written communication (1990).

[52] O. Williamson, *Markets and Hierarchies: Analysis and Antitrust Implications* (Free Press, New York, 1975).

[53] J. Ziman, Science in a "Steady State": the Research System in Transition, SPSG Concept No. 1, Science Policy Support Group, London, 1987.

[13]

National policies for technical change: Where are the increasing returns to economic research?

Keith Pavitt

This paper was presented at a colloquium entitled 'Science, Technology, and the Economy', organized by Ariel Pakes and Kenneth L. Sokoloff, held October 20-22, 1995, at the National Academy of Sciences in Irvine, CA.

ABSTRACT   Improvements over the past 30 years in statistical data, analysis, and related theory have strengthened the basis for science and technology policy by confirming the importance of technical change in national economic performance. But two important features of scientific and technological activities in the Organization for Economic Cooperation and Development countries are still not addressed adequately in mainstream economics: (*i*) the justification of public funding for basic research and (*ii*) persistent international differences in investment in research and development and related activities. In addition, one major gap is now emerging in our systems of empirical measurement – the development of software technology, especially in the service sector. There are therefore dangers of diminishing returns to the usefulness of economic research, which continues to rely completely on established theory and established statistical sources. Alternative propositions that deserve serious consideration are: (*i*) the economic usefulness of basic research is in the provision of (mainly tacit) skills rather than codified and applicable information; (*ii*) in developing and exploiting technological opportunities, institutional competencies are just as important as the incentive structures that they face; and (*iii*) software technology developed in traditional service sectors may now be a more important locus of technical change than software technology developed in 'high-tech' manufacturing.

From the classical writers of the 18th and 19th centuries to the growth accounting exercises of the 1950s and 1960s, the central importance of technical change to economic growth and welfare has been widely recognized. Since then, our under-standing – and consequent usefulness to policy makers – have been strengthened by systematic improvements in comprehensive statistics on the research and development (R&D) and other activities that generate knowledge for technical

Abbreviations: R&D, research and development; OECD, Organization for Economic Cooperation and Development.

change and by related econometric and theoretical analysis.

Of particular interest to national policy makers have been the growing number of studies showing that international differences in export and growth performance countries can be explained (among other things) by differences in investment 'intangible capital', whether measured in terms of education and skills (mainly for developing countries) or R&D activities (mainly for advanced countries). These studies have recently been reviewed by Fagerberg (1) and Krugman (2). Behind the broad agreement on the economic importance of technical change, both reveal fundamental disagreements in theory and method. In particular, they contrast the formalism and analytical tractability of mainstream neoclassical analysis with the realism and analytical complexity of the more dynamic evolutionary approach. Thus, Krugman concludes:

> Today it is normal for trade theorists to think of world trade as largely driven by technologial differences between countries; to think of technology as largely driven by cumulative processes of innovation and the diffusion of knowledge; to see a possible source of concern in the self-reinforcing character of technological advantage; and to argue that dynamic effects of technology on growth represent both the main gains from trade and the main costs of protection . . . the theory has become more exciting, more dynamic and much closer to the world view long held by insightful observers who were skeptical of the old conventional wisdom.
>
> Yet . . . the current mood in the field is one of at least mild discouragement. The reason is that the new approaches, even though they depend on very special models, are *too* flexible. Too many things can happen . . . a clever graduate student can produce a model to justify any policy. [ref. 2, p. 360.]

Fagerberg finds similar tensions among the new growth theorists:

> . . . technological progress is conceived either as a 'free good' (manna from heaven'), as a by-product (externality), or as a result of intentional R&D activities in private firms. All three perspectives have some merits. Basic research in universities and other public R&D institutions provides substantial inputs into the innovation process. Learning by doing, using interacting, etc., are important for technological progress. However . . . models that do not include the third source of technological progress (innovation . . . by intentional activities in private firms) overlook one of the most important sources of technological progress . . .
>
> . . . important differences remain . . . while formal theory still adopts the traditional neo-classical perspective as profit maximizers, endowed with perfect information and foresight, appreciative theorizing increasingly portrays firms as organizations characterized by different capabilities (including technology) and strategies, and operating under considerable uncertainty with respect to future technological trends . . . Although some formal theories now acknowledge the importance of firms for technological progress, these theories essentially treat technology as 'blueprints' and 'designs' that can be traded on markets. In contrast, appreciative theorizing often describes technology as organizationally embedded, tacit, cumulative in character, influenced by interaction between these firms and their environments, and geographically localized. [ref. 1, p. 1170.]

As a student of science and technology policy – and therefore unencumbered by an externally imposed need to relate my analyses to the assumptions and methods of mainstream neoclassical theory – I find what Krugman calls 'more exciting, more dynamic' theorizing and what Fagerberg calls 'appreciative' theorizing, far more useful in doing my job. More to the point of this paper, while the above differences have been largely irrelevant to past analyses of technology's economic importance, they are turning out to be critical in two important areas of policy for the future: the justification of public support for basic research and the determinants of the level of private support of R&D. They will therefore need to be addressed more explicitly in future. So, too, will the largely uncharted and unmeasured world of software technology.

The usefulness of basic research

The production of useful information?   In the past, the case for public policy for basic research has been strongly supported by economic analysis. Governments provide by far the largest proportion of the funding for such research in the Organization for Economic Cooperation and Development (OECD) countries. The well-known justification for such subsidy was provided by Nelson (3) and Arrow (4): the economically useful output of basic research is codified information, which has the property of a 'public good' in being costly to produce and virtually costless to transfer, use, and reuse. It is therefore economically efficient to make the results of basic research freely available to all potential users. But this reduces the incentive of private agents to fund it, since they cannot appropriate the economic benefits of its results; hence the need for public subsidy for basic research, the results of which are made public.

This formulation was very influential in the 1960s and 1970s, but began to fray at the edges in the 1980s. The analyses of Nelson and Arrow implicitly assumed a closed economy. In an increasingly open and interdependent world, the very public good characteristics that justify public subsidy to basic research also make its results available for use in any country, thereby creating a 'free rider' problem. In this context, Japanese firms in particular have been accused of dipping into the world's stock of freely available scientific knowledge, without adding much to it themselves.

But the main problem has been in the difficulty of measuring the national economics benefits (or 'spillovers') of national investments in basic research. Countries with the best record in basic research (United States and United Kingdom) have performed less well technologically and economically than Germany and Japan. This should be perplexing – even discouraging – to the new growth theorists who give central importance to policies to stimulate technological spillovers, where public support to basic research should therefore be one of the main policy instruments to promote technical change. Yet the experiences of Germany and Japan, especially when compared with the opposite experience of the United Kingdom, suggest that the causal linkages run the other way – not from basic research to technical change, but from technical change to basic research. In all three countries, trends in relative performance in basic research since World War II have lagged relative performance in technical change. This is not an original observation. More than one hundred years ago, de Tocqueville (5) and then Marx (6) saw that the technological dynamism of early capitalism would stimulate demand for basic research knowledge, as well as resources, techniques, and data for its execution.

At a more detailed level, it has also proved difficult to find convincing and comprehensive evidence of the direct technological benefit of the information provided by basic research. This is reflected in Table 1, which shows the frequency with which US patents granted in 1994 cite (i.e., are related to) other patents, and the frequency with which they cite science-refereed journals and other sources. In total, information from refereed journals provides only 7.2 per cent [= 0.9/(10.9 + 0.9 + 0.7), from last row of Table 1] of the information inputs into patented inventions, whereas academic research accounts for ≈17 per cent of all R&D in the United States and in the OECD as a whole. Since universities in the USA provide ≈70 per cent of refereed journal papers, academic research probably supplies less than a third of the information inputs into patented inventions than its share of total R&D would lead us to expect.

Furthermore, the direct economic benefits of the information provided by basic research are very unevenly spread amongst sectors, including among relatively R&D-intensive sectors. Table 1 shows that the intensity of use of published knowledge is particularly high in drugs, followed by other chemicals, while being virtually nonexistent in aircraft, motor vehicles, and nonelectrical machinery. Nearly half the citations journals are from chemicals, ≈37.5 per cent from electronic-related products and only just over 5 per cent from nonelectrical machinery and transportation. And in spite of this apparent lack of direct usefulness, many successful British firms recently advised the Government to continue to allow universities to concentrate on long-term academic research and training and to caution against diverting them to more immediately and obviously useful goals (7).

We also find that, in spite of the small direct impact on invention of published knowledge and contrary to the expectations of the mainstream theory, large firms in some sectors both undertake extensive amounts of basic research and then publish

Table 1   Citing patterns in US patents, 1994

| Manufacturing sector | No. of patents | No. of citations per patent to | | | Share of all citations to journals |
		Other patents	Science journals	Other	
Chemicals (less drugs)	10,592	9.8	2.5	1.2	29.1
Drugs	2,568	7.8	7.3	1.8	20.6
Instruments	14,950	11.8	1.0	0.7	16.3
Electronic equipment	16,108	8.8	0.7	0.6	12.2
Electrical equipment	6,631	10.0	0.6	0.6	4.4
Office and computing	5,501	10.0	0.7	1.0	4.3
Nonelectrical machinery	15,001	12.2	0.2	0.5	3.3
Rubber and miscellaneous plastic	4,344	12.4	0.4	0.6	1.9
Other	8,477	12.2	0.2	0.4	1.9
Metal products	6,645	11.6	0.2	0.4	1.5
Primary metals	918	10.5	0.8	0.7	1.0
Building materials	1,856	12.6	0.5	0.7	1.0
Food	596	15.1	1.3	1.6	0.9
Oil and gas	998	15.0	0.6	0.9	0.7
Motor vehicles and transportation	3,223	11.3	0.1	0.3	0.4
Textiles	567	12.4	0.3	0.8	0.2
Aircraft	905	11.6	0.1	0.3	0.1
Total	99,898	10.9	0.9	0.7	100.0

Data taken from D. Olivastro (CHI Research, Haddon Heights, NJ; personal communication).

the results. About 9 per cent of US journal publications come from firms. And Hicks et al. (8) have shown that large European and Japanese firms in the chemicals and electrical/electronic industries each publish >200 and sometimes up to 500 papers a year, which is as much as a medium-sized European or Japanese university.

The capacity to solve complex problems   Thus business practitioners persist in supporting both privately and publicly funded basic research, despite its apparently small direct contribution to inventive and innovative activities. The reason is that the benefits that they identify from public and corporate support for basic research are much broader than the 'information', 'discoveries', and 'ideas' that tend to be stressed by economists, sociologists, and academic scientists. Practitioners attach smaller importance to these contributions than to the provision of trained researchers, improved research techniques and instrumentation, background (i.e., tacit) knowledge, and membership of professional networks (see, in particular, refs. 9-14).

In general terms, basic research and related training improve corporate (and other) capacities to solve complex problems. According to one eminent engineer:

> . . . we construct and operate . . . systems based on prior experiences and we innovate in them by the human design feedback mode . . . first, we look at the system and ask ourselves 'How can we do it better?'; second, we make some change, and observe the system to see if our expectation of 'better' is fulfilled; third, we repeat this cycle of improvements over and over. This cyclic, human design feed back mode has also been called 'learning-by-doing', 'learning by using', 'trial and error', and even 'muddling through' or 'barefoot empiricism' . . . Human design processes can be quite rational or largely intuitive, but by whatever name, and however rational or intuitive . . . it is an important process not only in design but also in research, development, and technical and social innovations because it is often the only method available. [ref. 15, p. 63.]

Most of the contributions are person-embodied and institution-embodied tacit knowledge, rather than information-based codified knowledge. This explains why the benefits of basic research turn out to be localized rather than available indifferently to the whole world (8, 16, 17). For corporations, scientific publications are signals to academic researchers about fields of corporate interest in their (the academic researchers') tacit knowledge (18). And Japan has certainly not been a free rider on the world's basic research, since nearly all the R&D practitioners in their corporations were trained with Japanese resources in Japanese universities (19).

Why public subsidy?   These conclusions suggest that the justification for public subsidy for basic research, in terms of complete codification and nonappropriable nature of immediately applicable knowledge, is a weak one. The results of basic research are rarely immediately applicable, and making them so also increases their appropriable nature, since, in seeking potential applications, firms learn how to combine the results of basic research with other firm-specific assets, and this can rarely be imitated overnight, if only because of large components of tacit knowledge (20–22). In three other dimensions, the case for public subsidy is stronger.

The first was originally stressed strongly by Nelson (3); namely, the considerable uncertainties before the event in knowing if, when, and where the results of basic research might be applied. The probabilities of application will be greater with an open and flexible interface between basic research and application, which implies public subsidy for the former.

A second, and potentially new, justification grows out of the internationalization of the technological activities of large firms. Facilities for basic research and training can be considered as an increasingly important part of the infrastructure for downstream technological and production activities. Countries may therefore decide to subsidize them, to attract foreign firms or even to retain national ones.

The final and most important justification for public subsidy is training in research skills, since private firms cannot fully benefit from providing it when researchers, once trained, can and do move elsewhere. There is, in addition, the important insight of Dasgupta and David (23) that, since the results of basic research are public and those of applied research and development often are not, training through basic research enables more informed choices and recruitment into the technological research community.

Uneven technological development amongst countries

Evidence   Empirical studies have shown that technological activities financed by business firms largely determine the capacity of firms and countries both to exploit the benefits of local basic research and to imitate technological applications originally developed elsewhere (11, 24). Thus, although the output of R&D activities have some characteristics of a public good, they are certainly not a free good, since their application often requires further investments in technological application (to transform the results of basic research into innovations) or reverse engineering (to imitate a product already developed elsewhere). This helps explain why international differences in economic performance are partially explained by differences in proxy measures of investments in technological application, such as R&D expenditures, patenting, and skill levels.

Another important gap in our understanding is the persistent international differences in intangible investments in technological application. Even amongst the OECD countries, they are quite marked. Using census data, Table 2 shows that within Western Europe there are considerable differences in the level of training of the non-university-trained workforce. These broad statistical differences are confirmed by more detailed comparisons of educational attainment in specific subjects and their economic importance is confirmed by marked international differences in productivity and product quality (25). There is also partial evidence

Table 2   *Qualifications of the workforce in five European countries*

Level of qualification	Percentage of workforce				
	Britain*	Netherlands†	Germany‡	France*	Switzerland§
University degrees	10	8	11	7	11
Higher technician diplomas	7	19	7	7	9
Craft/lower technical diplomas	20	38	56	33	57
No vocational qualifications	63	35	26	53	23
Total	100	100	100	100	100

Data taken from ref. 25. Data shown are from the following years: *, 1988; †, 1989; ‡, 1987; and §, 1991.

that the United States resembles the United Kingdom, with a largely unqualified workforce, while Japan and the East Asian tigers resemble Germany and Switzerland (26).

In addition, OECD data show no signs of convergence among the member countries in the proportion of gross domestic produce spent on business-funded R&D activities. Japan, Germany, and some of its neighbors had already caught up with the US level in the early to mid-1970s (19). At least until 1989, they were forging ahead, which could have disquieting implications for future international patterns of economic growth, especially since there are also signs of the end of productivity convergence amongst the OECD countries (see, for example, ref. 27).

In spite of their major implications for both science and economic policies, relatively little attention has been paid to explaining these international differences. The conventional explanations are in terms of either macroeconomic conditions (e.g., Japan has an advantage over the United States in investment and R&D because of differences in the cost of capital) or in terms of market failures (e.g., given the lack of labor mobility, Japanese firms have greater incentives to invest in workforce training; see ref. 28).

Institutional failure   But while these factors may have some importance, they may not be the whole story. Some of the international differences have been long and persistent, and none more so (and none more studied) than the differences between the United Kingdom and Germany, which date back to at least the beginning of this century, and which have persisted through the various economic conditions associated with imperialism, Labour Party corporatism, and Thatcherite liberalism in the United Kingdom, and imperialism, republicanism (including the great inflation of 1924), nazism, and federalism in Germany (29). The differences in performance can be traced to persistent differences in institutions (30, 31), their incentive structures, and their associated competencies (i.e., tacit skills and routines) that change only slowly (if at all) in response to international differences in economic incentives.

One of the most persistent differences has been in the proportion of corporate resources spent on R&D and related activities. New light is now being thrown on this subject by improved international data on corporate R&D performance. Table 3 shows that, in spite of relatively high profit rates and low 'cost of funds', the major UK and US firms spend relatively low proportions of their sales on R&D. Similarly, despite higher cost of funds, Japanese firms spend higher shares of profits and sales on R&D than US firms. Preliminary results of regression analysis suggest that each firm's R&D/sales ratio is influenced significantly by its profits/sales ratio and by country-specific (i.e., institutional) effects. However, each firm's cost of funds/ profits ratio turns out not to be a significant influence, except for the subpopulation of US firms.

These differences cannot be explained away very easily. In a matched sample of firms of similar size in the United Kingdom and Germany, Mayer (33) and his colleagues found that, in the period from 1982 to 1988, the proportion of earnings paid out as dividends were 2 to 3 times as high in the UK firms. Tax differences could not explain the difference; indeed, retentions are particularly heavily discouraged in Germany. Nor could differences in inflation or in investments

Table 3   *Own R&D expenditures by world's 200 largest R&D spenders in 1994*

Country (n)	R&D as percentage of			Profits/sales, %	Cost of funds/profits, %
	Sales	Profits*	Costs of funds†		
Sweden (7)	9.2	73.4	194.3	12.5	37.8
Switzerland (7)	6.9	69.0	140.4	10.0	49.1
Netherlands (3)	5.6	103.8	201.0	5.4	51.6
Japan (60)	5.5	204.0	185.6	2.7	109.9
Germany (16)	4.9	149.0	202.9	3.2	73.4
France (18)	4.6	256.5	111.9	1.8	229.2
United States (67)	4.2	43.8	96.6	9.6	45.3
United Kingdom (12)	2.6	23.7	52.3	11.0	45.3
Italy (4)	2.3	N/A	34.0	N/A	N/A
Total (200)	4.7	72.1	119.1	6.5	63.1

Data taken from ref. 32. n, No. of firms; N/A, not applicable.
*Profits represent profits before tax, as disclosed in the accounts.
†Cost of funds represents (equity and preference dividends appropriated against current year profits) +
*   (interest servicing costs on debt) + (other financing contracts, such as finance leases).*

requirements explain it. Mayer attributes the differences to the structure of ownership and control. Ownership in the United Kingdom is dispersed, and control exerted through corporate takeovers. In Germany, ownership is concentrated in large corporate groupings, including the banks, and systems of control involve suppliers, purchasers, banks, and employees, as well as shareholders. On this basis, he concludes that the UK system has two drawbacks:

> [F]irst . . . the separation of ownership and control . . . makes equity finance expensive, which causes the level of dividends in the UK to be high and inflexible in relation to that in countries where investors are more closely involved. Second, the interests of other stakeholders are not included. This discourages their participation in corporate investment.
>   UK-style corporate ownership is therefore likely to be least well suited to co-operative activities that involve several different stakeholders, e.g. product development, the development of new markets, and specialized products that require skilled labour forces. [ref. 33, p. 191.]

I would only add that the UK financial system is likely to be more effective in the arm's-length evaluation of corporate R&D investments that are focused on visible, discrete projects that can be evaluated individually – for example, aircraft, oil fields, and pharmaceuticals. It will be less effective when corporate R&D consists of a continuous stream of projects and products, with strong learning linkages amongst them – for example, civilian electronics.

Similar (and independently derived) analyses have emerged in the USA, especially from a number of analysts of corporate behavior at Harvard Business School (34, 35). In addition to deficiencies in the financial system, they stress the importance of command and control systems installed by corporate managers. In particular, they point to the growing power of business school graduates, who are well trained to apply financial and organizational techniques, but have no knowledge of technology. They maximize their own advantage by installing decentralized systems of development, production, and marketing, with resource allocations and monitoring determined centrally by short-term financial criteria. These systems are intrinsically incapable of exploiting all the benefits of investments in technological activities, given their short-term performance horizons, their neglect of the intangible benefits in opening new technological options, and their inability to exploit opportunities that cut across established divisional boundaries. Managers with this type of competence therefore tend to underinvest in technological activities.

Institutions and changing technologies   But given the above deficiencies, how did the United States maintain its productivity advance over the OECD countries from 1870 to 1950? According to a recent paper by Abramovitz and David [ref. 36; similar arguments have been made by Freeman et al. (37), Nelson and Wright (38), and von Tunzelmann (39)], the nature of technical progress in this period was resource-intensive, capital-using, and scale-dependent – symbolized by the large-scale production of steel, oil, and the automobile. Unlike all other countries, the United States had a unique combination of the abundant natural resources, a large market, scarce labor, and financial resources best able to exploit this technological trajectory. These advantages began to be eroded after World War II, with new resource discoveries, the integration of national markets, and the improvements in transportation technologies. Furthermore, the nature and source of technology has been changing, with greater emphasis on intangible assets like training and R&D and lesser emphasis on economies of scale. Given these tendencies, Abramovitz and David foresee convergence amongst the OECD countries in future. The data in Tables 2 and 3 cast some doubt on this.

Is uneven technological development self-correcting?   But can we expect uneven international patterns of technological developments to be self-correcting in future? In an increasingly integrated world market, there are powerful pressures for the international diffusion of the best technological and related business practices through the international expansion of best practice firms, and also for imitation through learning and investment by laggard firms. But diffusion and imitation are not easy or automatic, for at least three sets of reasons.

First, technological (and related managerial) competencies, including imitative ones, take a long time to learn, and are specific to particular fields and to particular inducement mechanisms. For example, US strength in chemical engineering was strongly influenced by the opportunities for (and problems of) exploiting local petroleum resources (40). More generally, sectoral patterns of technological strength (and weakness) persist over periods of at least 20–30 years (19, 41).

Second, the location and rate of international diffusion and imitation of best practice depend on the cost and quality of the local labor force (among other things). With the growing internationalization of production, firms depend less on any specific labor market and are therefore less likely to commit resources in investment in local human capital. In other words, firms can adjust to local skill (or unskilled) endowments, rather than attempt to change them. National policies to develop human capital (including policies to encourage local firms to do so) therefore become of central importance.

Third, education and training systems change only slowly, and are subject to demands in addition to those of economic utility. In addition there may be self-reinforcing tendencies intrinsic in national systems of education, management, and finance. For example:

- The British and US structure of human capital, with well-qualified graduates and a poorly educated workforce, allows comparative advantage in sectors requiring this mix of competencies, like software, pharmaceuticals, and financial services. The dynamic success of these sectors in international markets reinforces demand for the same mix of competencies. In Germany, Japan and their neighboring countries, the dynamics will, on the contrary, reinforce demands in sectors using a skilled workforce.

- Decentralized corporate management systems based on financial controls breed managers in the same mold, whose competencies and systems of command and control are not adequate for the funding of continuous and complex technical change. Firms managed by these systems therefore tend to move out (or are forced out) of sectors requiring such technical change. See, for example, Greenen's ITT in the United States, and Weinstock's General Electric Company in the United Kingdom (35, 42).

- The British financial system develops and rewards short-term trading competencies in buying and selling corporate shares on the basis of expectations about yields, while the German system develops longer-term investment competencies in dealing with shares on the basis of expected growth. These competencies emerge from different systems of training and experience and are largely tacit. It is therefore difficult, costly, and time-consuming to change from one to the other. And there may be no incentive to do so, when satisfactory rates of return can be found in both activities.

Needless to say, these trends will be reinforced by explicit or implicit policy models that advocate 'sticking to existing comparative advantage', or 'reinforcing existing competencies'.

The measurement of software technology

The institutional and national characteristics required to exploit emerging technological opportunities depend on the nature and locus of these opportunities. Our apparatus for measuring and analysing technological activities is becoming obsolete, since the conventional R&D statistics do not deal adequately with software technology, to which we now turn.

There is no single satisfactory proxy measures for the activities generating technical change. The official R&D statistics are certainly a useful beginning, but systematic data on other measures show that they considerably underestimate both the innovations generated in firms with <1000 employees (where most firms do not have separately accountable R&D departments) and in mechanical technologies (the generation of which is dispersed across a wide variety of product groups; refs. 43 and 44).

A further source of inaccuracy is now emerging with the growth in importance of software technology, for the following reasons:

Table 4 *The growth of US science and engineering employment in life science, computing, and services*

Field	Ratio, no. of employees in 1992/no. of employees in 1980
All fields	1.44
Life sciences	3.12
Computer specialists	2.03
Manufacturing sectors	1.30
Nonmanufacturing sectors	1.69
Financial services	2.37
Computer services	4.10

Data taken from ref. 45.

Table 5 *Industries' percentages of business employment of scientists and engineers, 1992*

Field	Employment of scientists and engineers, % (computer specialists, %)
Manufacturing	48.1 (10.9)
Nonmanufacturing	51.9 (23.7)
Engineering services	9.1 (3.2)
Computer services	8.3 (51.8)
Financial services	6.1 (58.5)
Trade	5.2 (25.5)

Data taken from ref. 45.

- One revolutionary feature of software technology is that it increases the potential applications of technology, not only in the sphere of production, but also in the spheres of design, distribution, coordination, and control. As a consequence, the locus of technological change is no longer almost completely in the manufacturing sector, but also in services. In all OECD countries, a high share of installed computing capacity in the United States is in services, which have recently overtaken manufacturing as the main employers of scientists and engineers (see Tables 4 and 5).

- Established R&D surveys tend to neglect firms in the service sector. According to the official US survey, computer and engineering services accounted in 1991 for only 4.2 per cent of total company funded R&D compared with >8 per cent of science and engineering employment. The Canadian statistical survey has done better in 1995, ≈30 per cent of all measured business R&D was in services, of which ≈12 per cent was in trade and finance (46).

- This small presence of software in present surveys may also reflect the structural characteristics of software development. Like mechanical machinery, software can be considered as a capital good, in that the former processes materials into products and the latter processes information into services. Both are developed by user firms as part of complex products or production systems, as well as by small and specialized suppliers of machinery and applications software (for machinery, see ref. 47). As such, a high proportion of software development will be hidden in the R&D activities of firms making other products and in firms too small for the establishment of a conventional R&D department.

Conclusions

The unifying theme of this paper is that differences among economists about the nature, sources, and measurement of technical change will be of much greater relevance to policy formation in the future than they were in the past. These differences are at their most fundamental over the nature of useful technological knowledge, the functions of the business firm, and the location of the activities generating technological change. They are summarized, and their analytical and policy conclusions are contrasted, in Tables 6, 7, and 8. On the whole, the empirical evidence supports the assumptions underlying the right columns, rather than those on the left.

Basic research   The main economic value of basic research is not in the provision of codified information, but in the capacity to solve complex technological problems, involving tacit research skills, techniques, and instrumentation and membership in national and international research networks. Again, there is nothing original in this:

> [t]he responsibility for the creation of new scientific knowledge – and for most of its application – rests on that small body of men and women who understand the fundamental laws of nature and are skilled in the techniques of scientific research. [ref. 48, p. 7.]

Exclusive emphasis on the economic importance of codified-information:

- exaggerates the importance of the international free rider problem and encourages (ultimately self-defeating) techno-nationalism;
- reinforces a constricted view of the practical relevance of basic research by concentrating on direct (and more easily measurable) contributions, to the neglect of indirect ones;

Table 6   *Differing policies for basic research*

Subject	Assumptions on the nature of useful knowledge	
	Codified information	Tacit know-how
International free riders	Strengthen intellectual property rights; restrict international diffusion	Strengthen local and international networks
Japan's and Germany's better technological performance than United States and United Kingdom with less basic research	More spillovers by linking basic research to application	Increase business investment in technological activities
Small impact of basic research on patenting	Reduce public funding of basic research	Stress unmeasured benefits of basic research
Large business investment in published basic research	Public relations and conspicuous intellectual consumption	A necessary investment in signals to the academic research community

Table 7   *Differing policies for corporate technological activities*

	Assumptions on the functions of business firms	
Subject	Optimizing resource allocations based on market signals	Learning to do better and new things
Inadequate business investment in technology compared to foreign competition	R&D subsidies and tax incentives; reduce cost of capital; increase profits	Improve worker and manager skills; improve (through corporate governance) the evaluation of intangible competencies

- concentrates excessively on policies to promote externalities, to the neglect of policies to promote the demand for skills to solve complex technological problems (49, 50).

Uneven technological development   In this context, too little attention has been paid to the persistent international differences, even among the advanced OECD countries, in investments in R&D, skills, and other intangible capital to solve complex problems. Explanations in terms of macroeconomic policies and market failures are incomplete, since they concentrate entirely on incentives and ignore the competencies to respond to them. Observed 'inertia' in responding to incentives is not just a consequence of stupidity or self-interest, but also of cognitive limits on how quickly individuals and institutions can learn new competencies. Those adults who have tried to learn a foreign language from scratch will well understand the problem. Otherwise, the standard demonstration is to offer economists \$2 million to qualify as a surgeon within one year. (Some observers have been reluctant to make the reverse offer.)

These competencies are located not only in firms, but also in financial, educational, and managerial institutions. Institutional practices that lead to under- or mis-investment in technological and related competencies are not improved automatically through the workings of the market. Indeed, they may well be self-reinforcing (Table 7).

Software Technology   Although R&D statistics have been an invaluable source of information for policy debate, implementation, and analysis, they have always had a bias toward the technological activities of large firms compared with small ones and toward electrical and chemical technologies compared with mechanical engineering. The bias is now becoming even greater with the increasing

Table 8   *Differences in the measurement of technological activities*

	Assumptions on the nature of technological activities	
Subject	Formal R&D	Formal and informal R&D including software technology
The distribution of technological activities	Mainly in large firms, manufacturing, and electronics/chemicals/ transportation	Also in smaller firms in nonelectrical machinery and large and small firms in services

development of software technology in the service sector, while R&D surveys concentrate on manufacturing (Table 8).

As a consequence, statistical and econometric analysis will increasingly be based on incomplete and potentially misleading data. Perhaps more worrying, some important locations of rapid technological change will be missed or ignored. While we are bedazzled by the 'high-tech' activities of Seattle and Silicon Valley, the major technological revolution may well be happening among the distribution systems of the oldest and most venal of the capitalists: the money lenders (banks and other financial services), the grocers (supermarket chains), and the traders (textiles, clothing, and other consumer goods).

To conclude, if economic analysis is to continue to inform science and technology policy making, it must pay greater attention to the empirical evidence on the nature and locus of technology and the activities that generate it and spend more time collecting new and necessary statistics in addition to exploiting those that are already available. That the prevailing norms and incentive structure in the economics profession do not lend themselves easily to these requirements is a pity, just as much for the economists as for the policy makers, who will seek their advice and insights elsewhere.

This paper has benefited from comments on an earlier draft by Prof. Robert Evenson. It draws on the results of research undertaken in the ESRC (Economic and Social Research Council)-funded Centre for Science, Technology, Energy and the Environment Policy (STEEP) at the Science Policy Research Unit (SPRU), University of Sussex.

References

1. Fagerberg, J. (1994) *J. Econ. Lit.* **32**, 1147–75.
2. Krugman, P. (1995) in *Handbook of the Economics of Innovation and Technological Change*, ed. Stoneman, P. (Blackwell, Oxford), pp. 342–65.
3. Nelson, R. (1959) *J. Polit. Econ.* **67**, 297–306.
4. Arrow, K. (1962) in *The Rate and Direction of Inventinve Activity*, ed. Nelson, R. (Princeton Univ. Press, Princeton), pp. 609–25.
5. de Tocqueville, A. (1840) *Democracy in America* (Vintage Classic, New York), reprinted 1980.
6. Rosenberg, N. (1976) in *Perspectives on Technology* (Cambridge Univ. Press, Cambridge), pp. 126–38.
7. Lyall, K. (1993), M.Sc. dissertation (University of Sussex, Sussex, UK).
8. Hicks, D., Izard, P. and Martin, B. (1996) *Res. Policy* **23**, 359–78.
9. Brooks, H. (1994) *Res. Policy* **23**, 477–86.
10. Faulkner, W. and Senker, J. (1995) *Knowledge Frontiers* (Clarendon, Oxford).
11. Gibbons, M. and Johnston, R. (1974) *Res. Policy* **3**, 220–42.
12. Klevorick, A., Levin, R., Nelson, R. and Winter, S. (1995) *Res. Policy* **24**, 185–205.
13. Mansfield, E. (1995) *Rev. Econ. Stat.* **77**, 55–62.
14. Rosenberg, N. and Nelson, R. (1994) *Res. Policy* **23**, 324–48.
15. Kline, S. (1995) *Conceptual Foundations for Multi-Disciplinary Thinking* (Stanford Univ. Press, Stanford, CA).
16. Jaffe, A. (1989) *Am. Econ. Rev.* **79**, 957–70.
17. Narin, F. (1992) *CHI Res.* **1**, 1–2.
18. Hicks, D. (1995) *Ind. Corp. Change* **4**, 401–424.
19. Patel, P. and Pavitt, K. (1994) *Ind. Corp. Change* **3**, 759–87.
20. Galimberti, I. (1993) D. Phil. thesis (University of Sussex, Sussex, UK).
21. Miyazaki, K. (1995) *Building Competencies in the Firm: Lessons from Japanese and European Opto-Electronics* (Macmillan, Basingstoke, UK).
22. Sharp, M. (1991) in *Technology and Investment*, eds Deiaco, E., Hornell, E. and Vickery, G. (Pinter, London), pp. 93–114.
23. Dasgupta, P. and David, P. (1994) *Res. Policy* **23**, 487–521.

24. Cohen, W. and Levinthal, D. (1989) *Econ. J.* **99**, 569-96.
25. Prais, S. (1993) *Economic Performance and Education: The Nature of Britain's Deficiencies* (National Institute for Economic and Social Research, London), Discussion Paper 52.
26. Newton, K., de Broucker, P., McDougal, G., McMullen, K., Schweitzer, T. and Siedule, T. (1992) *Education and Training in Canada* (Canada Communication Group, Ottawa).
27. Soete, L. and Verspagen, B. (1993) in *Explaining Economic Growth*, eds Szirmai, A., van Ark, B. and Pilat, D. (Elsevier, Amsterdam).
28. Teece, D., ed. (1987) *The Competitive Challenge: Strategies for Industrial Innovation and Renewal* (Ballinger, Cambridge, MA).
29. Patel, P. and Pavitt, K. (1989) *Natl. Westminster Bank Q. Rev.* May, 27-42.
30. Keck, O. (1993) in *National Innovation Systems: A Comparative Analysis*, ed. Nelson, R. (Oxford Univ. Press, New York), pp. 115-57.
31. Walker, W. (1993) in *National Innovation Systems: A Comparative Analysis*, ed. Nelson, R. (Oxford Univ. Press, New York), pp. 158-91.
32. Company Reporting Ltd. (1995) *The 1995 UK R&D Scoreboard* (Company Reporting Ltd., Edinburgh).
33. Mayer, C. (1994) *Capital Markets and Corporate Performance*, eds Dimsdale, N. and Prevezer, M. (Clarendon, Oxford).
34. Abernathy, W. and Hayes, R. (1980) *Harvard Bus. Rev.* July/August, 67-77.
35. Chandler, A. (1992) *Ind. Corp. Change*, **1**, 263-84.
36. Abramovitz, M. and David, P. (1994) *Convergence and Deferred Catch-Up: Productivity Leadership and the Waning of American Exceptionalism* (Center for Economic Policy Research, Stanford, CA), CEPR Publication 401.
37. Freeman, C., Clark, J. and Soete, L. (1982) *Unemployment and Technical Innovation* (Pinter, London).
38. Nelson, R. and Wright, G. (1992) *J. Econ. Lit.* **30**, 1931-64.
39. von Tunzelmann, N. (1995) *Technology and Industrial Progress: the Foundation of Economic Growth* (Edward Elgar, Aldershot, UK).
40. Landau, R. and Rosenberg, N. (1992) in *Technology and the Wealth of Nations*, eds Rosenberg, N., Landau, R. and Mowery, D. (Stanford Univ. Press, Stanford, CA), pp. 73-119.
41. Archibugi, D. and Pianta, M. (1992) *The Technological Specialisation of Advanced Countries* (Kluwer Academic, Dordrecht, the Netherlands).
42. Anonymous (1995) *Economist* June 17, 86-92.
43. Pavitt, K., Robson, M. and Townsend, J. (1987) *J. Ind. Econ.* **35**, 297-316.
44. Patel, P. and Pavitt, K. (1995) in *Handbook of the Economics of Innovation and Technological Change*, ed. Stoneman, P. (Blackwell, Oxford), pp. 14-51.
45. National Science Board-National Science Foundation (1993) *Science and Engineering Indicators 1993* (US Government Printing Office, Washington, DC).
46. Statistics Canada (1996) *Service Bull.* **20**, 1-8.
47. Patel, P. and Pavitt, K. (1996) *Res. Policy* **23**, 533-46.
48. Bush, V. (1945) *Science and the Endless Frontier* (National Science Foundation, Washington, DC), reprinted 1960.
49. Mowery, D. (1983) *Policy Sci.* **16**, 27-43.
50. Metcalfe (1995) in *Handbook of the Economics of Innovation and Technological Change*, ed. Stoneman, P. (Blackwell, Oxford), pp. 409-512.

Name index

Abernathy, W. 37, 39, 114, 216
Abramovitz, M. 241
Acs, Z. 126
Allen, T. 7, 67
Altshuler, A. 182, 186–7
Andersen, E. 142
Anderson, P. 113
Ansoff, H. 27
Aoki, M. 116
Archibugi, D. 52, 74, 201
Arnold, E. 179, 186
Arrow, K. 78, 224, 235
Audretsch, D. 126
Averch, H. 224

Barclay, G. 8
Barras, R. 68, 84, 180
Barre, R. 72
Barton, J. 73
Bell, M. 12, 194
Bertin, G. 10, 130, 197
Bijker, W. 5
Bohlin, E. 94
Bosworth, D. 84
Brady, T. 56
Buer, T. 38, 39
Burns, T. 115
Bush, V. 231

Cantwell, J. 52, 72, 126, 135, 142, 194, 207
Carlsson, B. 71, 176
Carpenter, M. 6
Casson, M. 142
Chandler, A. 117, 120, 218–19
Chesbrough, H. 98
Christensen, C. 100, 115
Clark, K. 106, 108, 114, 115, 116, 207
Cohen, W. 73, 150
Cohendet, P. 111
Constant, E. 120
Cooper, A. 114
Cooper, R. 115
Corbett, J. 216
Curry, R. 183
Cusumano, M. 84

Dahlman, C. 198
Dasgupta, P. 5, 132, 150, 224, 231, 238
David, P. 5, 225, 238, 241

Davies, D. 8
Dodgson, M. 66
Dosi, G. 27, 34, 41, 43, 71, 72, 106, 114, 142, 189, 194
Dunning, J. 126, 142

Eliasson, G. 71, 186
Ergas, H. 194

Fagerberg, J. 126, 193, 194, 223, 234
Fleck, J. 56
Fottit, M. 8
Frame, J. 226, 229
Franko, L. 126, 194
Freeman, C. 5, 27, 34, 43, 62, 66, 112, 102–3, 126, 133, 169, 181, 183, 195, 226, 241
Frischtak, C. 129

Gaio, F. 11
Gambardella, A. 100, 107, 114
Geroski, P. 194
Gershuny, J. 27
Gibbons, M. 7
Gold, B. 22, 27, 43
Gönenc, R. 176, 179
Graham, M. 118
Granstrand, O. 52, 56, 74, 78, 94, 107, 119, 142, 143
Griliches, Z. 130, 195, 198

Hague, D. 8
Hakanson, L. 142
Hamel, G. 71, 76, 94, 97, 118
Hamilton, W. 117
Hanson, P. 10
Hare, P. 229
Harrison, T. 118
Hassid, J. 24, 39
Hayes, R. 216
Haywood, B. 185
Henderson, R. 115
Hicks, D. 104, 109, 229, 237
Hobday, M. 65
Hodson, C. 135
Hoffman, K. 179, 180
Hounshell, D. 115
Howells, J. 67
Hu, Y.-S. 216